Mineral Surfaces

Mineralogical Society Series

Series editor
Professor Geoffrey D. Price

The aim of the series is to provide up-to-date reviews through the selected but specialized contributions of leading experts. A particularly attractive feature of the series is the tutorial element, making it suitable for third year undergraduates and researchers of all levels not solely within the Earth Sciences. Each volume is purpose-designed, highly illustrated and serves as an excellent reference tool.

TITLES AVAILABLE

1 **Deformation Processes in Minerals, Ceramics and Rocks**
Edited by D.J. Barber and P.G. Meredith

2 **High-temperature Metamorphism and Crustal Anatexis**
Edited by J.R. Ashworth and M. Brown

3 **The Stability of Minerals**
Edited by Geoffrey D. Price and Nancy L. Ross

4 **Geochemistry of Clay-Pore Fluid Interactions**
Edited by D.A.C. Manning, P.L. Hall and C.R. Hughes

5 **Mineral Surfaces**
Edited by D.J. Vaughan and R.A.D. Pattrick

Mineral Surfaces

Edited by D.J. Vaughan and R.A.D. Pattrick
Department of Geology, University of Manchester, UK

Published in association with
The Mineralogical Society of Great Britain and Ireland

CHAPMAN & HALL

London · Glasgow · Weinheim · New York · Tokyo · Melbourne · Madras

Published by Chapman & Hall, 2–6 Boundary Row, London SE1 8HN, UK

Chapman & Hall, 2–6 Boundary Row, London SE1 8HN, UK

Blackie Academic & Professional, Wester Cleddens Road, Bishopbriggs, Glasgow G64 2NZ, UK

Chapman & Hall GmbH, Pappelallee 3, 69469 Weinheim, Germany

Chapman & Hall USA, One Penn Plaza, 41st Floor, New York NY 10119, USA

Chapman & Hall Japan, ITP-Japan, Kyowa Building, 3F, 2-2-1 Hirakawacho, Chiyoda-ku, Tokyo 102, Japan

Chapman & Hall Australia, Thomas Nelson Australia, 102 Dodds Street, South Melbourne, Victoria 3205, Australia

Chapman & Hall India, R. Seshadri, 32 Second Main Road, CIT East, Madras 600 035, India

First edition 1995

Typeset in 10/12 Times by Best-set Typesetter Ltd., Hong Kong
Printed in Great Britain at the University Press, Cambridge

ISBN 0 412 56340 1

A catalogue record for this book is available from the British Library

Library of Congress Catalog Card Number: 94-71811

∞ Printed on permanent acid-free text paper, manufactured in accordance with ANSI/NISO Z39.48-1992 and ANSI/NISO Z39.48-1984 (Permanence of Paper).

Contents

Contributors

G.E. Brown, Jr	Department of Geology, Stanford University, Stanford CA 94305, USA.
W.H. Casey	Land Air and Water Resources, University of California, Davis CA 95616, USA.
A. Dyer	Department of Chemistry and Applied Chemistry, University of Salford, Manchester M4 5WT, UK.
G.N. Greaves	SERC Daresbury Laboratory, Warrington WA4 4AD, UK.
M.F. Hochella, Jr	Department of Geological Sciences, Virginia Polytechnic Institute and State University, Blacksburg VA 24061, USA.
G.H. Kelsall	Department of Mineral Resources Engineering, Imperial College of Science, Technology and Medicine, London SW7 2BP, UK.
P.A. O'Day	Department of Geology and Geophysics, University of California, Berkeley CA 94720, USA.
G.A. Parks	Department of Applied Earth Sciences, Stanford University, Stanford CA 94305, USA.
R.A.D. Pattrick	Department of Geology, University of Manchester, Manchester M13 9PL, UK.
P.E. Richardson	Department of Mining and Minerals Engineering, Virginia Polytechnic Institute and State University, Blacksburg VA 24061, USA.
R.A. Schoonheydt	Centrum voor Oppervlaktechemie en Katalyse, Materials Research Centre, KU Leuven, 3001 Heverlee, Belgium.
J.A. Tossell	Department of Chemistry and Biochemistry, University of Maryland, College Park MD 20742, USA.
D.J. Vaughan	Department of Geology, University of Manchester, Manchester M13 9PL, UK.

Preface

In December 1992, the Mineralogical Society Winter Conference was held at the University of Manchester Geoscience Research Institute on the subject of 'Mineralogical Applications of Surface Science' and was preceded by a Short Course on the subject of 'Mineral Surfaces'. It is from the proceedings of the Short Course that this volume has been produced.

Financial support for the Short Course, defraying the expenses of invited speakers, student participants and publicity was received from the Mineralogical Society (including the Applied Mineralogy and Geochemistry Groups), the Royal Bank of Scotland, the Royal Society, the Natural Environment Research Council, the University of Manchester Research Support Fund and VG Scientific Instruments. These contributions are gratefully acknowledged.

In the preparation of this volume, we should like to thank Mrs Catherine Hardy for helping with all aspects of communication with authors and production of the manuscript, and Richard Hartley for help in the preparation of some of the figures.

CHAPTER ONE

Mineral surfaces: an overview

D.J. Vaughan

1.1 Introduction

The boundary is the best place for
acquiring knowledge

Paul Tillich,
Religiöse Verwirklichung (1929)

The above quotation, although it comes neither from mineralogy nor from geology but from theology, is a reminder that boundaries are the places where changes take place and it is here that processes of change can often be properly understood. In the case of minerals, the boundary is the surface of the mineral; a surface that in nature is also an interface, most commonly between the mineral and air, or the mineral and an aqueous fluid (or a melt).

Although the subdiscipline of **surface science** is well established within the major fields of physics and chemistry, in mineralogy (and related geosciences) the study of surfaces has only recently begun using modern methods. This is reflected, at least in part, by the relatively small number of texts and review articles dealing with this subject (Appendix 1.A). The situation is rapidly changing as a result of both the considerable advances taking place in techniques for the study of surfaces (both experimental methods and theoretical approaches) and the recognition of the relevance and potential applications of such studies in the earth, environmental and mineral sciences (and related technologies). In this introduction, I should like to draw attention to both technique development and applications, as well as introducing the chapters that follow in this book.

1.2 Surface science techniques

There are now a very large number of techniques that may be used to study solid surfaces. These techniques may be subdivided in a number of ways; for example, as to whether the technique yields information on the chemical composition of a surface (and if so, whether for a single monolayer of surface atoms, for several layers, or for many layers), whether the technique yields information on surface microtopography, or whether the information

Mineral Surfaces. Edited by D.J. Vaughan and R.A.D. Pattrick. Published in 1995 by Chapman & Hall, London. ISBN 0 412 56340 1

provided is actually about the 'geometric' arrangement of the atoms at the surface. An important distinction that also needs to be made is the distinction between techniques enabling surfaces to be studied *in situ* (e.g. with the solid examined while in contact with air or a fluid phase) and the *ex situ* techniques where the surface is most commonly studied in an ultra-high vacuum (UHV) environment. Both are important, because the former yields information closer to the 'real' world, whereas the latter can be used to study simpler and more controlled systems as a basis for understanding more complex ('real') systems. It is also true to say that some of the well-established techniques (such as LEED, SEM, XPS – Appendix 1.B) can now be thought of as the routine 'workhorse' methods of surface science, whereas there are many other methods that are highly specialized by comparison.

The techniques that are specifically mentioned in the text that follows this chapter are listed (with their acronyms) in Appendix 1.B. A text such as this is necessarily incomplete as regards coverage of all available methods, so that a list of other techniques used in surface science is given in Appendix 1.C (and this, again, also serves to provide a key to the plethora of acronyms in this field). Two other points need to be made in regard to surface science techniques. The first is that a very major advance took place in experimental surface science in the 1980s. This was the development of the technique known as scanning tunnelling microscopy (STM), followed closely by the related technique of atomic force microscopy (AFM). These techniques (which are now spawning many other related techniques) have enabled imaging of surfaces at atomic resolution, not only in UHV but also in air and under a layer of fluid. Using scanning tunnelling spectroscopy (STS), it is possible to identify individual atoms on a surface and monitor changes during reactions such as dissolution, oxidation and adsorption. The second point concerns advances in theory and the capabilities provided by modern computers in calculating geometric and electronic structures of surfaces and their chemical properties. As a detailed review of these developments is presented in Chapter 3, they will not be further discussed here.

1.3 Applications of mineral surface science

If we now turn to consider the relevance and the applications of mineral surface science, there are two broader areas of importance centred, first, on the earth and environmental sciences and, secondly, on aspects of applied mineralogy and mineral technology.

1.3.1 Applications in the earth and environmental sciences

The Earth and the 'environment' (generally taken to mean the surface and near-surface regions of the Earth where anthropogenic influences are important) are dynamic chemical systems. Almost all of the processes associated with the changes in these systems take place at surfaces or interfaces; these processes include the overall geochemical cycling of the elements, the special geochemical processes that are associated with concentration of particular elements to form ore deposits, and the dispersal into the environment of particular elements or compounds from (commonly man-made) sources of 'pollution'. The mineral surface is central to all of these systems because of the control exerted (ultimately at a molecular level) by such processes as mineral dissolution (including weathering and leaching), precipitation of minerals, and the sorption and desorption (i.e. the attachment and detachment) of chemical species on mineral surfaces. Some more specific examples will serve to illustrate the importance of mineral surfaces in this context.

(a) GEOCHEMICAL CYCLES OF THE ELEMENTS

The overall geochemical cycles of the elements have been a major subject of interest since the earliest days of geochemistry. If we consider sulfur as an example of an important non-metallic element, then the sulfur cycle can be represented very simplistically in the way shown in Figure 1.1. Here the fluxes between the various components of the system are shown in units of 10^9 kg/year. Both natural fluxes, such as the introduction of sulfur into the atmosphere from volcanic gases, weathering of sulfides yielding sulfur in soils, streams, rivers and the sea, and submarine volcanism introducing sulfur into the oceans, and the man-made introduction of sulfur into the atmosphere through the burning of fossil fuels and smelting of ores, are shown. In many cases, any real understanding of the processes (the mechanisms of the processes, their rates and the controls of those rates) must involve the study of mineral surfaces, whether it is the surfaces of sulfide minerals undergoing weathering or being precipitated via submarine volcanic exhalations, or the surfaces of other minerals reacting with sulfidic species in solution via sorption or dissolution reactions.

Metallic elements also provide many important examples of geochemical cycling, as in the case of cadmium, illustrated in Figure 1.2. Here, in a much reproduced diagram based on Brunner and Baccini (1981), are shown relative numbers representing reservoirs (numbers in boxes) and fluxes (numbers in ovals). As clearly shown by this diagram, the input of cadmium into the atmosphere from anthropogenic sources is more than five times that which is introduced naturally. Again, the role of mineral surfaces in regard to weathering and release of cadmium into the hydrosphere, and of sorption and desorption on mineral surfaces in the transport of cadmium, is of central importance.

(b) ELEMENT CONCENTRATION IN ORE FORMATION

The formation of metalliferous ore deposits is the most extreme natural example of selective concentration of the elements. In order to achieve

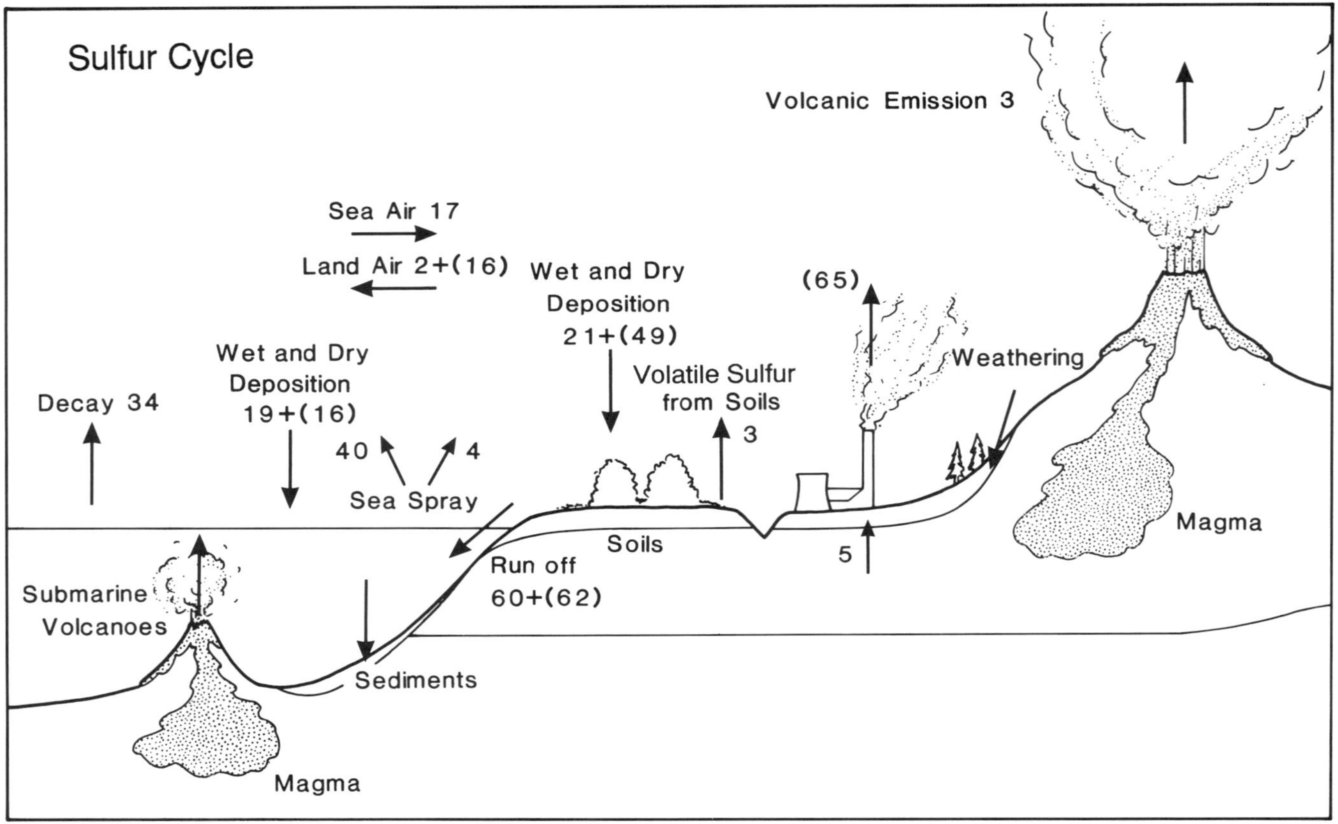

Figure 1.1 The geochemical cycle involving sulfur. (After a figure in *Our Future World*, published by the Natural Environment Research Council, 1989.)

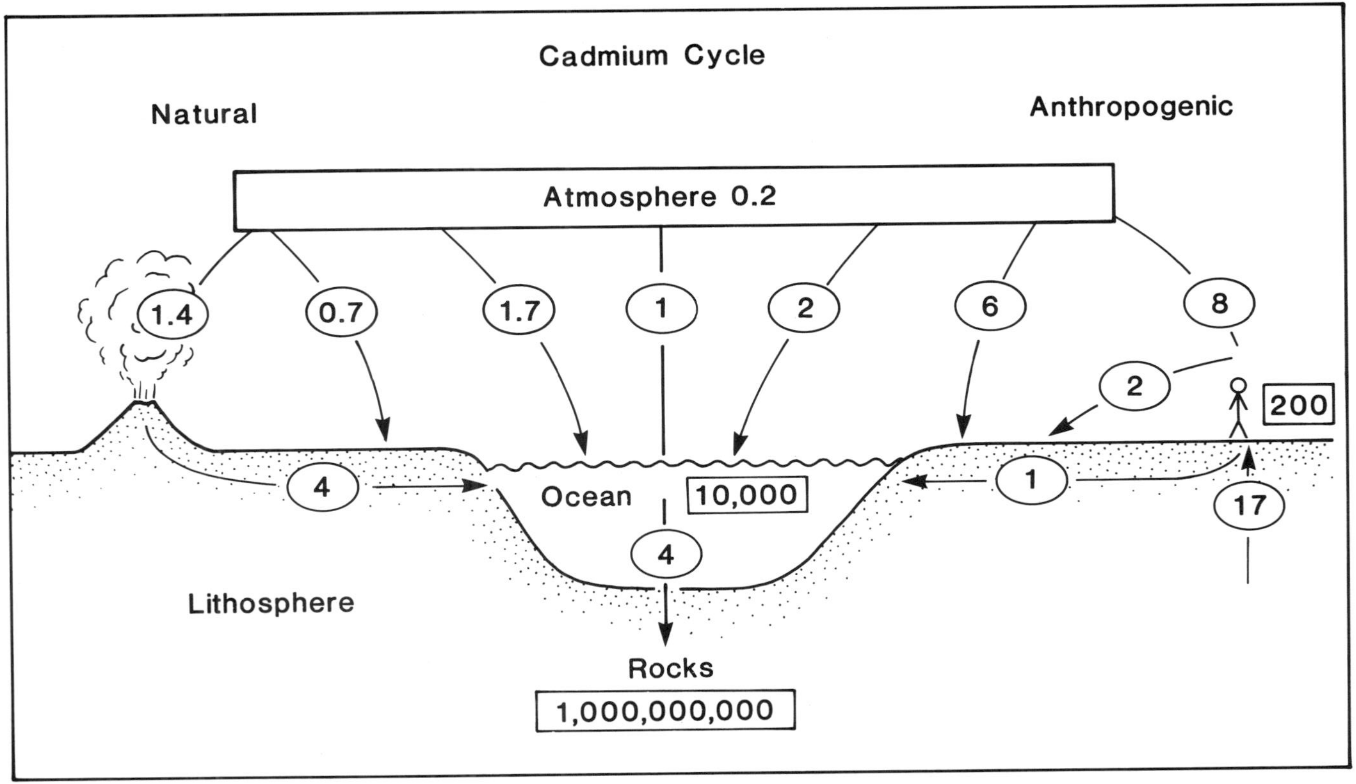

Figure 1.2 The geochemical cycle involving cadmium. (After Brunner and Baccini, 1981.)

concentrations that can be profitably mined, gold needs to be enriched above its average crustal abundance by a factor of ×250, tungsten by ×2000, lead by ×2400, chromium by ×4000, and mercury by ×12 400 (approximate figures; for further discussion see Craig, Vaughan and Skinner, 1988). Ore formation commonly involves three major stages: the release of the metal from a source (e.g. via leaching or weathering), transport of the metal, and deposition (or precipitation) as the ore mineral phases. Reactions and interactions at the mineral surface may be very important in all of these stages.

Clearly, weathering of rocks at the surface of the Earth can release metals into the hydrosphere and this may be an important source; more important in the case of many metals, however, is the passage of fluids through bodies of rock at elevated temperatures (hydrothermal leaching). For example, in volcanogenic sulfide deposits (Franklin, Lydon and Sangster, 1981), there is strong evidence that the metals to form the deposits were derived from the leaching of the underlying rocks by hydrothermal fluids. Studies of active hydrothermal systems, particularly those associated with mid-ocean ridges (Hekinian *et al.*, 1980), have done much to clarify the processes involved. Experimental studies (Seyfried, 1987) have also shown that seawater can very effectively leach and mobilize metals from the rocks that underly and surround mid-ocean ridges. Recent models (e.g. Bischoff and Rosenbauer, 1989) involve seawater circulation convection cells driven by the heat from the underlying magma chamber that can very effectively strip metals (Fe, Zn, Cu, etc.) from the rocks and create a hydrothermal fluid from which those metals can subsequently be precipitated on cooling or other interactions (as at the ‘black smoker’ hydrothermal vents). A very different environment in which hydrothermal fluids may be generated is that which forms on the burial and compaction of sediments and their trapped pore waters. In this situation, the trace amounts of metals (particularly Cu, Pb and Zn) contained within rock-forming silicates (pyroxenes, amphiboles, feldspars, sheet silicates, etc.) may be released when they dissolve in response to increasing temperature and changing redox conditions. Such processes are likely to be important in the formation of deposits found within (or flanking) sedimentary basins such as the so-called ‘sandstone-copper’ and ‘Mississippi Valley-type lead-zinc’ deposits (Sverjensky, 1986; Force, Eidel and Maynard, 1991). The release of metals through weathering and leaching reactions clearly involves processes at the mineral surface, processes that may result in complete dissolution of the mineral or its alteration and transformation by a selective removal of material.

Although the transport of metals from the source to the site of deposition commonly may be as dissolved ions or complexes in hydrothermal fluids, in some situations (particularly in the near-Earth-surface environment and hence at low temperatures) metals may have been transported through sorption on the surfaces of detrital minerals. Although such processes have been little studied, there is clearly the potential to transport substantial amounts of metals in this way, particularly on the surfaces of clays and the hydrated oxides of iron and aluminium (materials that are commonly very fine grained and have a large surface area with 'active' sites for metal sorption).

Precipitation of the metal from solution as ore mineral phases obviously involves fluid/mineral surface interactions. Such interactions may themselves be a concentration mechanism whereby a solution component that is below bulk saturation can 'precipitate' in the presence of a surface. An important example of this is the precipitation of precious metals on metal sulfide surfaces. Bancroft and Hyland (1990) have shown that this phenomenon is the result of adsorption – reduction reactions at the metal sulfide – aqueous solution interface, and other workers (e.g. Bakken *et al.*, 1989; Starling *et al.*, 1989) have found natural examples where such processes may have had a controlling influence in the formation of an ore deposit.

(c) ELEMENT DISPERSAL – ENVIRONMENTAL POLLUTION

In some senses, the release by humankind of metals or other elements into the environment is the reverse of a process like ore formation. From a highly concentrated source such as a smelter, mine dump or power generating facility, material is released into the atmosphere or hydrosphere and thereby enters the 'geochemical cycle'. Dumping of industrial waste from mining or manufacturing, the weathering of such materials, and the smelting of ores and burning of fossil fuels all contribute to this growing input. For example, it has been estimated that between 7000 and 70 000 metric tonnes of metals are released into the environment each year from mineral processing and smelting operations alone (Moore and Luoma, 1990).

As noted above (Figure 1.2), for a metal such as cadmium the contributions from anthropogenic sources to the global cycle far outweigh any natural input. Most of the cadmium released by human activity is deposited on the continents, where it enters various aquatic systems, within which its transport and distribution are critically influenced by interactions with such minerals as clays and the hydroxides of iron and manganese (Astruc, 1986; Bewers, Barry and MacGregor, 1987). There is considerable interest in the behaviour of cadmium because both the metal itself and solutions of its compounds are known to be highly toxic to humans, and because it has a strong tendency to accumulate in the food chain (Yasumura *et al.*, 1980).

Important sources of pollution that are now being intensively studied in various parts of the world are the former sites of metalliferous mining, particularly of metal sulfide ores. At such sites, formerly worked open pits, waste dumps and the fine-grained waste products of mineral processing (**tailings**) are exposed to the action of weathering. When metal sulfides react with oxygen-rich waters, very acid solutions are generated (e.g. Nordstrom, 1982) and the phenomenon known as 'acid mine drainage' itself causes

Figure 1.3 The old open pit mining area at Parys Mountain, Anglesey, Wales.

considerable environmental damage. However, such solutions are also very effective at breaking down metal-rich minerals and releasing the metals into groundwater and river systems. For example, the mining area known as Parys Mountain (Anglesey, North Wales) was the site of the largest copper mine in Europe in the eighteenth century, and extraction of copper, zinc and precious metals has been undertaken intermittently up until recent years. Current work at Manchester (S. Boult and R. Parkman, pers. comm., Figure 1.3) has shown that waters in the old open pits have pH values of ~2.3. Monitoring of pollutant levels in the only river draining the area, in the sediments of the estuary of that river and in shellfish in the estuary, show greatly enhanced levels of base metals (e.g. 22 000 ppm Cu and 14 000 ppm Zn in the estuarine sediments; R. Parkman and S. White, pers. comm.). Similar problems occur in the Clark Fork River Basin of western Montana, USA, but on a much larger scale. Here, over 125 years of metal mining and smelting has left 1600 km^2 of contaminated land with tailings and minewastes containing up to thousands of times the natural levels of As, Cd, Cu, Pb and Zn (see review by Moore and Luoma, 1990). In the Clark Fork drainage basin, elevated concentrations of toxic metals are found more than 560 km downstream from the primary source area.

The mechanisms and kinetics of breakdown of the sulfides in areas such as Parys Mountain and Clark Fork River, and of transport and subsequent fixation of the toxic metals from these sources, are actively being investigated. Research is also being undertaken on rehabilitation of such areas using a variety of approaches, including biotechnological methodologies. The role of the mineral surface and an understanding of processes at mineral surfaces are central to work in these fields.

1.3.2 Applications in industrial mineralogy and mineral technology

Some of the most important aspects of mineral surface science are associated with the processing of mined minerals, either to concentrate or to refine (or clean up) the products of mining so as to help in the extraction of certain elements (commonly metals) from them. Another important field of application arises when the minerals (or rocks) that are extracted (and, often, concentrated) are directly employed in industry. In the case of mineral products, this may be as catalysts or as sorbents in the chemical industry; in the case of rocks, it may be as building stones or as aggregate for construction purposes. These applications are further discussed below.

(a) ORE PROCESSING AND EXTRACTION

Metalliferous ores, following their mining, almost always require treatment to increase the percentage of metal and produce a concentrate suitable for smelting or some other metallurgical process (Wills, 1992). The most extensively used method of ore processing (beneficiation) is froth flotation, a method which makes use of the surface chemical properties of the minerals in a fine particle slurry of ore and associated gangue minerals. An understanding of mineral surface properties is, therefore, fundamental to the technology of flotation, a process employed in roughly 70% of all mineral processing operations and employed for the concentration of many non-metalliferous mining products (e.g. fluorite, coal) as well as metal ores.

Recent decades have seen a growth in the extraction of metals from their ores by dissolution in acid or alkali solutions (**leaching**). This is partly because leaching is less damaging to the environment than more traditional metal extraction processes (such as smelting) and can be applied to low-grade ores or even wastes (as in the **dump leaching** of sulfide ores using the sulfuric acid generated by sulfide oxidation). The dissolution of minerals in acid, alkali or strongly oxidizing solutions is obviously controlled by reactions taking place at the mineral surface (e.g. Bautista, 1984).

(b) MINERALS AS CATALYSTS AND SORBENTS

It has been suggested that 'the crust of planet Earth can be thought of as a catalytic bed of enormous diversity consisting of trillions of square kilometers of surface area' (Hochella and White, 1990). Minerals are not only effective as catalysts in nature but, along with their synthetic analogues, can be introduced into the laboratory or industrial processes. Particularly important in this respect are certain clay minerals and the zeolites; both offer a large amount of surface area on which reactions can take place, the latter through what can be regarded as 'internal' surfaces (Dyer, this volume). The same can be said of sorption processes, and in both cases, of course, the mineral surface plays the crucial role.

(c) WEATHERING OF MINERALS IN BUILDING STONES (AND RELATED MATERIALS)

The role of mineral surfaces in the weathering of rocks at a natural outcrop is the same when that rock has been quarried and cut for use as a building stone (or some kind of monument). Although the amounts of material subjected to breakdown in this way may be small compared with the weathering of rocks, the economic impact may be considerable. Many governments or international and private foundations are, therefore, spending considerable sums of money in attempts to prevent or slow down the deterioration of stone structures of cultural importance. The breakdown of rock materials in this way, which is often enhanced by anthropogenically derived pollutants, depends upon the surface properties of the constituent minerals of these rocks and building stones.

1.4 This book

The subject of mineral surfaces (and their interactions) is a vast field that overlaps with many others, as the above brief survey shows. A text such as this is necessarily illustrative rather than comprehensive in almost all respects.

The book begins with a chapter by Michael Hochella describing the tools of mineral surface science, the chemical and physical nature of mineral surfaces and the reactivity of mineral surfaces. This is followed by a discussion of theoretical approaches to mineral surfaces by Jack Tossell, in which models and methods for quantum mechanical calculation of the nature and properties of mineral surfaces are presented, with examples of state-of-the-art calculations. In Chapter 4, Neville Greaves gives an outline of synchrotron-based X-ray diffraction and spectroscopy techniques and their applications to surface mineralogy and related fields. Gordon Brown, George Parks and Peggy O'Day then discuss the application of these and related techniques and approaches to the study of sorption at mineral surfaces. The surface chemistry relevant to the dissolution of oxide and silicate minerals (and related materials) is outlined in Chapter 6 by Bill Casey. Chapters 7 and 8 deal with sulfide minerals, their electrochemistry and surface chemistry (Geoff Kelsall), and the surface chemistry relevant to flotation (Paul Richardson). Clay mineral surfaces are specifically treated in Chapter 9 by Robert Schoonheydt, and zeolite surfaces and reactivity in Chapter 10 by Alan Dyer. The applications of surface mineral science in mineral extraction and in catalysis are amongst the topics dealt with in more detail in these later chapters.

It is hoped that this text will prove of value to those engaged in teaching advanced undergraduate and graduate courses in the mineral sciences, that those actively involved in research on mineral surfaces will find it a useful summary of much current work, and that it will stimulate interest in this rapidly developing field within the earth sciences.

1.5 The future of mineral surface science

Mineral surface science is one of the most challenging new areas in the earth and environmental (as well as mineral) sciences. As will be seen from the chapters that follow this overview, experimental techniques and theoretical approaches are becoming available that should enable rapid advances to be made in this field. The specific challenges facing those involved in the study of mineral surfaces are:

1. the better description and characterization of surfaces and surface interactions at the molecular level;
2. the quantification of surface properties and processes occurring at surfaces; and
3. the modelling of surface properties and processes, and hence their prediction.

Beyond these specific goals are the challenges presented by incorporation of new models and concepts concerned with mineral surfaces at the molecular level into the overall understanding of Earth surface processes.

Appendix 1.A Mineral surface science publications

Books dealing with mineral surface science

Coyne, L.M., McKeever, S.W.S. and Blake, D.F. (eds) (1989) *Spectroscopic Characterization of Minerals and Their Surfaces*, American Chemical Society, Symp. Ser. 415.

Davis, J.A. and Hayes, K.F. (eds) (1986) *Geochemical Processes at Mineral Surfaces*, American Chemical Society, Washington, DC.

Hochella, M.F. Jr and White, A.F. (eds) (1990) *Mineral–Water Interface Geochemistry. Reviews in Mineralogy*, **23**, Mineralogical Society of America, Washington, DC.

Perry, D.L. (ed.) (1990) *Instrumental Surface Analysis of Geological Materials*, VCH Publishers, New York.

Stumm, W. (ed.) (1987) *Aquatic Surface Chemistry*, Wiley-Interscience, New York.

Stumm, W. (1992) *Chemistry of the Solid–Water Interface: Processes at the*

Mineral–Water and Particle–Water Interface in Natural Systems, J. Wiley and Sons, New York.

Review articles and chapters

Berry, F.J. (1985) Mineral surfaces and the chemical bond, in *Chemical Bonding and Spectroscopy in Mineral Chemistry* (eds F.J. Berry and D.J. Vaughan), Chapman & Hall, London.

Hochella, M.F. Jr (1988) Auger electron and X-ray photoelectron spectroscopies, in *Spectroscopic Methods in Mineralogy and Geology. Reviews in Mineralogy*, **18**, Mineralogical Society of America, Washington, DC.

Appendix 1.B Surface science techniques discussed or mentioned in this book

Techniques for determining surface microtopography

Scanning Electron Microscopy (SEM)
Transmission Electron Microscopy (TEM)
Scanning Tunnelling Microscopy (STM)
Atomic Force Microscopy (AFM)

Techniques for determining surface atomic structure

Low Energy Electron Diffraction (LEED)
Scanning Tunnelling Microscopy (STM)
Atomic Force Microscopy (AFM)
X-ray Standing Wave (XSW)

Techniques for determining surface chemical composition

X-ray Photoelectron Spectroscopy (XPS)
Auger Electron Spectroscopy (AES)
Ultraviolet Photoelectron Spectroscopy (UPS)
Scanning Auger Microscopy (SAM) or Auger Electron Microscopy (AEM)
Resonant Ionization Mass Spectroscopy (RIMS)
Rutherford Backscattering Spectroscopy (RBS)
Resonant Nuclear Reaction (RNR)
Surface Analysis by Laser Ionization (SALI)
Electron Tunnelling Spectroscopy (ETS) or Scanning Tunnelling Spectroscopy (STS)
Inverse Photoemission Spectroscopy (IPS)
X-ray Absorption Fine Structure (XAFS; also EXAFS and XANES)
Voltammetry (including Cyclic, and Linear Potential Sweep)

Rotary Ring Disc Electrodes (RRDE)
Fourier Transform Infrared Spectroscopy (FTIR)
Reflectance (optical) Spectroscopy
Nuclear Magnetic Resonance Spectroscopy (NMR, including MAS-NMR)

Appendix 1.C Other techniques employed in surface science

AED	Auger electron diffraction
AEPS	Auger-electron appearance potential spectroscopy
APS	Appearance potential spectroscopy
ASW	Acoustic surface-wave measurements
ATR	Attenuated total reflectance
BIS	Bremsstrahlung isochromat spectroscopy
CEMS	Conversion-electron Mössbauer spectroscopy
CIS	Characteristic isochromat spectroscopy
CL	Cathodoluminescence
COL	Colorimetry (IR, visible, UV, X-ray, and γ-ray absorption spectroscopy)
CPD	Contact potential difference (work function measurements)
DAPS	Disappearance potential spectroscopy
DTG	Differential thermal gravimetry
ΔH_{ABS}	Heat of absorption measurement
EELS	Electron energy loss spectroscopy
EHAS	Elastic helium atom scattering
EL	Electroluminescence
ELL	Ellipsometry
EMOSS	Emission Mössbauer spectroscopy
ENDOR	Electron nuclear double resonance
EPR	Electron paramagnetic resonance
ES	Emission spectroscopy
ESDI	Electron stimulated desorption of ions
ESDN	Electron stimulated desorption of neutrals
ESEEM	Electron spin echo envelope modulation
ESR or EPR	Electron spin resonance or electron paramagnetic resonance
EYEXAFS	Electron yield extended X-ray absorption fine structure
FD	Flash desorption
FDM	Field desorption microscopy
FDS	Field desorption spectroscopy
FEES	Field electron energy spectroscopy
FEM	Field emission microscopy
FIM	Field ion microscopy
FIM-APS	Field ion microscope atom probe spectroscopy
FYEXAFS	Fluorescence-yield extended X-ray absorption fine structure
FYXAS	Fluorescence-yield X-ray absorption spectroscopy

GDOS	Glow discharge optical spectroscopy
GIXS	Grazing-incidence X-ray scattering
HEED	High energy electron diffraction
HEIS	High energy ion scattering
HREELS	High resolution electron energy loss spectroscopy
IETS	Inelastic electron tunnelling spectroscopy
IIRS	Ion impact radiation spectroscopy
IIXS	Ion induced X-ray spectroscopy
ILS	Ionization loss spectroscopy
IMMA	Ion microprobe mass analysis
IMXA	Ion microprobe X-ray analysis
INS	Ion neutralization spectroscopy
IRAS	Infrared absorption spectroscopy
IRS	Internal reflectance spectroscopy
IS	Ionization spectroscopy
ISD	Ion stimulated desorption
ISN	Inelastic scattering of neutrons
ISS	Ion scattering spectroscopy
ITS	Inelastic tunnelling spectroscopy
LEIS	Low energy ion scattering
LMP	Laser microprobe
LS	Light scattering
MBRS	Molecular beam reactive scattering
MBSS	Molecular beam surface scattering
MOSS	Mössbauer spectroscopy
MS	Magnetic saturation
NEXAFS	Near edge X-ray absorption fine structure
NIRS	Neutral impact radiation spectroscopy
NS	Neutron scattering
PAS	Photoacoustic spectroscopy
PD	Photodesorption
PEM	Photoelectron microscopy
PES	Photoemission spectroscopy
PSD	Photon-stimulated desorption
ReflEXAFS	Reflection extended X-ray absorption fine structure
RHEED	Reflection high energy electron diffraction
σ(p)	Adsorption isotherm measurements
SC	Surface capacitance
SDMM	Scanning desorption molecule microscope
SEE	Secondary electron emission
SEELFS	Surface extended electron energy-loss fine structure
SERS	Surface enhanced raman spectroscopy
SEXAFS	Surface extended X-ray absorption fine structure
SFM	Scanning force microscopy

SI	Surface ionization
SIIMS	Secondary ion imaging mass spectroscopy
SIS	Surface infrared spectroscopy
SLEEP	Scanning low energy electron probe
SRS	Surface reflectance spectroscopy
STEM	Scanning transmission electron microscopy
SXAPS	Soft X-ray appearance potential spectroscopy
SXES	Soft X-ray emission spectroscopy
TE	Thermionic emission
TL	Thermoluminescence
XAS	X-ray absorption spectroscopy
XEM	Exoelectron microscopy
XES	Exoelectron spectroscopy
XRD	X-ray diffraction
XRSW	X-ray standing wave

References

Astruc, M. (1986) Evaluation of methods for the speciation of cadmium, in *Cadmium in the Environment* (eds H. Mislin and O. Ravera), Birkhauser Verlag, Basel, pp. 12–17.

Bakken, B.M., Hochella, M.F. Jr, Marshall, A.F. and Turner, A.M. (1989) High resolution microscopy of gold in unoxidized ore from the Carlin Mine, Nevada. *Econ. Geol.*, **84**, 171–9.

Bancroft, G.M. and Hyland, M.M. (1990) Spectroscopic studies of adsorption/reduction reactions of aqueous metal complexes on sulphide surfaces, in *Mineral–Water Interface Geochemistry* (eds M.F. Hochella, Jr and A.F. White), *Reviews in Mineralogy*, **23**, Mineralogical Society of America, Washington, DC, pp. 511–58.

Bautista, R.G. (ed.) (1984) *Hydrometallurgical Process Fundamentals*, NATO Conference Series (Series VI: Materials Science), Plenum Press, New York.

Bewers, J.M., Barry, P.J. and MacGregor, D.J. (1987) Distribution and cycling of cadmium in the environment, in *Cadmium in the Aquatic Environment* (eds J. Nriagu and J. Sprague), Wiley-Interscience, New York.

Bischoff, J.L. and Rosenbauer, R.J. (1989) Salinity variations in submarine hydrothermal systems by layered double-diffusive convection. *J. Geol.*, **97**, 613–23.

Brunner, P.H. and Baccini, P. (1981) Die Schermetalle, Sorgenkinde der Entsorgung? *Neue Zurcher Zeitung*, No. 70, 25 March.

Craig, J.R., Vaughan, D.J. and Skinner, B.J. (1988) *Resources of the Earth*, Prentice-Hall, New Jersey.

Force, E.R., Eidel, J. and Maynard, J.B. (eds) (1991) Sedimentary diagenetic mineral deposits: a basin analysis approach to mineral exploration. *Rev. Econ. Geol.*, **5**, 216 pp.

Franklin, J.M., Lydon, J.W. and Sangster, D.F. (1981) Volcanic-associated massive sulfide deposits. *Econ. Geol.*, 75th Anniv. vol., 485–627.

Hekinian, R., Faurier, M., Bischoff, J.L. *et al.* (1980) Sulfide deposits from the East Pacific Rise near 21°N. *Science*, **207**, 1433–44.

Hochella, M.F. Jr and White, A.F. (1990) Mineral–water interface geochemistry: an overview, in *Mineral–Water Interface Geochemistry* (eds M.F. Hochella, Jr and A.F. White), *Reviews in Mineralogy*, **23**, Mineralogical Society of America, Washington, DC, pp. 1–16.

Moore, J.N. and Luoma, S.N. (1990) Hazardous wastes from large-scale metal extraction: a case study. *Environ. Sci. Technol.*, **24**, 1278–85.

Nordstrom, D.K. (1982) Acid sulfate weathering. *Spec. Pub. No. 10. Soil Sci. Soc. Amer.*, Madison, Wisconsin, 37pp.

Seyfried, W.E. (1987) Experimental and theoretical constraints on hydrothermal alteration processes at mid-ocean ridges. *Ann. Rev. Earth Planet. Sci.*, **15**, 317–35.

Starling, A., Gilligan, J.M., Carter, A.H.C. *et al.* (1989) High temperature hydrothermal precipitation of precious metals on the surface of pyrite. *Nature*, **340**, 298–300.

Sverjensky, D.A. (1986) Genesis of Mississippi Valley lead zinc deposits. *Ann. Rev. Earth Planet. Sci.*, **14**, 177–99.

Wills, B.A. (1992) *Mineral Processing Technology*, 5th edn, Pergamon Press, Oxford.

Yasumura, S., Vartsky, D., Ellis, K.H. and Cohn, S.H. (1980) Cadmium in human beings, in *Cadmium in the Environment* (ed. J. Nriagu), Wiley-Interscience, New York, pp. 12–34.

CHAPTER TWO

Mineral surfaces: their characterization and their chemical, physical and reactive nature

M.F. Hochella, Jr

2.1 Introduction

The chemical reactivity of mineral surfaces has been studied for well over a century. Ebelmen (1847) and Daubree (1867) give particularly interesting accounts of some of the earliest mineral dissolution experiments. Liversidge (1893) gives an intriguing account of some of the first work on oxidation-reduction reactions at mineral–solution interfaces which was performed in the 1870s by Newbury, Skey and Wilkinson. In these early studies, and for a considerable time thereafter, the most fruitful way to study surfaces was by observing/measuring changes in the fluid (gas or liquid) composition in contact with the surface as a function of time, as well as to measure surface electrostatic charge. These methods were, and still are, very powerful and several sophisticated mineral–water interaction models with useful predictive capabilities have been developed from them (e.g. Davis and Kent, 1990; Parks, 1990; Stumm, 1992). To help to further advance this field of research, many surface-sensitive microscopic and spectroscopic techniques have been developed relatively recently. In addition, some techniques, generally used for bulk analysis, have been adapted for surface research. Most importantly, all of these methods have provided direct observations of either the mineral–water interface or the mineral surface before and after reaction. As a result of this, some previous models have been confirmed, while others have been discounted. In addition, one thing has become quite clear: mineral surfaces are complex in terms of atomic structure, composition, shape and, as a result, reactivity. Reactions that occur at complex geochemical interfaces proceed along a number of chemical fronts, and the most important of these (that which determines the overall rate) may be the fastest reaction, not the slowest. Whether one wishes to study mineral surfaces for the purposes of industrial heterogeneous catalysis, environmental geochemistry, or any number of other related fields, one should be cognizant of this complexity

Mineral Surfaces. Edited by D.J. Vaughan and R.A.D. Pattrick. Published in 1995 by Chapman & Hall, London. ISBN 0 412 56340 1

and of the tools, both theoretical and instrumental, needed for advanced characterization. Underlying this complexity, and perhaps the most important theme of this chapter, is the heterogeneous nature of mineral surfaces in terms of both chemistry and atomic structure. This heterogeneity is driven by the natural 'impurity' of minerals, as well as by the presence of complex microtopography (surface shape at the atomic/molecular scale). It is these two factors that are most often responsible for the heterogeneous reactive nature of mineral surfaces.

In this chapter, a detailed section is devoted to scanning probe microscopes and their application to mineral surface science, with specific emphasis on both vacuum- and air/liquid-based scanning tunnelling microscopy (STM) and electron tunnelling spectroscopy (ETS), as well as air/liquid-based atomic force microscopy (AFM; also called scanning force microscopy, SFM). In addition, many of the most important vacuum-based surface-sensitive techniques used to study mineral surfaces are briefly discussed. These include X-ray and ultraviolet photoelectron spectroscopy (XPS and UPS, respectively), Auger electron spectroscopy and scanning Auger microscopy (AES and SAM, respectively), secondary ion mass spectroscopy (SIMS) and related techniques, Rutherford backscattering spectroscopy (RBS) and resonant nuclear reaction (RNR) analysis, scanning and transmission electron microscopy (SEM and TEM, respectively), and low energy electron diffraction (LEED).

A general chemical and structural description of mineral surfaces is also given, based on the tools mentioned above and on the theories/methods needed to interpret the data from these tools. Finally, several specific examples of mineral surface reactivity are given, in some cases with observations down to the atomic level.

At the start, it is appropriate to answer one of the most often asked (and appropriate) questions in the application of vacuum-based instrumentation to mineral–water interaction chemistry. How can the information gained from surfaces characterized in ultra-high vacuum (UHV) be relevant to processes occurring at the mineral–water interface in nature? Certainly many of the most important surface-sensitive techniques must be performed in vacuum because of instrumental requirements. However, this is only one of the reasons for performing vacuum-based surface science for geochemical research, or any of the other fields in applied physics, materials science, chemistry and biochemistry that have major efforts in this regard. Vacuum environments offer a number of possibilities for fundamental scientific investigations relevant to non-vacuum conditions, but which are not possible in air or liquid:

- In UHV, a surface structure can be modified by the rehybridization (or redistribution) of dangling or other surface bonds leading to surface relaxations or reconstructions. The surface electronic properties are, therefore, intimately linked to the structural properties of the surface and its associated chemical properties. Various in-vacuum techniques can elucidate this connection between electronic structure and surface structure.

- Atomically clean surfaces can be prepared in vacuum to allow for the characterization of intrinsic surface properties, such as dangling bonds associated with coordinatively unsaturated surface atoms. Ultimately, both the local and long-range electronic structure of the surface can be investigated. It is this electronic structure that fundamentally controls the chemical reactivity of the surface.
- UHV provides for an environment in which the chemistry of a surface and surface reactants, as well as surface reaction kinetics, are relatively controllable. In addition, reactions can be started and stopped (by controlling reactant concentrations in vacuum) to allow for the systematic study of reaction progress.
- Practical application of quantum chemical theories, such as MO theory (Chapter 3), in the interpretation of the electronic structure and binding mechanisms of surface sites is strictly applicable only in UHV systems where 'chemical purity' is assured.
- The vacuum environment allows for the preparation/stabilization of reduced species on surfaces that may be unstable in air.

2.2 The tools of mineral surface science

2.2.1 Scanning tunnelling microscopy (STM)/electron tunnelling spectroscopy (ETS)

Despite the revolutionary contributions that STM/ETS has already made in other sciences (e.g. Güntherodt and Wiesendanger, 1992), its use has so far been rather limited in earth materials research. Therefore, a general review describing the capabilities of these methods is given here, including work involving surface atomic and electronic structure, followed by specific examples of their use in mineralogical and geochemical research.

Tunnelling techniques can be used for both imaging (STM) and spectroscopic (ETS) studies of surfaces. For the former, STMs can be used to scan areas as large as several thousands of square microns, and as small as several square nanometres. The application of STM to the study of surfaces over this wide range of magnifications is well established (e.g. Hansma *et al.*, 1988). However, perhaps the most dramatic scientific advance that this technology has provided is the ability to perform spectroscopic (ETS) studies on individual surface atoms (e.g. Avouris, 1990; Avouris and Lyo, 1991).

One of the current restrictions of STM/ETS is sample conductivity. Samples possessing bulk conductivity are always appropriate, but materials with relatively wide band gaps can also be tolerated if surface states provide adequate electronic paths near the Fermi level. Conventional STM/ETS on bulk insulators is not possible. It should be noted, however, that the future extension of tunnelling to insulating materials appears to be possible. Kochanski (1989) has reported a new method of operation described as nonlinear alternating current (AC) tunnelling microscopy; this technique has been used to observe nanometre-scale features on WSe_2 and CuO (both wide band gap semiconductors) and to stabilize an STM tip over Al_2O_3 (an insulator). Most recently, Stanick and Weiss (1992) have used this approach to image surfaces of insulating lead silicate glasses. While spectroscopic information has not yet been obtained, the frequency range of the alternating current used suggests that it may be possible to develop a local electron spin resonance (ESR) type probe using this approach.

(A) SURFACE ATOMIC STRUCTURE

Until the advent of STM, low energy electron diffraction (LEED) had been the primary method for determining surface atomic structure (section 2.2.11). However, from the first STM publications (Binnig and Rohrer, 1982; Binnig *et al.*, 1982a, b, 1983), it was apparent that tunnelling microscopy would be a revolutionary new surface structure tool. One of the most important applications of STM to date has been the determination of structures on reconstructed semiconductor surfaces. The most technologically important of these are reconstructed silicon surfaces (utilized in computer chip engineering and technology), which had been studied by LEED and other surface electron diffraction techniques for many years, and only partially understood. Two examples of such surfaces are the Si(001)-(2×1) and Si(111)-(7×7) (see also section 3.2). The latter reconstruction, obtained by heating an Si(111) surface to approximately 600 °C, is particularly complex, with the edges of the surface cell being seven times longer than the original surface cell repeats. Many efforts over the years failed to come up with a suitable model of this surface structure, and certainly none of the models could be proven. However, the first STM imaging studies of Si(111)-(7×7) led to a modified adatom model of this surface reconstruction (Binnig *et al.*, 1983). Further detailed STM studies later verified the dimer-adatom-stacking fault model of Takayanagi *et al.* (1985), which was proposed in the light of the original STM images and TEM/electron diffraction studies. In general, the interpretation of LEED results has now been dramatically facilitated by STM, which allows for the direct observation of surface atoms on an atomic scale (see below).

(B) SPECTROSCOPY AND ELECTRONIC STRUCTURE

Current–voltage (I–V) information measured with an STM, called electron tunnelling spectroscopy or ETS, is rich with information about surface electronic structure (e.g. Feenstra, Stroscio and Fein, 1987; Hamers, Avouris and Bozso, 1988; Hamers and Köhler, 1989; Avouris, 1990; Avouris and

Lyo, 1991). While the I–V curves are strongly dependent on the height of the tip above the surface and the tunnelling barrier height which depends on the electronic work functions of sample and tip, it has been demonstrated that reporting I–V data as a ratio of differential to total conductivity [$(dI/dV)/(I/V)$, equivalent to the function $d \ln I / d \ln V$], is approximately proportional to the surface density of electronic states near the Fermi level (Hamers, Tromp and Demuth, 1986; Feenstra, Stroscio and Fein, 1987; Hamers, Avouris and Bozso, 1988). When the sample is biased negatively relative to the tip, electrons tunnel out of occupied surface electronic states in the valence band of the sample, which gives a spectrum similar to that obtained by ultraviolet photoelectron spectroscopy (UPS); when the sample is biased positively relative to the tip, electrons tunnel into unoccupied surface electronic states in the conduction band of the sample, which gives a spectrum similar to that obtained by inverse photoemission spectroscopy (IPS). Electron tunnelling spectroscopy spectra can be collected in one of two ways. First, the tip can be positioned over a point of interest and the tunnelling current measured while the voltage is quickly ramped from some preset positive to negative (or negative to positive) voltage. Before the first I–V spectrum is run, the tip to sample distance is set by obtaining a preset tunnelling current chosen by the operator. If more than one voltage pass is desired for the purposes of averaging I–V spectra, a feedback loop is run to re-acquire the setpoint current in case the tip has drifted up or down. Lateral drift (thermal- as well as piezo-creep) may also be a problem, and little can be done except to minimize data collection time. The second method is known as current-imaging tunnelling spectroscopy (CITS) and, as the name implies, combines imaging and spectroscopy into one. In this case, the bias voltage is set to tunnel from, or into, a state of interest (as found by I–V spectra already collected), and an image is collected in the constant height mode to establish the spatial extent of such a state.

The unique contribution of ETS to the study of electronic properties is related to its high lateral spatial resolution. Whereas the conventional techniques for studying these properties (UPS, XPS, IPS, capacitance measurements, resistivity measurements, etc.) provide information averaged over a macroscopic portion of the sample or surface (10^{12} atomic sites or more), ETS can give electronic information on an atomic scale, essentially atom by atom. This was only possible from theoretical molecular orbital calculations before the development of STM (e.g. Munnix and Schmeits, 1983).

Before reviewing the STM work done on mineral surfaces, it is instructive to briefly describe a few of the remarkable breakthroughs made with tunnelling methods on important and well-studied, non-mineral surfaces. In particular, the work by Hamers and Köhler (1989) on the electronic properties of surface defects on silicon deserves special mention. Using the imaging capabilities of STM, these authors identified several types of characteristic defects on Si(001), including silicon divacancies (or dimer vacancies) and 'line vacancies' (or half-dimer vacancies) of up to four missing silicon atoms. The electronic properties of the different types of defects were measured by *I–V* spectroscopy and found to be radically different. The highest occupied molecular orbital (HOMO), or frontier orbital, associated with the dimer vacancies was found to be relatively deep in the valence band, while the line vacancies were found to have a high density of electronic states near the Fermi level. In addition, position dependent *I–V* measurements showed that the spatial extent of the electronic changes induced by the line vacancies was greater than a 10 Å radius, much larger than the geometric size of the defect. The electronic differences between dimer and line vacancy defects certainly must reflect variable reactivity characteristics on the surface. Measurements like these are some of the first steps in gaining a true fundamental understanding of the heterogeneous reactive nature of real surfaces.

Another very interesting example of ETS is provided by Avouris and Lyo (1991) who studied the adsorption mechanisms of H_2O and O_2 on an Si(111)-(7×7) reconstructed surface. During the STM imaging process, tunnelling spectra were measured at selected sites (e.g. where oxygen adsorbs). The STM images sampling the bottom of the conduction band (+2 V bias voltage) showed dark spots where adsorbed oxygen inhibits tunnelling because chemisorbed oxygen does not have states in this low-energy range. However, bright oxygen sites were also imaged. Measured tunnelling spectra taken over these bright spots were compared with calculated spectra based on a dimer adatom stack (DAS) model of the reconstructed Si(111) surface (Takayanagi *et al.*, 1985). Local density of states (LDOS) spectra of three different O–Si bonding mechanisms were calculated, using extended Hückel tight binding slab calculations (Lyo *et al.*, 1990). Only one of the configurations, where oxygen is inserted in an Si back-bond, shows the low-lying states on either side of the Fermi level (about −0.5 V and 0.5 V), which were observed in the tunnelling spectra and were responsible for the bright spots in the STM image. In this case, comparing LDOS calculations with tunnelling spectra not only elucidates the oxygen bonding configuration during the early stage of adsorption, but also provides information about the reaction mechanism.

(C) APPLICATION TO MINERAL SURFACES

A very important feature of STM in the study of mineral surface chemistry is its use as a spectroscopic tool. The results for defect characterization on Si(001) (Hamers and Köhler, 1989) have significant implications for surface reactions of mineral compounds such as oxides, where defects, especially oxygen vacancies (i.e. cation sites of reduced coordination), are often implicated as active sites in oxidation and reduction chemistry (i.e. part of a redox couple). Use of STM allows a direct spectroscopic characterization of the electronic properties of these atomic-scale point defects, and the range of their electronic influence over the surface. Also, imaging on the atomic scale provides a significant advantage over LEED in the study of structural characterization of semiconducting mineral surfaces. Oxides and sulfides, like many semiconductors, can undergo surface reconstructions when heated. These reconstructions occur to lower the surface energy that is common to many polar surfaces which have infinite Madelung energies in their unreconstructed and unrelaxed forms (Tasker, 1979). These reconstructions often exhibit complex low energy electron diffraction (LEED) patterns associated with thermal faceting (e.g. TiO_2(001) reconstructs to give {011} or {114} facets, depending on the annealing temperature; Firment, 1982). Other situations include reconstruction to form a coincident lattice at the surface because of changes in stoichiometry (e.g. the (4×1) reconstruction on SnO_2(110) is believed to be due to the formation of a reduced SnO(101) overlayer; Cox, Fryberger and Semancik, 1989). Scanning tunnelling microscopy would allow a direct determination of the surface structure in such circumstances.

To date, however, few studies of mineral surfaces by STM have been reported. There are reports of STM studies of several rutile (TiO_2) single crystal surfaces. For example, STM images of the rutile (001) surface in air have revealed the positions of surface titanium cations but not anions (Fan and Bard, 1990). Furthermore, this same surface has also been studied in air using tunnelling spectroscopy (Gilbert and Kennedy, 1989; Fan and Bard, 1990). A reduced rutile (110) surface has been imaged in UHV, and gives an indication of how complex the topography and electronic structure of non-stoichiometric mineral oxide surfaces can be (Rohrer, Henrich and Bonnell, 1990; Zhong, Vohs and Bonnell, 1992). A model involving ordered arrangements of crystallographic shear planes was proposed to explain the surface topography. Atomic resolution was obtained for the reduced (110) surface, although individual surface atoms were not imaged.

Recently, STM has been applied to certain minerals by earth scientists (Hochella *et al.*, 1989; Hochella, 1990; Eggleston and Hochella, 1990, 1991, 1992a, 1992b, 1993, 1994). This work has demonstrated that atomic resolution is possible on several common and geochemically important mineral surfaces. The use of tunnelling spectroscopy to study the band structure of clean and oxidized sulfide surfaces has also been demonstrated (Eggleston and Hochella, 1993). Identification of separate cation and anion features, distinguishable by differences in electronic structure, have also been accomplished (Eggleston and Hochella, 1992b).

Works by Hochella *et al.* (1989) and Eggleston and Hochella (1991, 1994) deal with the atomic identification of tunnelling 'peaks' in high-resolution images, a study of the oxidation of the galena (PbS) surface, and a study of the sorption/reduction of aqueous gold on galena, much of which is described in later sections of this chapter. The UPS-derived valence band electronic structure of PbS recorded in the literature (McFeely *et al.*, 1973;

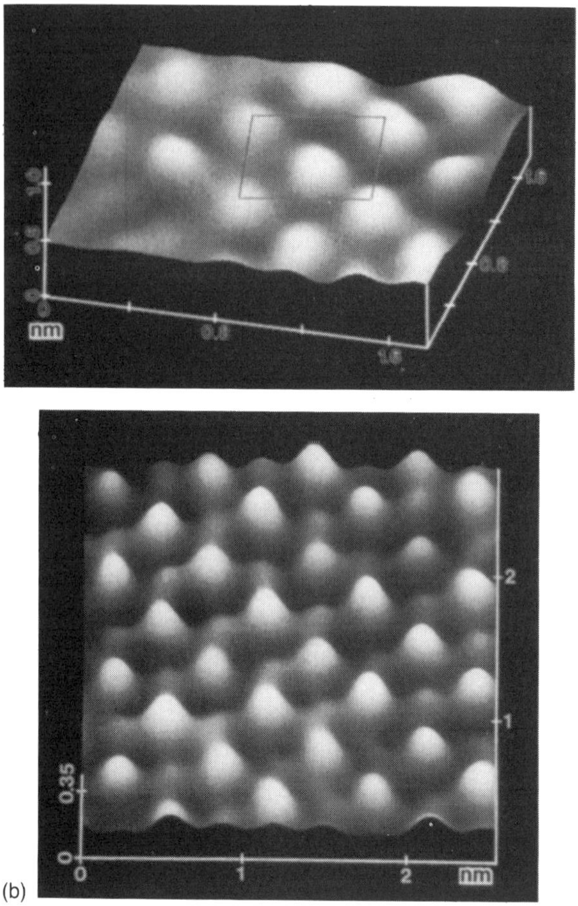

(a)

(b)

Figure 2.1 (a) STM image of the (001) surface of galena taken in the constant height mode at −600 mV sample bias and 2.9 nA tunnelling current. The peaks in the image are the tunnelling contribution for 3p electrons from individual sulfur atoms on the surface. The box outlines a surface unit cell, approximately 6 Å on an edge. (b) STM image of the same surface as in (a), but taken at +200 mV sample bias and 1.8 nA tunnelling current. The major peaks still have sulfur character, and the smaller peaks in between have lead character. All scales are in nanometres for both images. (From Eggleston and Hochella, 1990.)

Hagstrom and Fahlman, 1978; Evans and Raftery, 1982), as well as STM-derived valence and conduction band structure from our work (Eggleston and Hochella, 1994), provide interesting comparisons. The first populated states below the Fermi level are probably due to S 3p bands as seen in UPS spectra and also predicted from molecular orbital (MO) calculations (Hemstreet, 1975; Tossell and Vaughan, 1987). The STM spectra show that the first principal states in the valence band are near −400 mV, and in

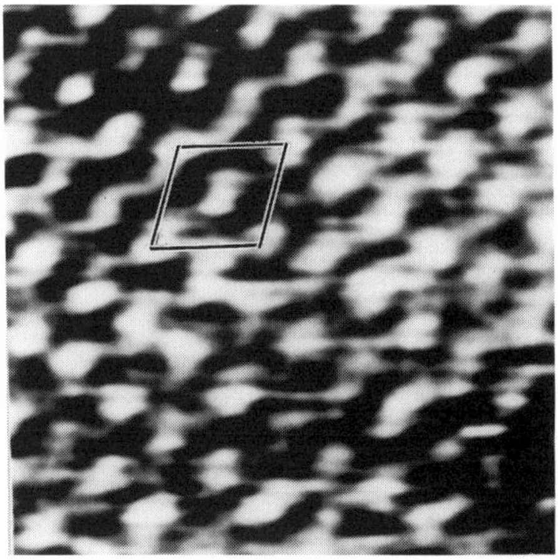

Figure 2.2 STM image of the (100) surface of pyrite taken in the constant current mode at a sample bias of −40 mV and 1.2 nA tunnelling current. The peaks are due to Fe 3p states at the top of the valence band. A unit cell, approximately 5.4 Å on a side, is shown, and the entire image is 31 × 31 Å. The cell is not square owing to instrument drift. (From Eggleston and Hochella, 1992a.)

keeping with the UPS and MO interpretations, this probably has S 3p character. Indeed, STM images taken with a similar bias generally show peaks that are consistent with S positions on the {100} surface (Figure 2.1(a)). Images taken at positive sample bias (Figure 2.1(b)) show peaks consistent with both S and Pb sites, reflecting the mixed S and Pb character at the bottom of the conduction band. Simultaneous image collection at both positive and negative bias shows that the dominant peaks of both images are in the same position, giving strong evidence that the dominant peaks in both images are, in fact, due to S.

Tunnelling microscopy of pyrite (FeS_2) is of particular interest because of its natural abundance. The study of the surface structure of pyrite {100} by Eggleston and Hochella (1992a) was performed in oil to avoid oxidation. Based on the molecular orbital calculations of Vaughan, Tossell and Johnson (1974) and Li *et al.* (1974), states in the top of the valence band have Fe 3d-like character (Figure 2.2), even though the surface is not in a vacuum (Jaegermann and Tributsch, 1988). Therefore, at small negative sample bias voltages, we were able to image Fe 3d surface states, in general agreement with the results of Fan and Bard (1991). One of our goals was to study the atomic/electronic structure between sites on terraces and those near steps. Figure 2.3 shows an image of an approximately 2.8 Å high step on the {100} surface, roughly half of the pyrite unit cell edge in this direction. The image near the step is clearly different from the regular arrangement of Fe contributions on terraces, as seen to the right of this image. In fact, some features

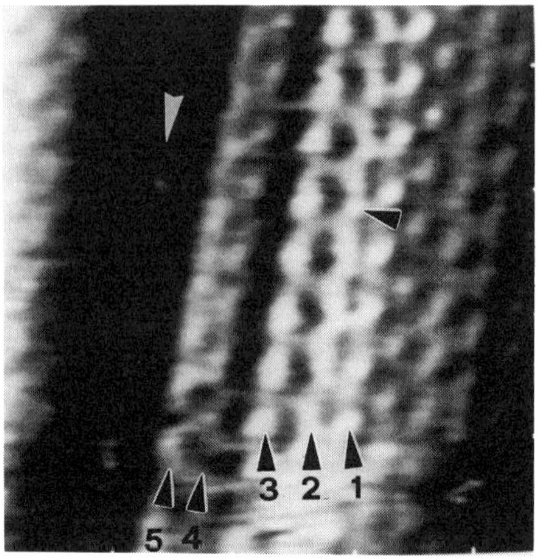

Figure 2.3 STM image of a step on a (100) pyrite surface, 44 × 44 Å, and taken under the same conditions used to collect the image in Figure 2.2. The step is 2.8 Å high, which is half of the pyrite unit cell edge. Rows 1–3, indicated by arrows, are brighter and spatially modified from terrace rows normally observed for imaging under these conditions (see rows to the right of row 1). Rows 4 and 5 are on the edge of the actual step. Unfortunately, the convolution of both atomic position and electronic structure information in this and other complex STM images does not allow for a unique interpretation of these data. More understanding will come when comparing these types of images with calculated model images. (From Eggleston and Hochella, 1992a.)

near the step might be attributed to sulfur sites. These perturbations could be caused by changes in electronic states due to band bending (i.e. alteration of occupied energy states at or close to the step relative to terraces), or due to atomic relaxation (shifting) at and near the step. Despite this ambiguity, Figure 2.3 is consistent with what we would expect theoretically (Zangwill, 1988); that is, there are both electronic and atomic perturbations at surface steps which vary significantly from terrace sites. It is also interesting to relate this work with that of Li *et al.* (1974) and Luther (1990), both of whom showed that the highest occupied molecular orbitals (HOMOs) on fresh {100} pyrite terraces were localized on Fe. This present study agrees with that, but also shows that HOMOs may be also localized on S sites near steps. It is interesting to speculate that the states imaged here with STM correspond to reactive frontier bands (somewhat equivalent to frontier orbitals), which play a major role in reactivity at these sites, and whose exact character depends on local surface structure and type of surface adsorbate.

A study of hematite {001} surfaces with STM has been particularly fruitful (Eggleston and Hochella, 1992b). Images collected at negative sample bias (and therefore sampling occupied states in the sample valence band) often show sites with 3 Å periodicity (Figure 2.4). This is as expected for a

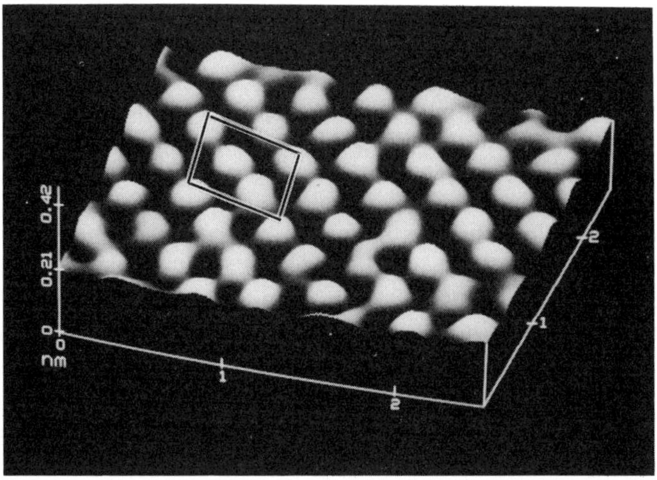

Figure 2.4 STM image of the (001) surface of hematite taken at −300 mV sample bias and 2.2 nA tunnelling current in the constant height mode. The peaks have 3 Å periodicity, as expected for the surface oxygen array. A surface unit cell is drawn in, approximately 5 Å on a side. All scales are in nanometres. (From Eggleston and Hochella, 1992b.)

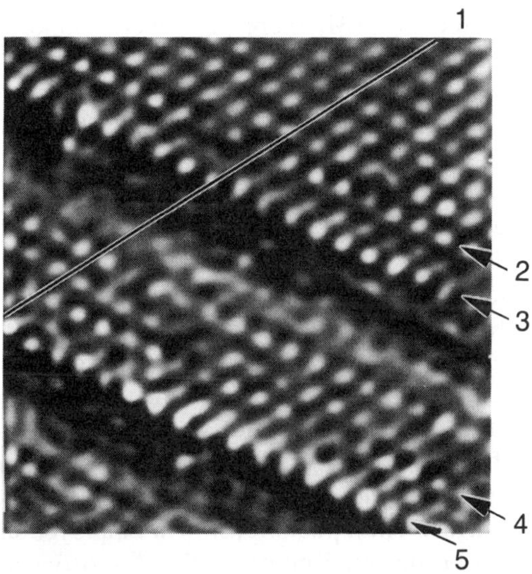

Figure 2.5 STM image of two monatomic steps on the hematite (001) surface taken at −300 mV sample bias and 1.5 nÅ tunnelling current in the constant height mode. The entire image is 54 × 54 Å. All peaks are due to O 2p-like states near the top of the valence band. The step edges (rows 3 and 5) have either a higher density of states in this energy range than oxygen a few rows away from the steps on terraces (to the right of rows 2 and 4), or these oxygens are physically higher compared with terrace oxygens owing to relaxation at this defect, or both. The line labelled 1 shows how a row of oxygens is offset as a step is crossed, exactly as one would expect from an ideal model of a stepped (001) hematite surface. (From Eggleston and Hochella 1992b.)

relaxed surface oxygen array, and is in agreement with previous LEED results (Hochella *et al.*, 1989). This also agrees with photoemission experiments, which suggest that the top of the valence band has, at least in part, oxygen character (for samples that have been exposed to air). In certain cases, Fe positions could be imaged at positive sample bias, though these images were variable and reflect the nature of the mixed states at the bottom of the hematite conduction band. In addition, STM clearly shows that these particular hematite surfaces have a high monolayer step density (in many places every 20–30 Å). It is evident that there is atomic/electronic reconstruction within two or three atomic dimensions on either side of every step (Figure 2.5), though unravelling the specific atomic position and electronic structure contribution is proving to be difficult and will require further study.

2.2.2 Atomic force microscopy (AFM)

Atomic force microscopy, also known as scanning force microscopy (SFM), is a recently developed technique first introduced by Binnig, Quate and Gerber (1986). It is a direct result of the invention of the STM. Although AFM is closely related to STM, there are several important fundamental differences. In particular, AFM is based on a 'mechanical' interaction between the sample and tip, as opposed to STM which is based on sample/tip electronic interaction. Therefore, insulators work with AFM just as well as conductors. Also in contrast to STM, the sample is typically rastered on the end of a single tube piezoelectric translator instead of the tip, and the tip is mounted on (or as part of) a flexible microcantilever which rides up and down as the sample moves under it. The most popular cantilever to date has been an Si_3N_4 triangular-shaped sheet, the flexible (thin) part being in the range of 100–200 μm long and approximately 0.5 μm thick; the tip on the end of this cantilever is formed by a crystalline termination of the silicon nitride which hangs below the cantilever plane. Many other cantilever/tip designs (both in terms of shape and material) are currently being tested. The vertical position of the cantilever/tip as it rides over the sample is monitored by means of a laser beam which reflects off the back of the cantilever and into a photodiode array. The optical lever arm employed allows deflections of the cantilever to be detected down to the Ångström level. Images are created by combining the scanner $x-y$ position (also controllable down to the Ångström level, as with STM) with the vertical deflection of the cantilever (called force, or constant height, mode) or the z-piezo response to keep the cantilever stationary (called height, or constant deflection, mode) (Figure 2.6). Exceptionally light tracking forces are desirable in order to minimize sample/tip distortion or damage. The minimum force obtainable for AFM performed in air is on the order of 10^{-7} N. This is due to adhesive surface tension forces between hydrated adventitious layers on both the sam-

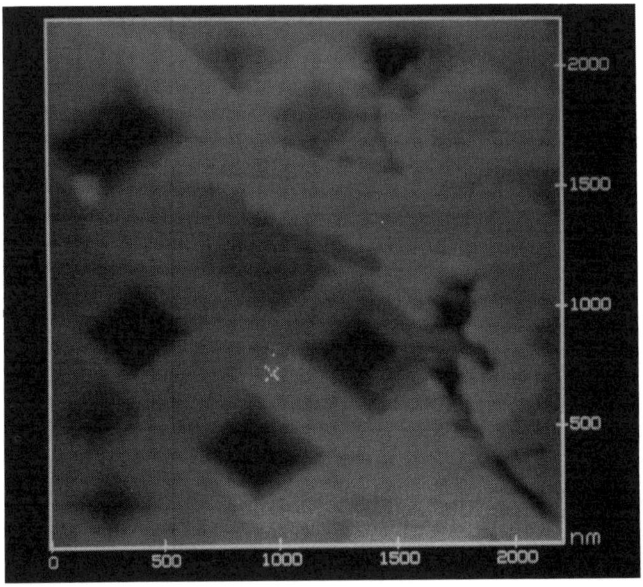

Figure 2.6 AFM image of a calcite rhombohedral cleavage surface after exposure to water for 20 s. The white cross is at the top of a 7 Å step, equivalent to 2 carbonate layers. Pits are up to 150 Å deep. (Image collected in height mode, constant deflection, by M.F. Hochella, Jr.)

ple and tip resulting from air exposure. Tracking forces in the 10^{-9}–10^{-10} N range can be relatively easily obtained in solution or in vacuum.

Although microtopography of surfaces can be observed down to the molecular scale (atomic scale in the *z* direction) with AFM, image details are dependent on the tip shape. The use of AFM to determine surface atomic structure is in a very early stage of development, and there is some controversy as to how an 'atomic' resolution image is created in the first place (e.g. Abraham and Batra, 1989). Tips are blunt relative to atomic dimensions, and the tracking forces would seem to exceed the structural integrity of an atomically sharp tip. Nevertheless, to date the AFM has demonstrated apparent atomic resolution on graphite (e.g. Binnig *et al.*, 1987; Albrecht and Quate, 1988), boron nitride (e.g. Albrecht and Quate, 1987; Binnig *et al.*, 1987) and a few insulating materials with simple surface structures (e.g. Hansma *et al.*, 1988). Until recently, despite numerous efforts and claims, AFM had not demonstrated true atomic resolution on more structurally complex surfaces such as albite (Hochella *et al.*, 1989; Drake and Hellmann, 1991), hematite (Johnsson, Eggleston and Hochella, 1991), calcite (Rachlin, Henderson and Goh, 1992; Dove and Hochella, 1993; Gratz, Hillner and Hansma, 1993), quartz (Gratz, Manne and Hansma, 1991), zeolites (MacDougall *et al.*, 1991) or several (001) surfaces of sheet silicates (Hartman *et al.*, 1990; Lindgreen *et al.*, 1991; Wicks,

Kjoller and Henderson, 1992). However, Ohnesorge and Binnig (1993) have demonstrated true atomic resolution on calcite using exceptionally small tip attractive force (10^{-11} N). It remains to be seen how attempts at atomic resolution imaging with AFM will proceed from here. Nevertheless, all of the above references are good examples of the applications of AFM to mineral surface science. It is the best technique for determining the shape of insulating surfaces down to at least molecular dimensions laterally and atomic dimensions vertically. The surface shape is a key to understanding aspects of mineral surface reactivity.

2.2.3 X-ray photoelectron spectroscopy (XPS)

X-ray photoelectron spectroscopy was the first vacuum-based surface chemical analysis technique to be developed (e.g. Siegbahn *et al.*, 1967). Because it is relatively straightforward to collect and interpret the data, and because it provides surface compositional, oxidation state and structural information, XPS has become the most widely used surface analytical tool today. Analysis depths range from a few to >100 Å, depending on the type of sample, the energy of the photoelectrons measured, and the instrumental set-up (e.g. Hochella, 1988; Hochella and Carim, 1988). Any surface that is compatible with moderate vacuums (as high as approximately 10^{-5} mbar) can be analysed, and semi-quantitative near-surface compositions can usually be calculated easily. The latest XPS technology provides for superior energy resolution (which provides for additional interpretation of oxidation and structural state information) and smaller analysis areas. Newly developed 'photoelectron microscopes' (some based on conventional hemispherical electron energy analysers with special entrance and exit optics, others on diverging magnetic fields and retarding field analysers – e.g. King, 1992) have already demonstrated 10 μm spatial resolution, with the potential for micron resolution. Commercially available systems now routinely analyse areas smaller than a few hundred microns in diameter.

In the XPS technique, soft X-rays strike a surface, ejecting photoelectrons from valence and core levels of the surface and near-surface atoms. An electrostatic electron energy analyser is used to measure accurately the kinetic energy, E_K, of the ejected photoelectrons. Photoelectrons that have not lost momentum on the way out of the sample have characteristic energies that can be used not only to identify the element from which they come, but also to determine the element's oxidation state and, in certain cases, structural details. X-ray photoelectron spectra are plotted as the electron binding energy E_B (the energy by which the electron was bound to its parent atom, which, simply speaking, is the incident photon energy minus E_K) versus signal intensity. The probability (cross-section) for photoelectron ejection from H and He atoms at the photon energies used for XPS are extremely low, and these are the only elements that cannot be detected with the technique.

Detailed reviews of XPS written from a mineralogical perspective are given in Hochella (1988) and Perry, Taylor and Wagner (1990); see also Tossell and Vaughan (1992). An example of the use of XPS in mineral surface research is given below in Figure 2.10 and associated text.

2.2.4 Auger electron spectroscopy (AES)/Scanning Auger Microscopy (SAM)

Electron-stimulated AES is the second most commonly used surface-sensitive spectroscopy after XPS, owing to its high spatial resolution capabilities. (The term SAM is often used to denote the spatial resolution capability of the technique, and is essentially synonymous with AES.) The surface sensitivity of the technique is equivalent to XPS as described above, although in many cases the Auger electrons that are most useful for analysis (i.e. from the most intense lines) have relatively low E_K and therefore originate from no deeper than a few monolayers beneath the surface of the sample. Analysis areas down to 300 Å in diameter are possible with commercially available equipment on non-insulators; sub-micron surface analysis has been demonstrated on insulators. Because insulators analysed with AES cannot be coated with a conductive film (as commonly used in SEM or electron microprobe analysis), owing to the surface-sensitive nature of the technique, charging problems may at first seem prohibitive. However, it is possible to balance the number of incoming (primary) electrons with the number of outgoing (primary backscattered, secondary and Auger) electrons, thus creating a charge neutral situation. Such a situation is possible when the primary beam voltage and the angle of incidence are carefully chosen (Hochella, Harris and Turner, 1986; Hochella, Turner and Harris, 1986). Another attribute of AES is the chemical state information that is available, much as for XPS. Unfortunately, AES is not often used for oxidation and structural state surface studies because of the ever-present possibility of electron beam damage, especially during high-resolution analysis when the beam current density can be very high. Another disadvantage of electron-stimulated Auger spectra is that they characteristically have a very high background (due to the number of backscattered electrons) and weak peak intensities. Therefore, it has become standard practice to differentiate the spectra electronically (with a lock-in amplifier) or mathematically (with a computer). The Auger peak position is conventionally taken at the minimum of the negative peak excursion.

Auger electrons may be generated when a core vacancy is created in an atom by any process (e.g. X-ray or electron bombardment). When this vacancy is filled by an electron from a higher level, the energy gained can be used to generate a photon (the basis of X-ray spectroscopy, such as X-ray fluorescence analysis), or this energy can be internally recombined to eject a still higher-lying electron, the so-called Auger electron. Note that owing to the number and configuration of electrons needed for Auger-generating transitions, H and He cannot be detected with AES. Lithium would not seem to be appropriate either, but 2s $\rightarrow$ 1s transitions have enough energy to eject valence electrons associated with lithium bonded to other atoms. As with XPS, if Auger electrons escape the material without undergoing an energy-losing collision event, they are characteristic of their parent element and their energy can be precisely measured by an electrostatic electron energy analyser.

It is the high spatial resolution, combined with the surface analytical capability of AES, that has permitted its use in distinguishing between true adsorption and surface precipitation in a number of cases. Precipitates, even those that are surface films as thin as a few monolayers and only a few hundred Ångströms across, can be detected. On the other hand, a distribution of sorbate in no more than monolayer thickness, which can be established by careful AES analysis, implies adsorption. It is not possible to distinguish between these two phenomena with indirect evidence, for example by following changes in solution composition (Sposito, 1986).

Reviews of Auger spectroscopy and microscopy applied specifically to geological materials can be found in Hochella, Harris and Turner (1986), Hochella, Turner and Harris (1986), Hochella (1988), Browning and Hochella (1990), and Mogk (1990); see also Tossell and Vaughan (1992). An actual example of the use of AES/SAM in mineral surface research can be found in section 2.3.1.

2.2.5 Secondary ion mass spectroscopy (SIMS)

Secondary ion mass spectroscopy has achieved popularity primarily due to the semiconductor industry, but its use in the earth sciences and other disciplines is increasing rapidly. In all forms of the technique, a mass spectroscopic analysis is performed on the surface of the sample, but because this is a destructive technique, depth profiles can be obtained several microns into the sample. The strengths of the technique lie in the fact that all elements can be analysed to trace levels (down to parts per billion in favourable cases), and isotopic ratios can be measured on the best instruments. Analysis with a few microns to sub-micron spatial resolution is possible, either by focusing the primary ion beam, or by utilizing spatially resolved ion optics which preserve the spatial integrity of the emitted ions throughout the flight path of the analyser. Insulating samples are often a challenge to analyse, but there are techniques available to overcome charging problems; quantitative analysis is possible, but generally difficult because of potentially complex primary ion–sample interactions.

The SIMS technique is performed by bombarding the surface to be analysed with an energetic ion beam (commonly used ions are Cs^+, O_2^+, and O^- at energies of 2–3 to usually no greater than 30 keV). The ion beam/surface interaction results in the ejection of elemental and molecular species from the top few layers of the surface, and <1% of these are charged species which can be extracted into the mass analyser, or spectrometer, for mass analysis. The mass analyser of the instrument can be a quadrupole, time-of-flight, or magnetic sector spectrometer.

Analysis via SIMS can be either the static or dynamic type. In static mode, primary ion beam current densities are extremely low ($<10^{-8}$ A cm^{-2}), allowing for analysis of (ideally) the top monolayer of atoms on a surface without its complete destruction. However, in the surface geochemical applications of SIMS, the dynamic mode is almost always used. Here, focused ion beams have current densities up into the A cm^{-2} range, and continuous analysis is achieved as the sample is sputtered. Elemental profiles to a depth of several microns are common, with a depth resolution of, generally, a few tens of Ångströms. Myhra *et al.* (1984) and Bancroft *et al.* (1987) used dynamic SIMS to study the leaching characteristics of titanite (sphene) and titanite composition glass as a potential host of radioactive waste. Nesbitt and Muir (1988) and Muir, Bancroft and Nesbitt (1989) were among the first to use dynamic SIMS to study the dissolution of common silicate minerals (mostly feldspars). In both groups of studies, depth profiles were obtained for all constituent elements (including H if present). These results, roughly in agreement with XPS/ion sputtering depth profiles, have been important in proving the formation of cation depletion layers and hydrogen infiltration during the dissolution of silicates.

Metson (1990) presents an excellent review of the SIMS technique from a geochemist's perspective, carefully explaining its advantages and disadvantages in dealing with earth materials. In addition, some of the most influential SIMS applications in the earth sciences to date are discussed.

2.2.6 *Surface analysis by laser ionization (SALI)*

Surface analysis by laser ionization has more similarities with SIMS than differences, although the differences make this a truly independent technique. The underlying concept behind SALI is to utilize the vast majority of the ions (99+%) sputtered from the sample surface that are neutral and cannot be mass analysed. Therefore, in this technique, a pulsed laser (passing just above and parallel to the sample surface) is used to ionize emitted neutral atoms or molecules for mass analysis. Depending on the wavelength and intensity of the laser, species coming off the surface can be universally ionized, thus reducing quantitative analysis problems, or they can be selectively ionized in order to simplify the mass spectra to an interpretable level. The former case has been used to analyse thin films of organic molecules on mineral surfaces (Tingle *et al.*, 1990; Tingle, Mathez and Hochella, 1991;

Tingle and Hochella, 1993; see also section 2.4.4). The latter case has been called resonant ionization mass spectroscopy or RIMS (e.g. Willis, Thonnard and Cole, 1989). This was the first technique available to systematically study the Re-Os isotopic system in rocks (e.g. Walker and Fassett, 1986; Walker *et al.*, 1989).

Whether universal or selective ionization SALI is used, the result is an extremely sensitive surface analytic technique, capable of obtaining analyses down to the part per million level on a single monolayer in the most ideal cases. The other major advantage of this 'post-ionization' technique is that the ionization step is completely separate from the desorption step. Therefore, the desorption step can be accomplished by heating, or by using an incident electron, ion, or laser beam. The desorption stimulus used can be chosen based on the type of material being analysed, the analysis desired, and the equipment available.

General descriptions of the SALI technique are given in Becker and Gillen (1984) and Schuhle, Pallix and Becker (1988). An example of the use of SALI in mineral surface research is given below in Figure 2.14 and associated text.

2.2.7 Rutherford backscattering (RBS)

Rutherford backscattering has been used only sparingly in mineral surface studies, in part because it requires a particle accelerator. Nevertheless, it is one of the few techniques that can produce non-destructive depth profiles. Disadvantages of the technique are that the results are difficult to interpret without some *a priori* knowledge of the surface chemistry of the sample, and also that the depth resolution of the elemental depth profiles is relatively poor, usually on the order of a few tens of Ångströms or perhaps more. Therefore, a compositional change within a few tens of Ångströms of the surface, for example, might go undetected.

This technique works by using a particle accelerator to generate a high-energy beam of helium (usually several MeV) which impacts on the surface of interest. The helium will backscatter from the target material as a convoluted function of the helium energy, the mass of the specific atom in the sample doing the backscattering, and the scattering angle. Elemental scattering cross-sections for He are known, so when the constituent elements and density of the target are known, a depth profile of these constituents in the near-surface of the sample can be established from the energy distribution of the backscattered He.

The RBS technique is described in detail in Chu, Mayer and Nicolet (1978), and examples of its use for the analysis of mineral surfaces can be found in Casey, Westrich and Arnold (1988) and in Casey *et al.* (1989). In the latter two references, Casey and others used RBS to characterize leached layers in labradorite feldspars that had been exposed to aqueous solutions of varying pHs. Under acidic conditions (pH 1–3), RBS clearly shows surface layers up to several hundred Ångströms thick that were depleted in sodium, calcium, and aluminum. Elastic recoil detection (ERD) analysis shows that there is insufficient hydrogen in this leached layer for charge balance. This suggests that the silica-rich surface of the labradorite has at least partially repolymerized, forming Si—O—Si bridging bonds and expelling hydrogen.

2.2.8 *Resonant nuclear reaction (RNR) analysis*

Although RNR analysis has not generally been used to study the near-surface chemistry of minerals, it has found a limited but important application in geochemistry. It has been used to depth profile for hydrogen in the near-surface of minerals. This has been particularly useful in the study of mineral dissolution and weathering processes (Petit *et al.*, 1987, 1989; Petit, Dran and Della Mea, 1990). This is despite the fact that, like RBS, the depth resolution is no better than several tens of Ångströms. Nevertheless, these studies were the first to show that silicates (specifically feldspars, olivines, and pyroxenes) can become protonated and/or hydrated to depths of several hundred Ångströms during dissolution reactions.

Hydrogen depth profiles can be obtained by taking advantage of the nuclear reaction $^{15}N + {}^{1}H \rightarrow {}^{12}C + {}^{4}He + 4.43\,MeV\,\gamma$ (where γ represents a gamma-ray emission) which can be induced by bombarding the sample with high energy ^{15}N ions. As ^{15}N penetrates the sample to deeper and deeper levels, it loses energy at a predictable rate. When the ^{15}N energy reaches exactly 6.385 MeV (the resonant energy in this case), there is a dramatic increase in the probability of ^{15}N interaction with hydrogen, and the above reaction will proceed. At ^{15}N energies higher or lower than this, the reaction probability drops off sharply. Therefore, if a ^{15}N energy of greater than the resonant energy for this reaction is used, the nitrogen will get to a certain depth before it obtains the resonant energy, at which point easily detectable 4.43 MeV gamma-rays will be produced, depending on the amount of H at that depth in the sample. In general, depth profiles are produced by varying the energy of the incoming ions and measuring the intensities of the characteristic gamma-rays which are emitted. From this, hydrogen depth profiles are constructed.

2.2.9 *Scanning electron microscopy (SEM)*

Scanning electron microscopes (SEMs) have been in common use since commercial introduction in 1965 and, with their relative affordability use of SEM is considered almost as routine as optical examination with a light microscope. Quite simply, SEM is probably the most universally useful technique available for the study of grain shapes and the physical characteristics of solid surfaces. It is easy to use, it has a large depth of field (similar to the human eye), and it has an enormous magnification range, capable of image resolution from light microscope capabilities down into the neighbourhood of 10–20 Å. The most advanced SEMs do not depend on thermionic emission (e.g. from a hot tungsten filament) for an electron source, but utilize field emission sources which make use of a strong electrostatic field gradient to pull electrons from a sharp, unheated metallic (usually tungsten) tip. The result is a highly focused, high-flux beam even at low accelerating voltages. Because of favourable beam characteristics even at low voltages, imaging insulating surfaces without a conductive coating is possible. In general, performing SEM without a sample coating has become more and more important as resolution capabilities have improved. At ultra-high magnifications, the surface texture of the conductive film, if present, may be observed and misinterpreted as the underlying sample surface.

Scanning electron microscopy is performed with a highly focused electron beam which is rastered over a sample surface with an accelerating potential of typically between 3 and 20 keV. These electrons interact with the sample, producing backscattered primary electrons, secondary electrons (those originating in the sample), and Auger electrons. The electron detector on the SEM is set up to measure the flux of low-energy secondary electrons (typically with energies less than 50 eV), which generally originate from the valence and conduction bands of near-surface atoms. At low beam accelerating voltages (0.5–3.0 keV), these secondary electrons are, for the most part, generated very close to the point of impact of the beam, allowing for potentially higher image resolution. Secondary electron generation is primarily a function of composition and the angle at which the incident beam strikes the sample (more secondary electrons are produced at glancing angles). This latter phenomenon and the shadowing effects on rough surfaces are together responsible for generating contrast in a typical SEM image. Lateral dimensions can be directly measured off a SEM photomicrograph. However, unlike AFM (section 2.2.2), SEM does not produce any direct height or vertical relief information.

The application of SEM to geological materials, especially when combined with an energy-dispersive X-ray spectrometer (for elemental information) and a backscatter detector (giving image contrast, even on a flat surface, as a function of average atomic number) may seem quite obvious. When backed by other systematic studies, SEM observations can be relatively diagnostic. Take for example the examination of clay minerals, which have particle (grain) sizes generally less than 2 μm. Henning and Storr (1986) have provided a large collection of SEM images of well-characterized clays. The study of an unknown clay with SEM goes a long way towards its identification and characterization when using such a reference. In association with this example, SEM is also invaluable in gaining an understanding of the distribution of clays in soils and rock, which may be critical in porosity and permeability studies.

Comprehensive reviews of SEM and related techniques can be found in the books by Goldstein *et al.* (1981) and Newbury *et al.* (1985). More recently, a review of SEM along with applications to geological materials has been written by Blake (1990).

2.2.10 Transmission electron microscopy (TEM) and transmission electron diffraction (TED)

Transmission electron microscopy is generally thought of as a 'bulk' electron diffraction and imaging technique. In this case, the bulk is in the form of a thin film (at least several hundred Ångströms thick) that high energy electrons (up to a few hundred thousand volts in energy for conventional microscopes) can pass through easily. However, standard TEM can also be used to study surface shape up to ultra-high magnifications by means of what has come to be known as the TEM decoration technique. The technique was developed by Bethge and Krohn (1965) and Krohn and Bethge (1977) as a way to use the TEM to routinely study surface morphology. The technique works by vapour-depositing a thin film of gold on to the surface of interest, which has been cleaned and heated. Next, the surface is carbon-coated. Finally, the original specimen is etched away with an acid, leaving the thin carbon–gold film, which is effectively a molecular-scale mould of the original surface and amenable to TEM imaging. An example of this technique, used to image molecular-scale growth spirals on kaolin group minerals, can be found in Sunagawa and Koshino (1975).

Transmission electron diffraction has also been used to study surface atomic structure. The majority of the electron diffraction originates from the bulk of the film, but if the surfaces of the film are reconstructed relative to the bulk, weak diffraction spots will be observable originating from these surfaces and they can be used to help 'solve' the surface structure. Its most famous use in this regard was the solution of the 7×7 reconstruction of the Si(111) surface (section 2.2.1). Takayanagi *et al.* (1985) first proposed what is now generally agreed as the correct model (the dimer-adatom-stacking fault model) for this reconstruction using TED. The STM images were important in verifying the model.

2.2.11 Low energy electron diffraction (LEED)

Before the advent of STM, LEED was by far the most popular generally applicable tool for determining the atomic structure of surfaces. It can be used on both conducting and insulating surfaces, it is sensitive to the first few monolayers only, and it can be used in a qualitative or quantitative sense (see below). In this technique, a low energy beam of electrons (usually between 30–200 eV) impinges on (and is perpendicular to) a surface. Primary electrons which are elastically backscattered (that is backscattered with the same energy as the primary beam) will undergo Bragg diffraction if the surface is atomically ordered (i.e. crystalline). The spatial distribution of these backscattered/diffracted electrons is viewed on a hemispherical fluorescent screen situated over the sample, where retarding grids repel inelastically backscattered (lower energy) electrons to keep the background in the diffraction pattern low. For qualitative LEED, this pattern is photographed and measured, from which the unit cell size and shape can be determined. In quantitative LEED, the diffracted electron beam intensities must be measured as the primary beam energy is varied. With this, it is possible to solve the atomic structure of the surface by coming up with the model that most closely accounts for the change in diffracted peak intensities versus primary beam energy. However, this is generally a complex and not always unambiguous task, especially for large surface cells.

An example of the use of qualitative LEED in mineral surface research can be found in Hochella *et al.* (1989). In this study, LEED was performed on the (001) surface of hematite. The spot pattern and spacings are those expected for a surface that has an Fe and O arrangement that is fundamentally the same as that found in the bulk (001) plane. In other words, LEED shows that this surface is not reconstructed. If relaxation of the surface atoms has occurred (this is most likely to occur perpendicular to the surface, but is also possible in the plane of the surface), the movement has been such that the surface unit cell dimensions match equivalent cell distances in the bulk. In fact, STM shows that the surface oxygens have relaxed on the (001) of hematite relative to the bulk such that they are all equidistant from one another (Eggleston and Hochella, 1992b), although, as determined by LEED, the surface cell dimensions do not change. Other examples of the application of qualitative LEED to mineral surface structure analysis can be found in Hochella *et al.* (1989, 1990) and Hochella (1990). A full description of surface structure analysis using LEED can be found in Clarke (1985) and Van Hove, Weinberg and Chan (1986).

2.3 The chemical and physical nature of mineral surfaces

The chemical properties (reactivity) of a bulk phase depend on its composition and atomic structure. The chemical properties of a surface depend on its composition, atomic structure and its microtopography (i.e. the morphology on the general dimensional scale of the chemical interactions that will take place there, usually from a few to perhaps a few tens of Ångströms). Surface atomic structure and microtopography are not entirely independent, but the distinction is important, and they are considered separately below.

2.3.1 Composition

Mineral surface compositions are rarely representative of the bulk. Only 'unreactive' surfaces (such as a gold surface), or one created by fracturing in ultra-high vacuum, have the potential to be clean. Otherwise, mineral surfaces exposed to air or an aqueous solution immediately take up hydrogen, oxygen, usually carbon, and sometimes nitrogen (from O_2, H_2O, CO_2, hydrocarbons, and N_2 and/or NO_x in the air or solution). This is generally known as 'adventitious' material. Often, adventitious molecules in the first layer on the surface (the first monolayer) dissociate during adsorption. For example, infrared and ultraviolet photoelectron spectroscopies show that the first monolayer of water molecules on an oxide surface dissociate (Henrich, 1985; Kurtz and Henrich, 1987; Parks, 1990; and references therein). For example, for an Si—O—Si linkage on a silica surface, this results in the formation of surface silonal groups: $Si{-}O{-}Si + H_2O \rightarrow 2(Si{-}OH)$. The build-up of these and other adventitious materials 'passivates' the clean, highly reactive surface very quickly. Excluding gross contamination, adventitious materials of the type mentioned above may not obtain a thickness of more than a few to several Ångströms. Generally, in this case, only the first one or two monolayers of adventitious material are considered to be important to the reactive nature of the mineral surface because these atoms/molecules are interacting directly with the top layer of atoms on the mineral.

Besides the ubiquitous nature of adventitious influences on mineral surfaces, compositional modification can also result from sorption of other species, or the desorption of components in the near-surface region of the mineral. These sorption and desorption processes may affect the top layer of atoms, but they also have the potential to modify the mineral to a considerable depth, owing to near-surface solid state diffusion.

Perhaps the most studied aspect of mineral surface modification is the sorption of species from aqueous solutions (e.g. Brown, 1990; Davis and Kent, 1990; Schindler, 1990). The opposite process has also been widely studied. Chemical weathering of minerals generally results in a modification of the surface composition in the form of a leached layer. The thickness of this layer can be a few Ångströms, or it can be several hundreds to thousands of Ångströms thick, or even deeper (Petit *et al.*, 1987; Schott and Petit, 1987; Casey, Westrich and Arnold, 1988; Casey *et al.*, 1989; Mogk and Locke, 1988; Nesbitt and Muir, 1988; Hellmann *et al.*, 1990).

One aspect of mineral surface composition analysis that is often neglected is lateral heterogeneity. This situation can develop through either sorption or desorption processes. Evidence, both direct and circumstantial, for surface compositional heterogeneity due to sorption can be found throughout the literature (two examples, galena oxidation and gold sorption, and uranium sorption at a sulfide–water interface, are given in section 2.4). It is clear from these types of studies that various sites on a surface have different reaction potentials with, for example, species in aqueous solution. Such arguments are not as obvious for dissolution phenomena, but high-resolution surface analysis of dissolving mineral surfaces show that dramatic heterogeneities can be present even in these cases. For example, we have mapped the surface Ca concentration (depth of analysis approximately 10 Å) on a relatively flat plagioclase surface after dissolution under hydrothermal conditions (DI water at 350 °C for 57 days) using AES/SAM with a lateral resolution of 0.25 μm (Hochella, 1990). While a quantitative assessment is difficult, some areas show only slight leaching of Ca, while others have very little Ca left in the near-surface region. At least in this case, the style of dissolution varies, depending on a number of factors, including the variations in surface composition of the pre-dissolved surface, the starting microtopography of the surface, and the variation of the flow of fluids over different parts of the surface. All these must be considered in a complete dissolution model of any mineral.

2.3.2 *Microtopography*

The appearance of a surface as seen with the unaided eye can be highly deceiving. For example, the (001) surface of specular hematite or galena may appear to have mirror-like flatness (which is enhanced by their metallic lustre), yet on the molecular level, these surfaces may be very rough. Surfaces may even appear flat when viewed using an optical microscope or an SEM, yet they may still have considerable roughness, with consequences for the reactivity of the surface. The most useful tools for characterizing microtopography are the scanning probe microscopes (sections 2.2.1, 2.2.2). In general, it is becoming more and more apparent that natural mineral surfaces have very high roughness factors (Sunagawa, 1987; Hochella, 1990). Even cleavage planes described as perfect may have atomic steps every 20–30 Å (e.g. Eggleston and Hochella, 1992b).

Models for the microtopography of surfaces have been developed from many studies over the last few decades (e.g. Somorjai, 1981). The general

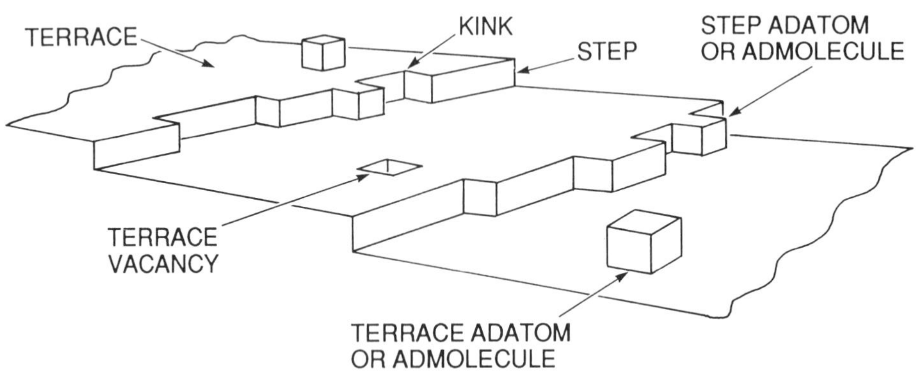

Figure 2.7 Schematic model of the microtopography of a solid surface. Blocks may represent a single atom or some atomic group or molecule. Features depicted are briefly defined in the text.

model of surface shape that has been developed from sorption, growth, dissolution and heterogeneous catalysis studies includes flat areas, called terraces, which are separated by steps (Figure 2.7). Steps can be from one to many atomic layers high. Kink sites appear where a step changes direction twice in a short distance along itself. An atom added to a step or a terrace is called a step or terrace adatom, respectively. Likewise, an added molecule is call an admolecule. An atomic- or molecular-sized hole in a terrace is called a vacancy. It is interesting to note that microtopography studied with phase contrast optical microscopy, TEM, STM and AFM have generally supported this model. However, STM and AFM are adding tremendously to our knowledge of surface shape by revealing the frequency and spatial distribution of these important features.

Atoms at the base of steps or underlying terrace vacancies have a higher coordination (with neighbouring surface and near-surface atoms) than any other surface site. Terrace atoms have lower coordination. At the top edge of a step, the number of nearest-neighbours is reduced again, and atoms at the outer corners of kink sites have even fewer neighbours. Terrace adatoms generally have the fewest surface nearest-neighbours of all the different surface sites. A number of vacuum and non-vacuum based studies have generally shown that surface sites with lower coordination (i.e. lower number of nearest neighbour atoms in the substrate) are the most chemically reactive (e.g. Christmann and Ertl, 1976; Iwasawa *et al.*, 1976; Somorjai, Van Hove and Bent, 1988). Unfortunately, none of these detailed studies has been performed on a mineral surface. However, this factor of the influence of surface coordination on the reactivity of a site is still emphasized in geochemical modelling studies of fluid–rock interaction (e.g. Dibble and Tiller, 1981; Steefel and Van Cappellen, 1990).

2.3.3 Atomic structure

It is typical for the atoms on surfaces, and sometimes those as deep as several layers down, to assume positions different from their equilibrium positions in the bulk. These structural differences, if they exist, can be subtle or dramatic. In general, mineral surfaces will not be reconstructed unless they have been heated to a few hundred degrees or more. For a surface exposed simply by fracture, the surface atomic structure is likely to be only relaxed (not reconstructed) from an equivalent plane in the bulk, especially if it is a low-symmetry structure (e.g. a feldspar surface; Hochella *et al.*, 1990). Higher symmetry structures (e.g. the rocksalt structure of galena) may show no lateral relaxation at all (Hochella *et al.*, 1989). However, in all cases, some relaxation perpendicular to the surface would be expected. Specifically, most clean surfaces (i.e. in ultra-high vacuum) relax inward, toward the bulk, with the spacing between the first and second atomic layers being reduced by up to 15% (Somorjai, 1981, 1990, and references therein). By eliminating the dangling bonds, these surface layers will again expand, sometimes exceeding what they are in the bulk, depending on what is sorbed to the surface.

Another generally neglected aspect of mineral surface chemistry is the structural effect that sorbates have on a mineral surface. It is generally known in the vacuum-based surface science literature (e.g. Somorjai, 1990, and references therein) that substrate surface atoms that are nearest to the sorbate may shift in position to better accommodate the sorbed species. This generally occurs only when the sorbate–surface interaction is strong, i.e. the sorbate is chemically reactive with the surface and strong bonding occurs. Nevertheless, it is dangerous to think only of the sorbate as being dynamic, and the substrate as being static.

2.4 Reactivity of mineral surfaces

In the following sections, examples of mineral surface reactivity from various experimental geochemical systems are given. In the first three examples, techniques that allow for high spatial resolution surface analysis/observation show the chemical heterogeneity that we suggest is the rule rather than the exception in mineral surface reactivity. High spatial resolution techniques are not utilized in the last example given below; nevertheless, this is an excellent example of the heterogeneous catalytic nature of mineral surfaces.

2.4.1 Galena oxidation and gold sorption

There is a substantial amount of interest in the fundamental oxidation processes of sulfides and the consequences of these reactions in terms of acid mine drainage and aqueous metal concentrations in natural waters (e.g. Nordstrom, 1982; Morse *et al.*, 1987), ore deposit formation (e.g. Starling *et al.*, 1989; Bakken, 1990; Bancroft and Hyland, 1990, and references therein; Bakken, Fleming and Hochella, 1991; Eggleston and Hochella, 1991), and ore processing/mineral beneficiation (e.g. Chapter 8 and references therein). Galena is a particularly convenient sulfide to study, particularly with STM, because it is relatively slow to oxidize (not as air-sensitive as pyrite), it has perfect cleavage, providing relatively flat surfaces of known orientation, and it has a simple crystal structure (NaCl-type). Eggleston and Hochella (1990, 1991, 1993) use STM to propose atomistic models for the oxidation of this surface and the sorption/reduction of aqueous gold. A discussion of the oxidation and sorption studies follows.

When a surface sulfur oxidizes, its tunnelling image is extinguished (under the instrumental conditions of our scans); that is, the peak due to the surface sulfur becomes what looks like a hole in the image (Figure 2.8). Therefore, it is easy to watch the progression of the oxidation reaction on the surface. The first thing that is readily apparent is that surface S sites do

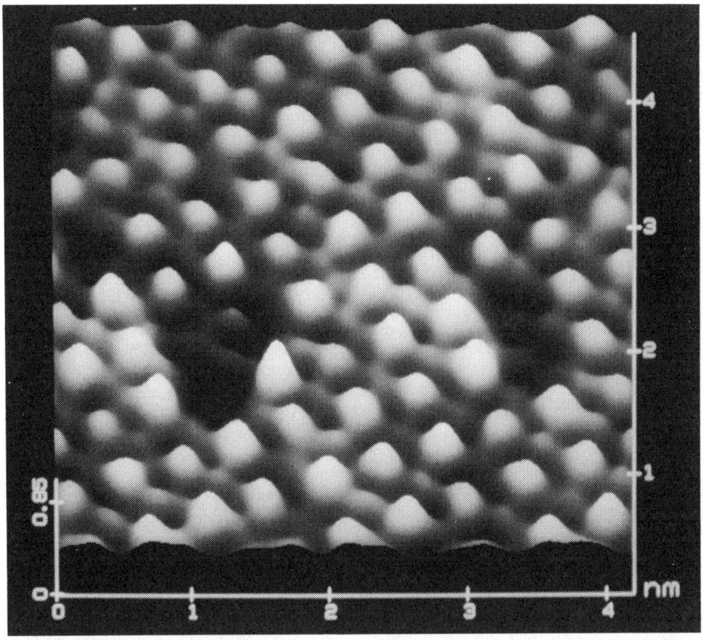

Figure 2.8 STM image of the (001) surface of galena taken after exposure to water for 1 min. The conditions of image collection were the same as in Figure 2.1(b). The high peaks represent surface sulfur, and the low bumps in between lead atoms. The apparent vacancies are oxidized sulfur sites; the tunnelling current to these sites is very low. (Image collected by C.M. Eggleston.)

not randomly oxidize, but that oxidized regions tend to nucleate and grow. Secondly, the boundaries of the oxidized patches are almost always along [110] directions. We believe that this is due to the fact that an unoxidized S across a [110] front has only one nearest-neighbour oxidized sulfur, while an unoxidized S across a [100] front, for example, has two nearest-neighbour oxidized sulfurs. Perhaps due to this difference, [100] fronts are fast-moving and obliterate themselves; [110] fronts are slow-moving and become dominant. This is a particularly pleasing example of the power of STM imaging, allowing us the freedom to bypass potentially ambiguous models and obtain results quickly and directly.

For the Au sorption studies, both fresh and pre-oxidized galena surfaces were exposed to 3 p.p.m. Au(III) solutions (added as $KAuCl_4$) for several seconds to several days (Eggleston and Hochella, 1991, 1993). Sorbed gold (in the form of minute islands) were readily apparent in STM images (Figure 2.9(a)). The presence of metallic gold was confirmed by the collection of tunnelling spectra on the islands; these spectra matched those taken on gold foil (Eggleston and Hochella, 1993). This has also been confirmed by XPS

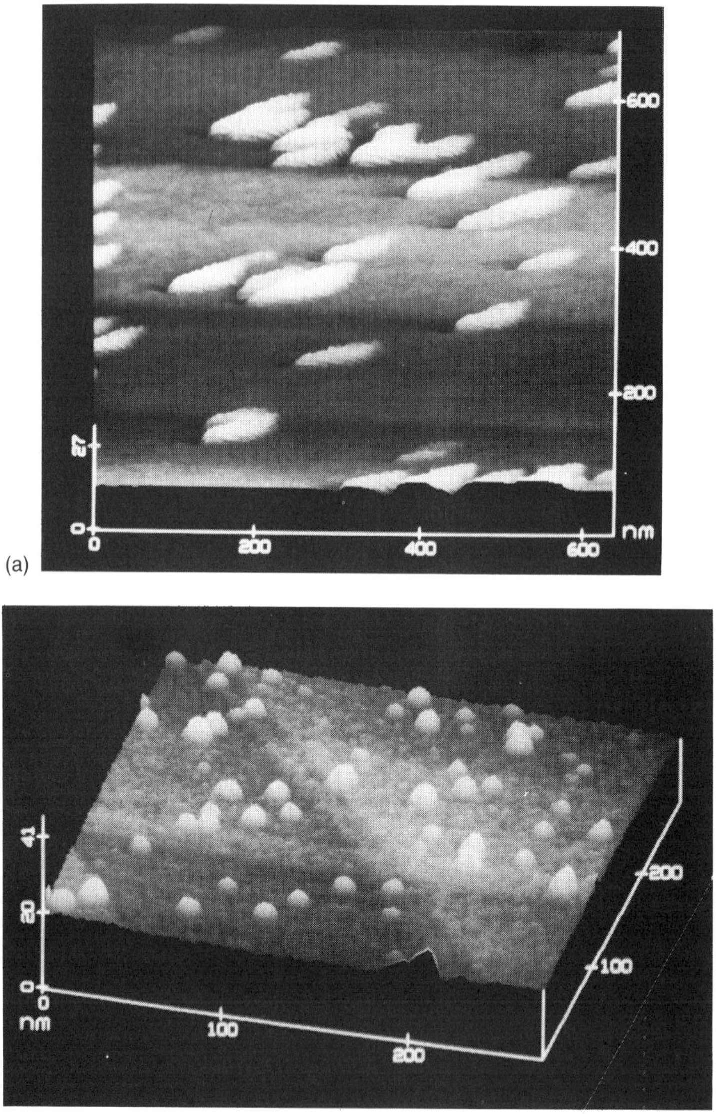

(a)

(b)

Figure 2.9 (a) STM image of gold islands growing on a galena (001) surface in the [110] direction. This galena sample was in contact with a 3 p.p.m. Au(III)-chloride solution for 1.5 h. Image collected in constant current mode at +200 mV sample bias (by C.M. Eggleston). (b) STM image of gold islands on a pre-oxidized galena surface. Otherwise, the imaging conditions and the sample handling were the same as in (a).

work (Bancroft and Hyland, 1990). Atomic resolution imaging on the gold islands, as well as on the galena just away from an island, shows that there is no obvious structural registry between the two. Instead, we have concluded that there is a direct coupling of gold reduction and sulfide oxidation that controls initial island growth on unoxidized surfaces (the islands have boundaries in [110] directions; see above). To further support this conclusion, on pre-oxidized surfaces, gold islands grow more slowly by a factor of 3 and show no preferred growth orientation (Figure 2.9(b)).

2.4.2 Uranium at the sulfide–water interface

Work reported in Wersin *et al.* (1994), and summarized here, deals with the sorption/reduction of uranyl ($U(VI)_{aq}$) on pyrite and galena surfaces. The understanding of the mobility of uranyl in groundwater is important in understanding certain types of uranium ore deposits and in long-term safety aspects of high-level radioactive waste disposal. In particular, the mobility of uranium depends critically on immobilization processes such as sorption to solid particles and reduction of U(VI) to U(IV) (e.g. Morse and Choppin, 1991). Although certain bacteria have been shown to be effective uranium reducers (e.g. Lovley *et al.*, 1991), sulfide surfaces also provide a convenient redox pair and no experimental work has been available in this area until now.

Sorption experiments were performed with powders and single crystals of galena and pyrite at room temperature, pHs generally from 5 to 6, U concentrations up to 1000 p.p.m., and exposure times from a few hours up to 9 days. Great care was taken to perform all experiments under anoxic conditions. Under the assumption of homogeneous uptake of U on the sulfide surfaces, both solution and XPS data independently yield estimates of 1–2 monolayers of U coverage on the fine powders ($<40\,\mu m$). However, AES/SAM results prove that this assumption is not valid. Inhomogeneous distribution of the U is observed in the case of both pyrite and galena. In addition, after careful calibration and standardization of our XPS instrument, the U4f spin-orbit split photopeak pair clearly indicates a mixed U(VI)–U(IV) surface complex/oxide, probably similar to U_3O_8 (Figure 2.10). Supporting evidence comes from FTIR, used in the diffuse reflectance mode, which shows the presence of a mixed oxidation state. Finally, experiments designed to study the kinetics of this process show that the oxidation step is rapid on both galena and pyrite, but that the sorption process is slow.

Specifically for galena, XPS shows that sulfur on the surface is oxidized to a polysulfide, similar to that seen for gold sorption/reduction on galena (Bancroft and Hyland, 1990). Auger electron spectroscopy/scanning Auger spectroscopy shows that flat cleavage surfaces of galena sorb little or no uranium, but on rough surfaces, 500 Å wide acicular uranium oxide precipitates abound. For pyrite, at a pH of 5.2–5.5, the reaction products

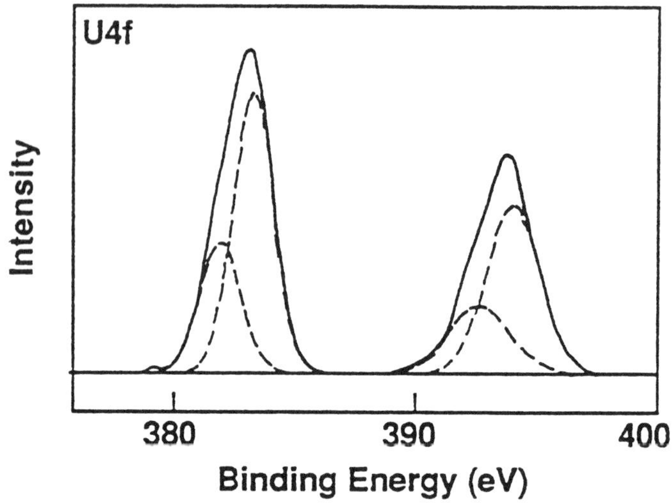

Figure 2.10 Uranium 4f XPS spectrum of a galena sample exposed to 2 mM uranyl nitrate solution. The main peak (solid curve) at lower binding energy is the U $4f_{7/2}$ line, while the higher binding energy peak is the U $4f_{5/2}$ line. Peak components (dashed curves) are for U(IV) (to the low binding energy side of each peak) and U(VI) (to the high binding energy side of each peak). (From Wersin *et al.*, 1994.)

observed at the surface are very similar to those observed on galena. Examination with AES/SAM reveals that these oxidation-reduction products are inhomogeneously distributed over the surface, with certain areas of the pyrite surface seemingly completely unreacted. At higher pHs (around 6), XPS and AES/SAM indicate both polysulfide formation and Fe(III) oxide formation, and XPS and FTIR indicate a higher U(VI) component compared to lower pH samples. It is clear that at this point two U-species are present, an unreduced U(VI) oxide associated with the secondary Fe(III) oxide on the pyrite surface, and a mixed U(IV)–U(VI) oxide associated with a surface Fe-polysulfide.

In summary, it is clear that sulfide minerals are efficient scavengers of soluble uranyl. Further, sulfide surfaces act as reductants, resulting in precipitation of insoluble reduced uranium oxides, except at higher pH (>6) for pyrite. In this latter case, unreduced uranyl is sorbed by Fe(III) oxides forming on the oxidizing pyrite surface. This is important in natural systems; reduced U is not significantly affected by changes in pH, whereas unreduced U(VI) is strongly dependent on pH variations and can be easily re-released.

2.4.3 Calcite precipitation and growth, and inhibition of growth

Because the kinetics of carbonate precipitation, growth and dissolution are fundamental to our understanding of many natural and anthropogenic geochemical systems, there are now numerous rate equations describing these phenomena for a wide range of conditions (e.g. for nucleation, Bischoff, 1968; Goodarz-nia and Motamedi, 1980; Reddy, 1986; Kralj, Brecevic and Nielsen, 1990; for growth, Nancollas and Reddy, 1971; Berner, 1975; Reddy, Plummer and Busenberg, 1981; Mucci and Morse, 1983; Inskeep and Bloom, 1985; Busenberg and Plummer, 1986; Compton and Daly, 1987; Reddy, 1988). An examination of the studies reporting carbonate growth kinetics shows that the proposed mechanisms and rate laws are constructed largely from experiments which monitor changes in the solution chemistry with time, occasionally in conjunction with examination of the initial/final run products by microscopic and spectroscopic methods. It has recently become possible to observe directly reaction processes occurring at a mineral–solution interface in real time using AFM (e.g. Hillner *et al.*, 1992; Gratz, Hillner and Hansma, 1993). We have attempted to improve our knowledge of calcite precipitation and growth by using AFM to observe near-equilibrium calcite growth processes in solutions of known composition and saturation state in real time (Dove, Hochella and Reeder, 1992; Dove and Hochella, 1993). Interestingly, we observe that when solution saturation with respect to calcite is >1–2, precipitation always begins with the formation of surface nuclei (Figure 2.11). These nuclei spread and eventually coalesce. Under the conditions of our experiments, only after nearly 2 h is there a transition to a mechanism resembling spiral growth. Steps (along which growth occurs) are typically between 1 and 3 calcium carbonate monolayers in height.

The influence of phosphate was also examined and observations

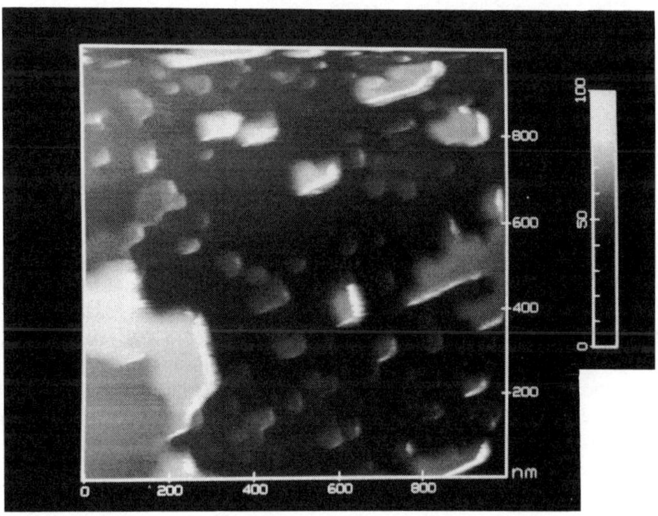

Figure 2.11 AFM image of the birth and spread of calcite nuclei on a calcite crystal surface taken in solution. Height of nuclei range from 6 to 30 nm. See text for additional information. (From Dove *et al.*, 1992.)

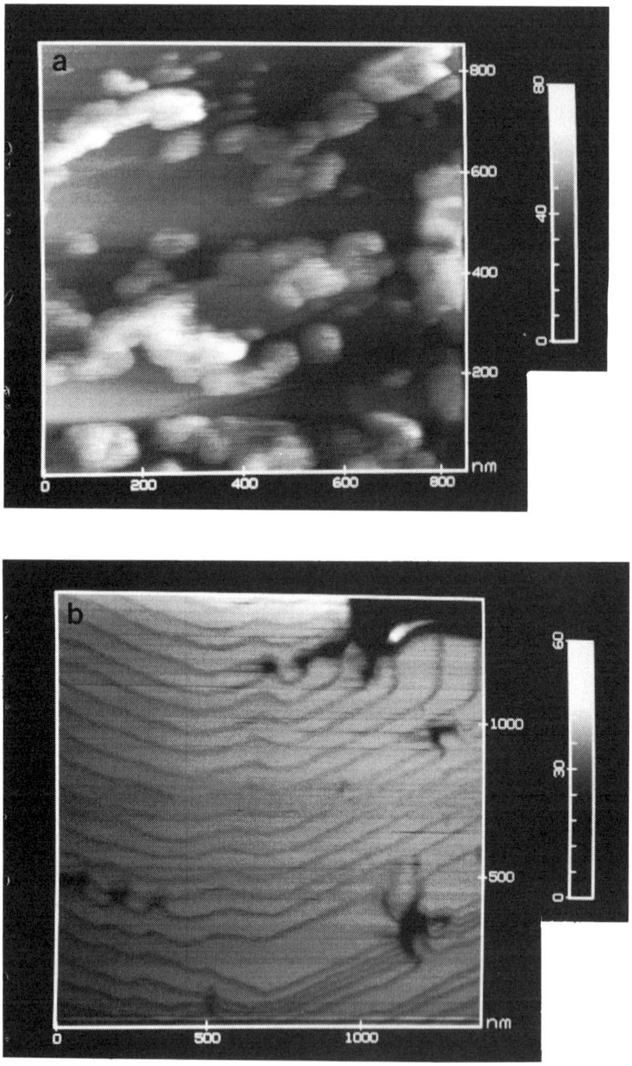

Figure 2.12 AFM image of phosphate poisoning of calcite growth taken in solution. (a) Anhedral nuclei caused by the introduction of phosphate during early stages of growth. (b) Layer growth of phosphate-free system shows normal smooth and relatively straight steps. Step relief is generally no more than a few carbonate units in height (6–10 Å). (c) After introduction of phosphate to the growth solution in (b), step advance is dramatically reduced, steps become jagged, and eventually growth stops all together. See text for additional information. (From Dove *et al.* 1992.)

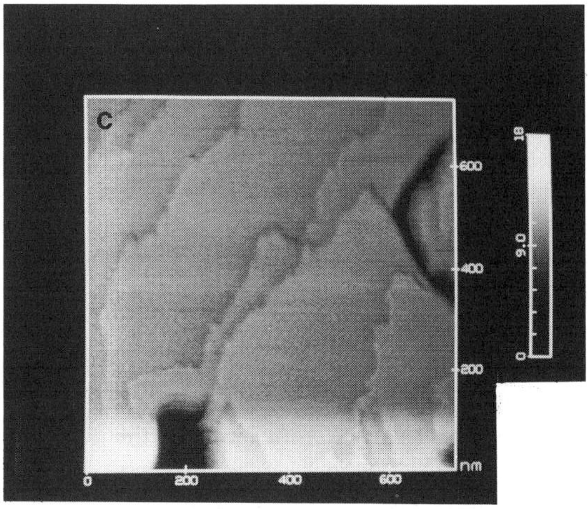

Figure 2.12 *Continued*

suggested two inhibition mechanisms, depending on surface history. Phosphate (6–10 μmol PO_4) introduced during the nucleation stage results in the formation of nuclei with amorphous shapes (Figure 2.12(a)). Phosphates introduced during layer growth disrupt the relatively straight steps produced during PO_4-free growth (Figure 2.12(b)). The steps become jagged and then cease to advance (Figure 2.12(c)). It is most likely that PO_4 preferentially sorbs to edges and steps, poisoning growth locations on both incipient nuclei and established crystals.

A comparison of our observations with kinetic interpretations reported by solution chemistry studies of calcite precipitation suggests that some rate experiments may include the early influence of a rapid nonlinear growth phase caused by surface nucleation. If true, the analysis of data and interpretations of growth mechanisms may be affected and this suggests that rates of stable calcite growth in natural systems may be slower than experimentally determined values would predict.

2.4.4 Heterogeneous catalysis on mineral surfaces

Examples of mineral surface catalytic activity in nature are difficult to document and can be quite speculative. By their very nature, catalytically active surfaces should promote reactions, but not in themselves change. As a result, this subject is only rarely discussed in the geochemical literature, yet it probably has great geochemical consequence (Coyne and McKeever, 1989, and references therein). Judging from catalytic investigations by chemists, chemical engineers and material scientists, many mineral surfaces have either a proven or a potential capability of possessing relevant catalytic activity (King and Woodruff, 1984; Bradley, Gattuso and Bertolacini, 1989; Richardson, 1989). The following is an example of natural catalytic activity, in this case a mineral surface which actively participates in the generation of species that would otherwise not have been present.

Studies by Tingle *et al.* (1990) and by Tingle, Mathez and Hochella (1991) have investigated trace amounts of organic molecules that have been found in thin films (<3–4 nm thick as shown by XPS-based sputter depth profiling) of carbonaceous matter on microcrack surfaces within mantle xenoliths from Hualalai Volcano, Hawaii, and San Carlos, Arizona, basalt flows (Figure 2.13), as well as from the Jagersfontein kimberlite. There is considerable evidence that the organics represent neither field nor laboratory contamination. Fresh olivine surfaces exposed to laboratory air and handling do not have these films, and they are not generated in the UHV instrumentation used for SALI measurement. We consider field contamination to be unlikely, based on our inability to 'contaminate' interior microcrack surfaces by soaking crystals in solutions containing high concentrations of certain organic markers for several months. On the other hand, these films and organics do not appear to be from the deep earth (upper mantle in this case), because healed microcracks, formed and healed in the upper mantle or lower crust, do not have these films. Therefore, we are able to establish that these films and their associated organics are most likely formed sometime during eruption and cooling. It is also interesting to note that these films do not form on basaltic glass surfaces recovered from an active lava lake. Therefore, we have suggested that the formation of this reduced carbonaceous matter occurs by heterogeneous 'catalytic' reaction of volcanic gases (H_2O and CO_2, and some reduced gases such as H_2, CH_4 and CO) on fresh, chemically active crack surfaces produced by thermal stresses during rapid cooling events (Mathez, 1987). The organics, which are thermally desorbed by heating in the SALI instrument used for these analyses, may represent pyrolysis products of macromolecules in the films and include a number of alkenes (butene, propene, etc.), alkyl fragments such as butyl and propyl, and aromatic compounds such as benzene and toluene, among other compounds (Figure 2.14). Low energy electron diffraction shows that the films are amorphous.

In our most recent study (Tingle and Hochella, 1993), we have proposed the hypothesis that these films are the result of simple volcanic gases reacting with fresh olivine surfaces at high temperature. This is based on experimentally generating films which are very similar to those found in nature. Our experiments involved the exposure of fresh {010} surfaces of San Carlos olivine to C—O—H gases generated by the decomposition of oxalic acid and oxalic acid dihydrate at between 400 and 800 °C. The reactivity of fresh olivine surfaces was assured by generating crack surfaces with thermal shocking during the actual run in sealed capsules. Under these conditions, carbonaceous films form readily on these surfaces and achieve thicknesses of

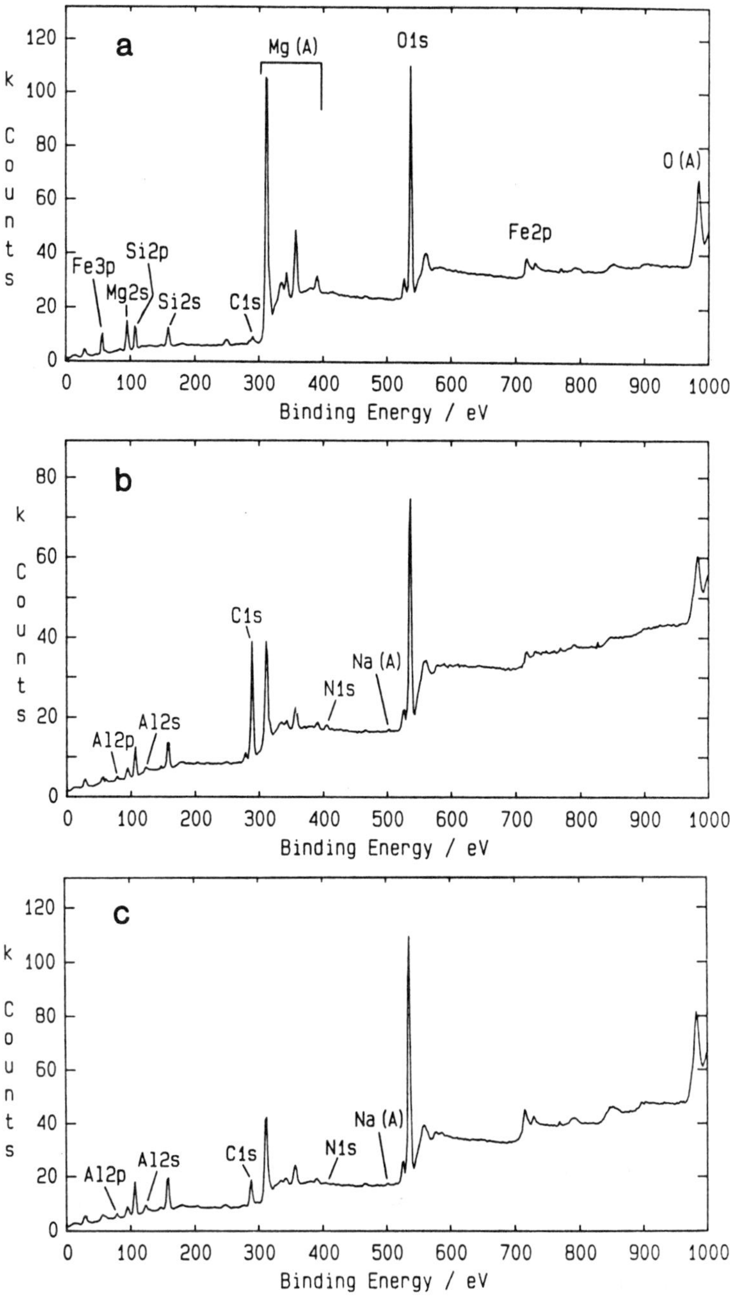

Figure 2.13 XPS spectra of San Carlos olivine surfaces. (a) (010) cleaved surface from flawless single crystal. All olivine components are clearly present, and a minor amount of surface carbon is present (C1s peak at 286 eV) owing to the exposure of this surface to air for a short time. (b) Olivine microcrack surface with carbonaceous film. Note C1s intensity and relative reduction of the other lines. (c) Same surface as in (b) after sputtering with Ar ions at 1 kV for 1 min. Large reduction of C1s intensity indicates that film is no more than a few nanometers thick. (From Tingle *et al.*, 1990.)

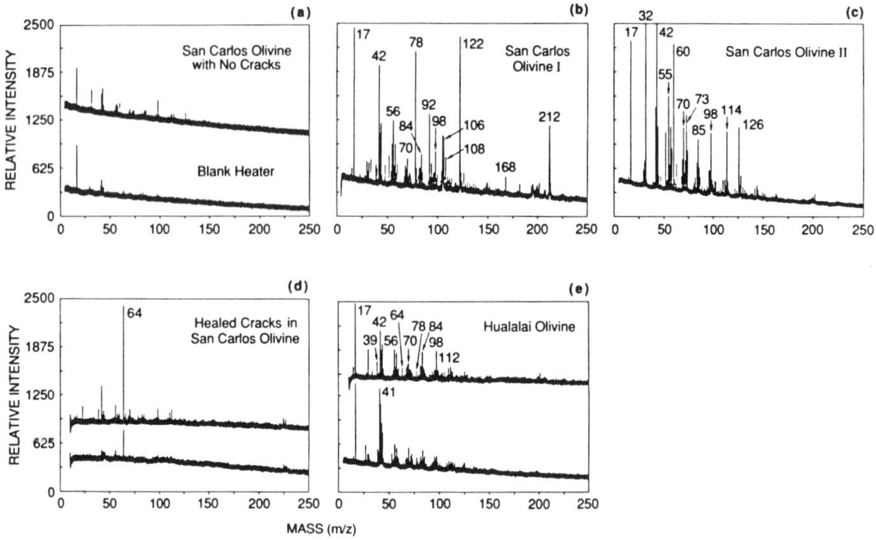

Figure 2.14 Temperature-integrated (24–700 °C) thermal desorption SALI spectra (relative intensity vs. mass). (a) Blank heater strip (used as a control) and surface equivalent to that analysed in Figure 2.13(a). (b) and (c) Typical spectra obtained from carbonaceous films from San Carlos, Arizona, xenoliths. (d) Spectrum obtained from healed crack as described in text. (e) Carbonaceous films in xenoliths from Hualalai Volcano, Hawaii. (From Tingle *et al.*, 1990.)

a few nanometres, comparable with those observed in natural samples. At relatively oxidizing conditions, we found that the film consisted of principally C—C and C—H bonded species with smaller amounts of C—O bonded species. These carbon speciations, and the ones mentioned below, were determined by C1s photoelectron spectra (XPS). At relatively reducing conditions, the carbonaceous films consist of C—C/C—H, C—O, and metal—C (carbide) species. Although we analysed these films with the ultra-sensitive SALI technique, we failed to detect organic compounds in these films, leaving open the question of abiotic synthesis of organic matter on crack surfaces in basalts. Nevertheless, it is clear from this work that crack surfaces in olivine (and probably other minerals) are capable of stabilizing compounds that otherwise would not be present in cooling lava.

2.5 Conclusions

The four examples briefly described above should give the reader some idea of current mineral surface geochemical research. In each case, the reactions and reactivity are found to be relatively complex, especially when viewed from the molecular and atomic levels. Unfortunately, despite significant recent advances in these areas of research, we still have little idea of the fundamental mechanisms which drive such surface and interface reactions. For example, during the time that a species is attached to a specific surface site, how is the sorbed molecule, and the site on which it sits, electronically modified? What is the molecular orientation of attachment, and what are the relative probabilities that this sorbed species will dissociate, migrate to another site, diffuse into the near-surface, or desorb? What specific affinities do various sorbates have for the different surface sites available, and can these be measured directly? How and at what rate do surface species diffuse and combine with other species to form products? What are the actual physiochemical steps in a heterogeneous catalysis reaction at a mineral–fluid interface, and how can we recognize these reactions in nature? Answering these kinds of questions is one way to obtain exceptional insight into the workings of natural mineral–fluid interfaces. In the process, as we probe more thoroughly and characterize these systems, the specific contributions to measurable thermodynamic quantities (e.g. surface-free energies) at interfaces will become more apparent.

Our science must go well beyond a knowledge of the electrostatic nature of surfaces and the development of surface complexation theory (for sorption reactions) and cation exchange and leach-layer formation (for desorption reactions). Ultimately, knowledge of surfaces and the chemical interactions that occur on them must come from a direct and detailed knowledge of chemical bonding at the interface.

Acknowledgements

The author is very grateful to Jodi Junta, David Cox, Udo Becker, and Carrick Eggleston for assistance in preparing this chapter. Financial support from the National Science Foundation (EAR-9105000), the Petroleum Research Fund, administered by the American Chemical Society (PRF 22892-AC5,2), Virginia Polytechnic Institute and State University, and the Center for Materials Research at Stanford University is also greatly appreciated.

References

Abraham, F.F. and Batra, I.P. (1989) Theoretical interpretation of atomic force microscope images of graphite. *Surface Sci.*, **209**, L125–32.

Albrecht, T.R. and Quate, C.F. (1987) Atomic resolution imaging of a nonconductor by atomic force microscopy. *J. Appl. Phys.*, **62**, 2599–602.

Albrecht, T.R. and Quate, C.F. (1988) Atomic resolution with the atomic force microscope on conductors and nonconductors. *J. Vac. Sci. Tech.*, **6**, 271–4.

Avouris, P. (1990) Atom-resolved surface chemistry using the scanning tunneling microscope. *J. Phys. Chem.*, **94**, 2246–56.

Avouris, P. and Lyo, I. (1991) Probing and inducing surface chemistry with the STM: the reactions of Si(111)-7×7 with H_2O and O_2. *Surface Sci.*, **242**, 1–11.

Bakken, B.M. (1990) Gold mineralization, wall-rock alteration, and the geochemical evolution of the hydrothermal system in the main orebody, Carlin mine, Nevada. PhD Disseration, Stanford University.

Bakken, B.M., Fleming, R.H. and Hochella, M.F., Jr (1991) High-resolution microscopy of auriferous pyrite from the Post Deposit, Carlin District, Nevada, in *Process Mineralogy XI – Characterization of Metallurgical and Recyclable Products* (eds D.M. Hausen, W. Petruk, R.D. Hagni and A. Vassiliou), Minerals, Metals and Materials Society, Warrendale, PA, pp. 13–23.

Bancroft, G.M. and Hyland, M.M. (1990) Spectroscopic studies of adsorption/reduction reactions of aqueous metal complexes on sulphide surfaces, in *Mineral–Water Interface Geochemistry* (eds M.F. Hochella, Jr and A.F. White), *Reviews in Mineralogy*, **23**, Mineralogical Society of America, Washington, DC, pp. 511–58.

Bancroft, G.M., Metson, J.B., Kresovic, R.A. and Nesbitt, H.W. (1987) Leaching studies of natural and synthetic titanites using secondary ion mass spectroscopy. *Geochim. Cosmochim. Acta*, **51**, 911–18.

Becker, C.H. and Gillen, K.T. (1984) Surface analysis by non-resonant multiphoton ionization of desorbed or sputtered species. *Anal. Chem.*, **56**, 1671–4.

Berner, R.A. (1975) The role of magnesium in the crystal growth of calcite and aragonite from seawater. *Geochim. Cosmochim. Acta*, **39**, 489–504.

Bethge, H. and Krohn, M. (1965) Wachstumsvorgang auf NaCl-kristallen nach Anlosung, in *Adsorption et Croissance Cristalline*. CNRS, Paris, pp. 389–406.

Binnig, G. and Rohrer, H. (1982) Scanning tunneling microscopy. *Helv. Phys. Acta*, **55**, 726–35.

Binnig, G., Quate, C.F. and Gerber, Ch. (1986) Atomic force microscope. *Phys. Rev. Lett.*, **56**, 930–3.

Binnig, G., Rohrer, H., Gerber, Ch. and Weibel, E. (1982a) Tunneling through a controllable vacuum gap. *Appl. Phys. Lett.*, **40**, 178–80.

Binnig, G., Rohrer, H., Gerber, Ch. and Weibel, E. (1982b) Surface studies by scanning tunneling microscopy. *Appl. Phys. Lett.*, **49**, 57–61.

Binnig, G., Rohrer, H., Gerber, Ch. and Weibel, E. (1983) 7×7 reconstruction on Si(111) resolved in real space. *Phys. Rev. Lett.*, **50**, 120–3.

Binnig, G., Gerber, Ch., Stoll, E., Albrecht, T.R. and Quate, C.F. (1987) Atomic resolution with atomic force microscope. *Surface Sci.*, **189/190**, 1–6.

Bischoff, J.L. (1968) Kinetics of calcite nucleation: magnesium ion inhibition and ionic strength catalysis. *J. Geophy. Res.*, **73**, 3315–22.

Blake, D.F. (1990) Scanning electron microscopy, in *Instrumental Surface Analysis of Geologic Materials* (ed. D.L. Perry), VCH Publishers, New York, pp. 11–44.

Bradley, S.A., Gattuso, M.J. and Bertolacini, R.J. (1989) *Characterization and Catalyst Development*, *ACS Symposium Series* 411. American Chemical Society, Washington, DC.

Brown, G.E. Jr (1990) Spectroscopic studies of chemisorption reaction mechanisms at oxide–water interfaces, in *Mineral–Water Interface Geochemistry* (eds M.F. Hochella and A.F. White) *Reviews in Mineralogy*, **23**, Mineralogical Society of America, Washington, DC, pp. 309–64.

Browning, R. and Hochella, M.F. Jr (1990) Auger electron spectroscopy and microscopy, in *Instrumental Surface Analysis of Geologic Materials* (ed. D.L. Perry), VCH Publishers, New York, pp. 87–120.

Busenburg, E. and Plummer, N.L. (1986) A comparative study of the dissolution and crystal growth kinetics of calcite and aragonite, in *Studies in Diagenesis* (ed. F.A. Mumpton), *USGS Bull.*, 1578, pp. 139–68.

Casey, W.H., Westrich, H.R. and Arnold, G.W. (1988) Surface chemistry of labradorite feldspar reacted with aqueous solutions at pH = 2, 3, and 12. *Geochim. Cosmochim. Acta*, **52**, 2795–807.

Casey, W.H., Westrich, H.R., Arnold, G.W. and Banfield, J.F. (1989) The surface chemistry of dissolving labradorite feldspar. *Geochim. Cosmochim. Acta*, **53**, 821–32.

Christmann, K. and Ertl, G. (1976) Interaction of hydrogen with Pt(111): the role of atomic steps. *Surface Sci.*, **60**, 365–84.

Chu, W.K., Mayer, J.W. and Nicolet, M.A. (1978) *Backscattering Spectroscopy*, Academic Press, New York.

Clarke, L.J. (1985) *Surface Crystallography, an Introduction to Low Energy Electron Diffraction*, John Wiley & Sons, New York.

Compton, R.G. and Daly, P.J. (1987) The dissolution/precipitation kinetics of calcium carbonate: an assessment of various kinetic equations using a rotating disk method. *J. Coll. Int. Sci.*, **115**, 493–8.

Cox, D.F., Fryberger, T.B. and Semancik, S. (1989) Surface reconstruction of oxygen deficient SnO_2(110). *Surface Sci.*, **224**, 121–42.

Coyne, L.M. and McKeever, S.W.S. (1989) Spectroscopic characterization of minerals and their surfaces, an overview, in *Spectroscopic Characterization of Minerals and Their Surfaces* (eds L.M. Coyne, S.W.S. McKeever and D.F. Blake), *ACS Symposium Series* 415, American Chemical Society, Washington, DC, pp. 1–29.

Daubree, M. (1867) Experiences sur les decompositions chimiques provoquees par les actions mecaniques dans divers mineraux tels que le feldspath. *Comptes Rendus*, **51**, 339–45.

Davis, J.A. and Kent, D.B. (1990) Surface complexation modeling in aqueous geochemistry, in *Mineral–Water Interface Geochemistry* (eds M.F. Hochella, Jr and A.F. White), *Reviews in Mineralogy*, **23**. Mineralogical Society of America, Washington, DC, pp. 177–260.

Dibble, W.E. Jr and Tiller, W.A. (1981) Non-equilibrium water/rock interactions – I Model for interface-controlled reactions. *Geochim. Cosmochim. Acta*, **45**, 79–92.

Dove, P.M. and Hochella, M.F. Jr (1993) Calcite precipitation mechanisms and inhibition by orthophospate: in situ observations by scanning force microscopy. *Geochim. Cosmochim. Acta*, **57**, 705–14.

Dove, P.M., Hochella, M.F. Jr and Reeder, R.J. (1992) In situ investigation of near-equilibrium calcite precipitation by atomic force microscopy, in *Water–Rock Interaction* (eds Y. Kharaka and A. Maest), Balkema, Rotterdam, pp. 141–4.

Drake, B. and Hellmann, R. (1991) Atomic force microscopy of the albite (010) surface. *Amer. Mineral.*, **76**, 1773–6.

Ebelmen, J.J. (1847) Recherches sur la decomposition des roches. *Annales des Mines*, **12**, 627–54.

Eggleston, C.M. and Hochella, M.F. Jr (1990) Scanning tunneling microscopy of sulfide surfaces. *Geochim. Cosmochim. Acta*, **54**, 1511–17.

Eggleston, C.M. and Hochella, M.F. Jr (1991) Scanning tunneling microscopy of galena (100) surface oxidation and sorption of aqueous gold. *Science*, **254**, 983–6.

Eggleston, C.M. and Hochella, M.F. Jr (1992a) Scanning tunneling microscopy of pyrite {100}: surface structure and step reconstruction. *Amer. Mineral.*, **77**, 221–4.

Eggleston, C.M. and Hochella, M.F. Jr (1992b) The structure of hematite {001} surfaces by scanning tunneling microscopy: image interpretation, surface relaxation, and step structure. *Amer. Mineral.*, **77**, 911–22.

Eggleston, C.M. and Hochella, M.F. Jr (1993) Tunneling spectroscopy applied to PbS(001) surfaces: fresh surfaces, oxidation, and sorption of aqueous Au. *Amer. Mineral.*, **78**, 877–83.

Eggleston, C.M. and Hochella, M.F. Jr (1994) Atomic and electronic structure of PbS{100} surfaces and chemisorption-oxidation reactions, in *The Environmental Chemistry of Sulfide Oxidation*, *ACS Symposium Series*, **550** (eds C.N. Alpers and D.W. Blowes), American Chemical Society, Washington, DC, pp. 201–22.

Evans, S. and Raftery, E. (1982) Electron spectroscopic studies of galena and its oxidation by microwave-generated species and by air. *J. Chem. Soc., Faraday Transactions 1*, **78**, 3545–60.

Fan, F. and Bard, A.J. (1990) Scanning tunneling microscopy and tunneling spectroscopy of the n-TiO_2(001) surface. *J. Phys. Chem.*, **94**, 3761–6.
Fan, F. and Bard, A.J. (1991) Scanning tunneling microscopy and tunneling spectroscopy of n-type iron pyrite (n-FeS_2) single crystals. *J. Phys. Chem.*, **95**, 1969–76.
Feenstra, R.M., Stroscio, J.A. and Fein, A.P. (1987) Tunneling spectroscopy of the Si(111) 2×1 surface. *Surface Sci.*, **181**, 295–306.
Firment, L.E. (1982) Thermal faceting of the rutile TiO_2(001) surface. *Surface Sci.*, **116**, 205–16.

Gilbert, S.E. and Kennedy, J.H. (1989) Observation of hysteresis induced by the tip field in scanning tunneling microscope spectroscopic probing of TiO_2 and α-Fe_2O_3 single-crystal surfaces. *Langmuir*, **5**, 1412–15.
Goldstein, F.I., Newbury, D.E., Echlin, P. *et al.* (1981) *Scanning Electron Microscopy and X-ray Microanalysis*. Plenum Press, New York.
Goodarz-nia, I. and Motamedi, M. (1980) Kinetics of calcium carbonate crystallization from aqueous solutions. *J. Crystal Growth*, **48**, 125–31.
Gratz, A.J., Hillner, P.A. and Hansma, P.K. (1993) Step dynamics and spiral growth on calcite. *Geochim. Cosmochim. Acta*, **57**, 491–5.
Gratz, A.J., Manne, S. and Hansma, P.K. (1991) Atomic force microscopy of atomic-scale ledges and etch pits formed during dissolution of quartz. *Science*, **251**, 1343–6.
Güntherodt, H.-J. and Wiesendanger, R. (eds) (1992) *Scanning Tunneling Microscopy I: Applications to Clean and Adsorbate-covered Surfaces*. Springer-Verlag, Berlin.

Hagstrom, A.L. and Fahlman, A. (1978) The interaction of oxygen and the lead chalcogenides at room temperature studied by photoelectron spectroscopy. *Appl. Surf. Sci.*, **1**, 455–70.
Hamers, R.J. and Köhler, U.K. (1989) Determination of the local electronic structure of atomic-size defects on Si(001) by tunneling spectroscopy. *J. Vac. Sci. Technol. A*, **7**, 2854–9.
Hamers, R.J., Avouris, Ph. and Bozso, F. (1988) A scanning tunneling microscopy study of the reaction of Si(001)-(2×1) with NH_3. *J. Vac. Sci. Technol. A*, **6**, 508–11.
Hamers, R.J., Tromp, R.M. and Demuth, J.E. (1986) Surface electronic structure of Si(111)-(7×7) resolved in real space. *Phys. Rev. Lett.*, **56**, 1972–5.
Hansma, P.K., Elings, V.B., Marti, O. and Bracker, C.E. (1988) Scanning tunneling microscopy and atomic force microscopy: application to biology and technology. *Science*, **242**, 209–16.
Hartman, H., Sposito, G., Yang, A. *et al.* (1990) Molecular-scale imaging of clay mineral surfaces with the atomic force microscope. *Clays Clay Miner.*, **38**, 337–42.
Hellmann, R., Eggleston, C.M., Hochella, M.F. Jr and Crerar, D.A. (1990) Formation of leached layers on albite during dissolution under hydrothermal conditions. *Geochim. Cosmochim. Acta*, **54**, 1267–81.
Hemstreet, L.A. (1975) Cluster calculations of the effects of single vacancies during dissolution under hydrothermal conditions. *Geochim. Cosmochim. Acta*, **11**, 2260–70.
Henning, K.-H. and Storr, M. (1986) *Electron Micrographs (TEM, SEM) of Clays and Clay Minerals*. Akademie-Verlag, Berlin.
Henrich, V.E. (1985) The surfaces of metal oxides. *Rep. Prog. Phys.*, **48**, 1481–541.
Hillner, P.E., Gratz, A.J., Manne, S. and Hansma, P.K. (1992) Atomic-scale imaging of calcite growth and dissolution in real-time. *Geology*, **20**, 359–62.
Hochella, M.F. Jr (1988) Auger electron and X-ray photoelectron spectroscopies, in *Spectroscopic Methods in Mineralogy and Geology* (ed. F.C. Hawthorne), *Reviews in Mineralogy*, **18**, Mineralogical Society of America, Washington, DC, pp. 573–637.
Hochella, M.F. Jr (1990) Atomic structure, microtopography, composition, and reactivity of mineral surfaces, in *Mineral–Water Interface Geochemistry* (eds M.F. Hochella, Jr and A.F. White), *Reviews in Mineralogy*, **23**, Mineralogical Society of America, Washington, DC, pp. 87–132.

Hochella, M.F. Jr and Carim, A.F. (1988) A reassessment of electron escape depths in silicon and thermally grown silicon dioxide thin films. *Surface Sci.*, **197**, L260–8.

Hochella, M.F. Jr, Harris, D.W. and Turner, A.M. (1986) Scanning Auger microscopy as a high resolution microprobe for geologic materials. *Amer. Mineral.*, **71**, 1247–57.

Hochella, M.F. Jr, Turner, A.M. and Harris, D.W. (1986) High resolution scanning microscopy of mineral surfaces. *Scan. Electron Micros.*, 1986, pp. 337–49.

Hochella, M.F. Jr, Eggleston, C.M., Elings, V.B. *et al.* (1989) Mineralogy in two dimensions: scanning tunneling microscopy of semiconducting minerals with implications for geochemical reactivity. *Amer. Mineral.*, **74**, 1235–48.

Hochella, M.F. Jr, Eggleston, C.M., Elings, V.B. and Thompson, M.S. (1990) Atomic structure and morphology of the albite{010} surface: an atomic-force microscope and electron diffraction study. *Amer. Mineral.*, **75**, 723–30.

Inskeep, W.P. and Bloom, P.R. (1985) An evaluation of rate equations for calcite precipitation kinetics at pCO_2 less than 0.01 atm and pH greater than 8. *Geochim. Cosmochim. Acta*, **49**, 2165–80.

Iwasawa, Y., Mason, R., Textor, M. and Somorjai, G.A. (1976) The reactions of carbon monoxide at coordinatively unsaturated sites on a platinum surface. *Chem. Phys. Lett.*, **44**, 468–70.

Jaegermann, W. and Tributsch, H. (1988) Interfacial properties of semi-conducting transition metal chalcogenides. *Prog. Surf. Sci.*, **29**, 1–167.

Johnsson, P.A., Eggleston, C.M. and Hochella, M.F. Jr (1991) Imaging molecular-scale structure and microtopography of hematite with the atomic force microscope. *Amer. Mineral.*, **76**, 1442–5.

King, D. and Woodruff, D.P. (1984) *The Chemical Physics of Solid Surfaces and Heterogeneous Catalysis*, Elsevier, New York.

King, P. (1992) *Photoelectron microscopy*. PhD Dissertation, Stanford University.

Kochanski, G.P. (1989) Nonlinear alternating-current tunneling microscopy. *Phys. Rev. Lett.*, **62**, 2285–8.

Kralj, D., Brecevic, L. and Nielsen, A.E. (1990) Vaterite growth and dissolution in aqueous solution. I. Kinetics of crystal growth. *J. Crystal Growth*, **104**, 793–800.

Krohn, M. and Bethge, H. (1977) Step structures of real surfaces, in *Current Topics in Materials Science*, Vol. 2 (ed. E. Kaldis), North-Holland, Amsterdam, pp. 141–61.

Kurtz, R.L. and Henrich, V.E. (1987) Surface electronic structure and chemisorption on corundum transition metal oxides: α-Fe_2O_3. *Phys. Rev. B*, **36**, 3413–21.

Li, E.K., Johnson, K.H., Eastman, D.E. and Freeouf, J.L. (1974) Localized and bandlike valence-electron states in FeS_2 and NiS_2. *Phys. Rev. Lett.*, **32**, 470–2.

Lindgreen, H., Garnaes, J., Hansen, P.L. *et al.* (1991) Ultrafine particles of North Sea illite/smectite clay minerals investigated by STM and AFM. *Amer. Mineral.*, **76**, 1218–22.

Liversidge, A. (1893) On the origin of gold nuggets. *J. Roy. Soc. New South Wales*, **27**, 303–43.

Lovley, D., Phillips, E.J.P., Gorby, Y.A. and Landa, E.R. (1991) Microbial reduction of uranium. *Nature*, **350**, 413–16.

Luther, G.W. (1990) The frontier-molecular orbital theory approach in geochemical processes, in *Aquatic Chemical Kinetics*. (ed. Werner Stumm), John Wiley and Sons, New York, pp. 173–98.

Lyo, I., Avouris, P., Schubert, B. and Hoffmann, R. (1990) Elucidation of the intial stages of the oxidation of Si(111) using scanning tunneling microscopy and spectroscopy. *J. Phys. Chem.*, **94**, 4400–3.

MacDougall, J.E., Cox, S.D., Stucky, G.D. *et al.* (1991) Molecular resolution of zeolite surfaces as imaged by atomic force microscopy. *Zeolites*, **11**, 429–33.

Mathez, E.A. (1987) Carbonaceous matter in mantle xenoliths: composition and relevance to the isotopes. *Geochim. Cosmochim. Acta*, **51**, 2339–47.

McFeely, F.R., Kowalczyk, L., Ley, L. *et al.* (1973) High-resolution X-ray photoemission spectra of PbS, PbSe, and PbTe valence bands. *Phys. Rev. B*, **7**, 5228–36.

Metson, J. (1990) Secondary ion mass spectometry, in *Instrumental Surface Analysis of Geologic Materials* (ed. D.L. Perry), VCH Publishers, New York, pp. 311–52.

Mogk, D.W. (1990) Application of Auger electron spectroscopy to studies of chemical weathering, in *Reviews of Geophysics*, **28**, American Geophysical Union, Washington, DC, pp. 337–56.

Mogk, D.W. and Locke, W.W. (1988) Application of Auger electron spectroscopy (AES) to naturally weathered hornblende. *Geochim. Cosmochim. Acta*, **52**, 2537–42.

Morse, J.W. and Choppin, G.R. (1991) The chemistry of transuranic elements in natural waters. *Reviews in Aquatic Science*, **4**, 1–22.

Morse, J.W., Millero, F.J., Cornwell, J.C. and Rickard, D. (1987) The chemistry of the hydrogen sulfide and iron sulfide systems in natural waters. *Earth Sci. Rev.*, **24**, 1–42.

Mucci, A. and Morse, J.W. (1983) The incorporation of Mg^{2+} and Sr^{2+} into calcite overgrowths: influences of growth rate and solution composition. *Geochim. Cosmochim. Acta*, **47**, 217–33.

Muir, I.J., Bancroft, G.M. and Nesbitt, H.W. (1989) Characterization of altered labradorite surface by SIMS and XPS. *Geochim. Cosmochim. Acta*, **53**, 1235–41.

Munnix, S. and Schmeits, M. (1983) Electronic structure of tin dioxide surfaces. *Phys. Rev. B*, **27**, 7624–35.

Myhra, S., Savage, D., Atkinson, A. and Riviere, J.C. (1984) Surface modification of some titanite minerals subjected to hydrothermal chemical attack. *Amer. Mineral.*, **69**, 902–9.

Nancollas, G.H. and Reddy, M.M. (1971) The crystallization of calcium carbonate II. Calcite growth mechanism. *J. Colloid Interface Sci.*, **37**, 824–30.

Nesbitt, H.W. and Muir, I.J. (1988) SIMS depth profiles of weathered plagioclase and processes affecting dissolved Al and Si in some acidic soil solutions. *Nature*, **334**, 336–8.

Newbury, D.E., Joy, D.C., Echlin, P. *et al.* (1985) *Advanced Scanning Electron Microscopy and X-ray Microanalysis*. Plenum Press, New York.

Nordstrom, D.K. (1982) *Acid sulfate weathering*. Special Publication no. 10, Soil Science Society of America, Madison, WI.

Ohnesorge, F. and Binnig, G. (1993) True atomic resolution by atomic force microscopy through repulsive and attractive forces. *Science*, **260**, 1451–6.

Parks, G.A. (1990) Surface energy and adsorption at mineral/water interfaces: an introduction, in *Mineral–Water Interface Geochemistry* (eds M.F. Hochella, Jr and A.F. White), *Reviews in Mineralogy*, **23**. Mineralogical Society of America, Washington, DC, pp. 133–175.

Perry, D.L., Taylor, J.A. and Wagner, C.D. (1990) X-ray induced photoelectron and Auger spectroscopy, in *Instrumental Surface Analysis of Geologic Materials* (ed. D.L. Perry), VCH Publishers, New York, pp. 45–86.

Petit, J.-C., Dran, J.-C. and Della Mea, G. (1990) Energetic ion beam analysis in the earth sciences. *Nature*, **344**, 621–6.

Petit, J.-C., Della Mea, G., Dran J.-C., Schott, J. and Berner, R.A. (1987) Mechanism of diopside dissolution from hydrogen depth profiling. *Nature*, **325**, 705–7.

Petit, J.-C., Dran, J.-C., Paccagnella, A. and Della Mea, G. (1989) Structural dependence of crystalline silicate hydration during aqueous dissolution. *Earth Planet. Sci. Lett.*, **93**, 292–8.

Rachlin, A.L., Henderson, G.S. and Goh, M.C. (1992) An atomic force microscope (AFM) study of the calcite cleavage plane: image averaging in Fourier space. *Amer. Mineral.*, **77**, 904–10.

Reddy, M.M. (1986) Effect of magnesium ions on calcium carbonate nucleation and crystal growth in dilute aqueous solutions at 25 °C, in *Studies in Diagenesis* (ed. F.A. Mumpton), *USGS Bulletin* 1578, pp. 169–82.

Reddy, M.M. (1988) Physical-chemical mechanisms that affect regulation of crystallization, in *Chemical Aspects of Regulation of Mineralization* (eds C.S. Sikes and A.P. Wheeler), University of Southern Alabama Publication Services, pp. 3–8.

Reddy, M.M., Plummer, L.N. and Busenberg, E. (1981) Crystal growth of calcite from calcium bicarbonate solutions at constant P_{CO_2} and 25 °C: a test of a calcite dissolution model. *Geochim. Cosmochim. Acta*, **45**, 1281–9.

Richardson, J.T. (1989) *Principles of Catalyst Development*. Plenum Press, New York.

Rohrer, G.S., Henrich, V.E. and Bonnell, D.A. (1990) Structure of the reduced TiO_2(110) surface determined by scanning tunneling microscopy. *Science*, **250**, 1239–41.

Schindler, P.W. (1990) Co-adsorption of metal ions and organic ligands: formation of ternary surface complexes, in *Mineral–Water Interface Geochemistry* (eds M.F. Hochella, Jr and A.F. White), *Reviews in Mineralogy*, **23**. Mineralogical Society of America, Washington, DC, pp. 281–307.

Schott, J. and Petit, J.-C. (1987) New evidence for the mechanisms of dissolution of silicate minerals, in *Aquatic Surface Chemistry* (ed. W. Stumm), Wiley Interscience, New York, pp. 293–315.

Schuhle, U., Pallix, J.B. and Becker, C.H. (1988) Sensitive mass spectrometry of molecular adsorbates by stimulated desorption and single photon ionization. *J. Amer. Chem. Soc.*, **110**, 2323–7.

Siegbahn, K., Nordling, C.N., Fahlman, A. *et al.* (1967) *ESCA: Atomic, Molecular, and Solid State Structure Studied by Means of Electron Spectroscopy*, Almqvist and Wiksells, Uppsala, Sweden.

Somorjai, G.A. (1981) *Chemistry in two dimensions: Surfaces*. Cornell University Press, Ithaca, NY.

Somorjai, G.A. (1990) Modern concepts in surface science and heterogeneous catalysis. *J. Phys. Chem.*, **94**, 1013–23.

Somorjai, G.A., Van Hove, M.A. and Bent, B.J. (1988) Organic monolayers on transition-metal surfaces. The catalytically important sites. *J. Phys. Chem.*, **92**, 973–8.

Sposito, G. (1986) Distinguishing adsorption from surface precipitation, in *Geochemical Processes at Mineral Surfaces* (eds J.A. Davis and K.F. Hayes), *ACS Symposium Series* 323. American Chemical Society, Washington, DC, pp. 217–28.

Stanick, S.J. and Weiss, P.S. (1992) Paper presented at the 39th Symposium of the American Vacuum Society, Chicago.

Starling, A., Gilligan, J.M., Carter, A.H.C., Foster, R.P. and Sanders, R.A. (1989) High temperature hydrothermal precipitation of precious metals on the surface of pyrite. *Nature*, **340**, 298–300.

Steefel, C.I. and Van Cappellen, P. (1990) A new kinetic approach to modeling water–rock interaction: the role of nucleation, precursors, and Ostwald ripening. *Geochim. Cosmochim. Acta*, **54**, 2657–77.

Stumm, W. (1992) *Chemistry of the Solid–Water Interface*. John Wiley & Sons, New York.

Sunagawa, I. (1987) Surface microtopography of crystal faces, in *Morphology of Crystals* (ed. I. Sunagawa), Terra Scientific Publishing, Tokyo, pp. 323–65.

Sunagawa, I. and Koshino, Y. (1975) Growth spirals on kaolin group minerals. *Amer. Mineral.*, **60**, 407–12.

Takayanagi, K., Nanishiro, Y., Takahashi, M., Motoyoshi, H. and Yagi, K. (1985) Structural analysis of Si(111)-7×7 by UHV-transmission electron diffraction and microscopy. *J. Vac. Sci. Technol. A*, **3**, 1502–6.

Tasker, P.W. (1979) The stability of ionic crystal surfaces. *J. Phys. C: Solid State Phys.*, **12**, 4977–84.

Tingle, T.N. and Hochella, M.F. Jr (1993) Formation of reduced carbonaceous matter in basalts and xenoliths: reaction of C—O—H gases on olivine crack surfaces. *Geochim. Cosmochim. Acta*, **57**, 3245–9.

Tingle, T.N., Mathez, E.A. and Hochella, M.F. Jr (1991) Carbonaceous matter in peridotites and basalt studied by XPS, SALI, and LEED. *Geochim. Cosmochim. Acta*, **55**, 1345–52.

Tingle, T.N., Hochella, M.F. Jr, Becker, C.H. and Mulhotra, R. (1990) Organic compounds on crack surfaces in olivine from San Carlos, Arizona and Hualalai Volcano, Hawaii. *Geochim. Cosmochim. Acta*, **54**, 477–85.

Tossell, J.A. and Vaughan, D.J. (1987) Electronic structure and the chemical reactivity of the surface of galena. *Can. Mineral.*, **25**, 381–92.

Tossell, J.A. and Vaughan, D.J. (1992) *Theoretical Geochemistry: Application of Quantum Mechanics in the Earth and Mineral Sciences*. Oxford University Press, Oxford.

Van Hove, M.A., Weinberg, W.H. and Chan, C.M. (1986) *Low-energy Electron Diffraction: Experiment, Theory and Surface Structure Determination*. Springer-Verlag, Berlin.

Vaughan, D.J., Tossell, J.A. and Johnson, K.H. (1974) The bonding of ferrous iron to sulfur and oxygen in tetrahedral coordination: a comparative study using SCF Xα scattered wave molecular orbital calculations. *Geochim. Cosmochim. Acta*, **50**, 1509–20.

Walker, R.J. and Fassett, J.D. (1986) Isotopic measurement of subnanogram quantities of rhenium and osmium by resonance ionization mass spectrometry. *Anal. Chem.*, **58**, 2923–7.

Walker, R.J., Shirley, S.B., Hanson, G.N., Rajamani, V. and Horan, M.F. (1989) Re-Os, Rb-Sr, and O isotopic systematics of the Archean Kolar schist belt, Karnataka, India. *Geochim. Cosmochim. Acta*, **53**, 3005–13.

Wersin, P., Hochella, M.F. Jr, Persson, P. *et al.* (1994) Interaction between aqueous U(VI) and sulfide minerals: spectroscopic evidence of sorption and reduction. *Geochim. Cosmochim. Acta* (in press).

Wicks, F.J., Kjoller, K. and Henderson, G.S. (1992) Imaging the hydroxyl surface of lizardite at atomic resolution with the atomic force microscope. *Can. Mineral.*, **30**, 83–91.

Willis, R.D., Thonnard, N. and Cole, D.R. (1989) *Resonance ionization spectroscopy and its potential application in geosciences. USGS Prof. Paper* 1890, pp. 117–28.

Zangwill, A. (1988) *Physics at Surfaces*. Cambridge University Press, Cambridge.

Zhong, Q., Vohs, J.M. and Bonnell, D.A. (1992) Effect of reduction on the topographic and electronic structure of TiO_2(110) surfaces. *Surface Sci.*, **274**, 35–43.

CHAPTER THREE

Mineral surfaces: theoretical approaches

J.A. Tossell

3.1 Introduction

In this chapter some recent applications of 'state-of-the-art' quantum mechanical methods to problems in surface science will be reviewed. Since relatively little work has been done on actual minerals, many of the examples will be from other inorganic materials. The different quantum mechanical methods, their accuracy and their range of application will also be described, along with both infinite lattice and localized models for surfaces, showing how the two can be meshed. Both the positive and the negative aspects of the great advances in computer technology that have made possible accurate calculations on real materials within the past two decades will be described. It is important to realize that all the quantum mechanical methods currently applicable to large molecules, surfaces and solids involve numerous approximations in their search for solutions to Schrödinger's equation, the fundamental equation describing the properties of atoms, molecules and solids. It is therefore unreasonable to expect calculations exactly to match experimental data. On the other hand, the errors caused by various approximations are often systematic and well understood, so the nature and magnitude of discrepancies between calculation and experiment can often be anticipated. More importantly, quantum mechanics provides a vehicle to synthesize the different properties of a material into a unified and understandable whole.

3.2 Overview of recent state-of-the-art calculations

What is the present state of the art in the quantum mechanical computation of surface properties – i.e. what are the best calculations we can do and how do they compare with experiment? It is worth while to consider the quality of agreement between calculation and experiment before discussing the various methods themselves. Let us first consider some recent theoretical studies of the complex and extensively studied (7×7) reconstruction of the (111) surface of Si explained using a dimer-adatom-stacking fault (DAS) model. Several studies on this surface structure are described in Srivastava and Jones (1992). Figure 3.1, taken from Brommer *et al.* (1992), shows this

Mineral Surfaces. Edited by D.J. Vaughan and R.A.D. Pattrick. Published in 1995 by Chapman & Hall, London. ISBN 0 412 56340 1

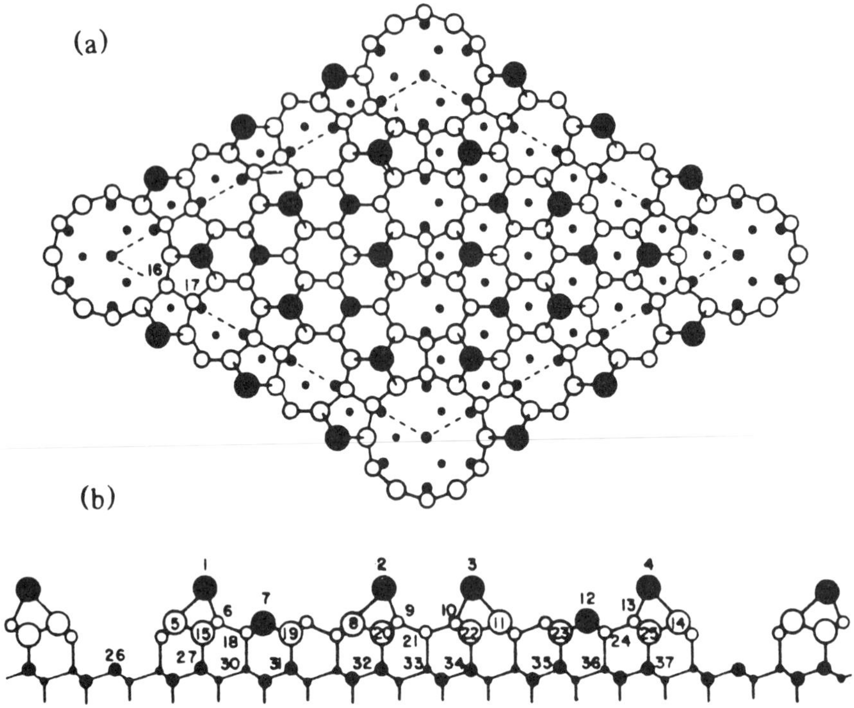

Figure 3.1 (a) Top view of the DAS model for the Si (111)- (7×7) surface reconstruction. The (7×7) unit cell is outlined by the broken lines; (b) side view – the numbers label the 37 unique atoms in the four-layer slab. (From Brommer *et al.*, 1992.)

surface schematically. Ancilotto *et al.* (1990), Brommer *et al.* (1992) and Stich *et al.* (1992) have studied this surface using very high level and computationally demanding methods based on a local-density-functional formalism (which will be described in more detail later in this chapter). The conclusions of these studies were that the DAS (7×7) reconstruction is, in fact, more stable than several alternative simpler reconstructions (in agreement with experiment), that the calculated relaxed geometries are in good agreement with most of the experimental data, that the scanning tunnelling spectroscopy (STS) images for occupied and unoccupied states agree with experiment, and that the surface phonon spectrum can be definitively assigned using the calculations.

Given the enormous computational resources employed in these studies (as suggested by the phrase 'a challenge for massively parallel computation' in the title of Brommer *et al.*, 1992), one might ask 'Why this surface?' Clearly Si is a material of enormous technological importance and the (7×7) reconstruction of the (111) surface is extreme in its effects upon bond distances and angles, and of dramatic complexity. The extreme difficulty of the problem of describing this surface also attracted researchers with state-of-the-art computational capabilities. One should not expect this much computational effort to be applied to most mineral surfaces unless a really critical or perplexing experimental problem emerges. Of course, computations that are extremely demanding now may well be routine in a decade's time.

It is also important to note that semi-empirical methods can be usefully applied to surface problems of this complexity. Qian and Chadi (1987) used a semi-empirical, parameterized version of the tight-binding band method to determine energies of atomic arrangements for the DAS reconstruction of Si(111), obtaining results in reasonable agreement with experiment and with the later, more accurate first-principles calculations.

A mineral of technological interest which has been extensively studied theoretically is molybdenite, MoS_2. Coley, Goddard and Baldeschwieler (1991) have calculated the STS images of this mineral employing Hartree–Fock (HF) and HF plus configuration interaction (CI) techniques using a cluster model (Figure 3.2) consisting of seven Mo atoms, six S atoms and 18 point charges (see section 3.6 for a brief description of HF theory and a reference to CI techniques). To evaluate STS images we must first obtain an accurate description of the energies for removal or addition of electrons to the system. This requires a cluster of adequate size and composition and an adequate theoretical model for the ionization or electron addition process, as is discussed later. One must then evaluate the electron tunnelling probability between the sample and the scanning tunnelling microscope (STM) tip, which may be approximated in the formalism of Tersoff and Hamann (1985) in terms of the electron density at a given energy and spatial position or, more correctly, in terms of integrals of the Hamiltonian or total energy

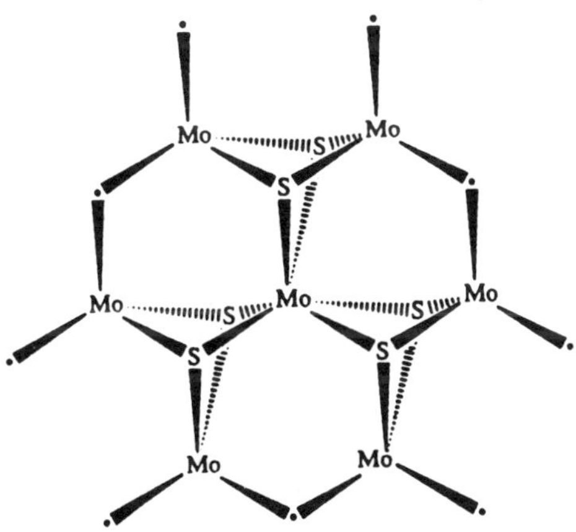

Figure 3.2 The $Mo_7S_6PC_{18}$ cluster model used to describe the scanning tunnelling spectroscopy images of MoS_2. (From Coley, Goddard and Baldeschwieler, 1991.)

operator between inital and final states as described in Coley, Goddard and Baldeschwieler (1991). However, even using their highly accurate HF + CI approach, Coley, Goddard and Baldeschwieler (1991) could not definitively assign the atoms in the experimental STS images from the computational results; they discussed the need for both better calculations and additional experiments.

A ground-breaking study on the reactivity of simple mineral surfaces was performed by McCarthy and Hess (1992) and McCarthy *et al.* (1993). In their first paper these authors used the periodic HF or HF band method to calculate bond distances and energetics for several different coverages of Cl_2 on a periodic slab model of the MgO(001) surface, finding that Cl_2 bonded preferentially at the O position. They also presented electron density maps and total and atom projected densities of states for both the MgO surface and the Cl_2/MgO adsorption complex and plots of electrostatic potentials produced by the electron density distributions of the surface and the adsorbate molecule. The adsorption of Cl_2 at the O site was explained by the electrostatic interaction of the quadrupole moment of O_2 and the gradient of the electric field at the surface. Since the computational difficulty of a HF band calculation is considerably greater than that of a molecular HF calculation and since analytical gradient geometry optimization procedures are not yet available for periodic HF calculations, the geometry optimization was incomplete, with the MgO slab geometry held fixed. Although these results may be changed somewhat by further geometry optimization and other refinements, they represent a pioneering application of such techniques to mineral reactivity. McCarthy *et al.* (1993) added a correction for the effects of electron correlation based on density functional theory which increased the magnitudes of the adsorption energies without changing the preferred mode of adsorption.

3.3 Models for mineral surfaces: infinite repeating unit vs. local cluster

Basically, there are two ways to model the surface of a solid. In the periodic model one exploits the two-dimensional periodicity of the infinite system, using either a supercell model consisting of alternating bulk material and vacuum layers, or a slab model with a vacuum at the top and bottom. A general supercell model as described by Srivastava and Jones (1992) is shown in Figure 3.3. The accuracy of the model will, of course, generally increase with the slab width. A **three-layer** slab used to describe one of the surfaces of Li_2O in Lichanot *et al.* (1991) is shown in Figure 3.4. The alternative molecular cluster approach is to choose a small cluster of atoms that have the most important characteristics of the real surface. A number of different models for Zn on the surface of solid ZnS are considered below. The cluster model is often used in studies of chemisorption, e.g. of small molecules on metals, as in the recent study of Bagus and Pacchioni (1992) of

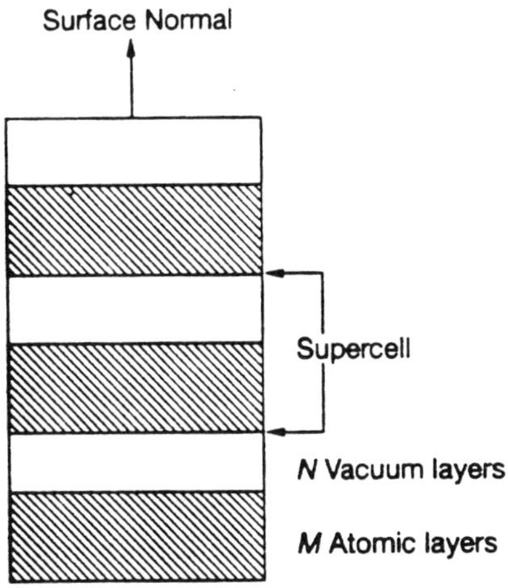

Figure 3.3 General supercell model for a surface. (From Srivastava and Jones, 1992.)

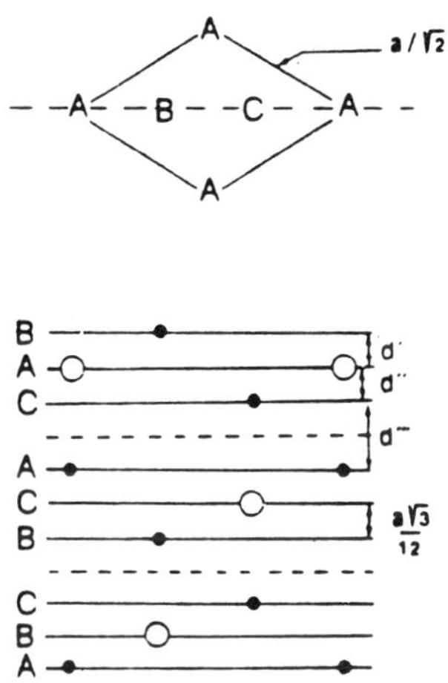

Figure 3.4 Slab model for the (111) surface of Li_2O from Lichanot *et al.* (1991), showing the surface unit cell and a vertical section of a three-layer slab. Lithium and oxygen atoms are marked by filled and open circles, respectively.

CO adsorbed on Li, Na and Cu metals, which describes the metal surface in terms of a 14 atom cluster. Some advantages of the periodic model are clear – it can properly describe both short- and long-range aspects of the surface structure and no artefacts are introduced through the presence of dangling bonds or *ad hoc* saturations of them. A proper energetic coupling is also possible between local and distant geometrical parameters, e.g. surface and bulk bond distances. There are also many important experimental techniques that probe aspects of structure that are long range with respect to atomic dimensions, e.g. low energy electron diffraction (LEED) and STM/STS. The disadvantage of using the periodic approach is that it is computationally very difficult, even if the two-dimensional symmetry is exploited as much as possible. A cluster approach is generally much easier computationally, allowing studies to be performed at a higher level of quantum mechanical rigour. Many properties of materials (e.g. core excitation spectra such as X-ray absorption near-edge spectra, XANES) are also best described in terms of localized spectral excitations, which can be quite easy to treat within a cluster approach and much more difficult within a delocalized scheme, which may require a much larger supercell with periodic excitations. Another advantage of the cluster approach is that it makes more explicit the similarity of properties of the same local species in different environments, e.g. $Si(OM)_4$ group in gas phase molecules, solutions and solids.

3.4 Quantum mechanical methods: HF and local density functional theory

An excellent short review of methods in molecular quantum mechanics, focusing on HF and HF plus configuration interaction (CI) but including many-body perturbation theory, local density functional theory and the more general density functional theory (DFT) as well, is given in Simons (1991). For more details one should consult quantum chemistry texts such as Levine (1991) and Szabo and Ostlund (1989), articles in the solid-state physics literature like Srivistava and Weaire (1987) and texts in the geochemical literature such as Tossell and Vaughan (1992). Some salient points of the two most popular approaches – the HF method and the local density functional (LDF) method are described below. Both methods are of the so-called one-electron or independent electron type in that they approximate the exact wave-function for the many-electron system as a product of wave-functions for the individual electrons. For materials of the complexity of minerals and their surfaces such an **independent electron approach** will be necessary for some time in the future. None the less, since each electron feels the averaged charge density of all the other electrons in both the HF and the LDF methods, the phrase ‘independent electron’ is perhaps a misnomer. Although the effect of the instantaneous correlation of electron motions, which are ignored in these methods, can certainly be substantial, the surprising thing is that the HF and LDF independent electron methods work as well as they do. For minerals and their surfaces the present state of the art in quantum mechanical computation employs either LDF theory or HF theory (perhaps with some correlation correction through either many-body perturbation theory or some density functional approach). Therefore, significant discussion of more accurate methods, such as CI, which are currently applicable only to fairly small molecules, will not be included.

The HF method involves as its basic components:

- an orbital approximation to the many-electron wave-function;
- orbitals which are determined by the variational method, minimizing the total energy with respect to the form of the orbitals;
- the self-consistent-field method, in which a starting approximate electron density is refined until it is self-consistent, i.e. no longer changing;
- the exchange energy between electrons is non-local, expressed in terms of electron–electron repulsion integrals;
- the molecular orbitals (MO) are expanded in a basis set of simpler functions, usually Gaussian type orbitals (GTO).

The most serious deficiencies of the HF method arise from basis set truncation error, owing to use of an incomplete basis set and the neglect of electron correlation (for electrons of different spin). Improvements upon the HF method can utilize:

- many-body perturbation theory;
- the multi-configuration SCF (self-consistent field) method (and its many variants);
- the CI method.

The advantages of the HF method are:

- it has been thoroughly tested for molecules and its capabilities are well understood;
- the necessary computer software is readily available for structure, energy and property evaluation;
- the results compare better with experiment than might be expected for many important properties;
- the errors due to basis set truncation and neglect of correlation are fairly systematic.

Its disadvantages are:

- it scales poorly with the size of system, with the difficulty of the computation increasing at about n^4, where n is the number of orbitals in the basis set;
- many spectral energies are not described well in terms of the ground state orbital energies of HF theory;
- basis set truncation and correlation errors are often fairly large.

Some of the basic equations of HF theory (taken mostly from Levine, 1991) include:

- the time-independent Schrödinger equation:

$$H\psi(q) = E\psi(q) \tag{3.1}$$

- the electronic and nuclear Schrödinger equations, after applying the Born–Oppenheimer approximation (which separates electronic and nuclear motion)

$$H_e\psi_e = E_e\psi_e \tag{3.2}$$

$$H_N\psi_N = E_N\psi_N \tag{3.3}$$

- the nonrelativistic electronic Hamiltonian

$$\hat{H}_e = \sum_A \sum_{B>A} \frac{Z_A Z_B}{R_{AB}} + \sum_i \left(\frac{\nabla_i^2}{2} - \sum_A \frac{Z_A}{r_{iA}}\right) + \sum_i \sum_{j>i} r_{ij}^{-1} \tag{3.4}$$

- the total energy as an expectation value of the Hamiltonian

$$E = \int \psi^* \hat{H} \psi \, d\tau \tag{3.5}$$

- the wave-function for an n-electron system in the independent electron approximation, written as a Slater determinant of one-electron wave-functions or orbitals (the independent electron approximation)

$$\Phi_i = \frac{1}{\sqrt{n!}} \begin{vmatrix} \phi_1(1) & \phi_1(2) & \ldots & \phi_1(n) \\ \phi_2(1) & \phi_2(2) & \ldots & \phi_2(n) \\ \phi_n(1) & \phi_n(2) & \ldots & \phi_n(n) \end{vmatrix} \tag{3.6}$$

- the HF equation, obtained by minimizing the energy of the independent electron wave-function using the variation principle

$$\hat{F}(1)\phi_i(1) = \varepsilon_i\phi_i(1), \qquad \text{where} \qquad i = 1, \ldots, n \tag{3.7}$$

- the Fock Hamiltonian operator

$$\hat{F}(1) = -\frac{1}{2\nabla_1^2} - \sum_A \frac{Z_A}{r_{1A}} + \sum_{j=1}^{n/2} [2J_i(1) - K_j(1)] \tag{3.8}$$

- expansion of the MOs in a basis set, as suggested by Roothaan (1951):

$$\phi_i = \sum_k c_{ki}\chi_k \tag{3.9}$$

- the Hartree–Fock–Roothaan (HFR) equation in determinantal form

$$\det(F_{jk} - \varepsilon_i S_{jk}) = 0 \tag{3.10}$$

Since construction of the Fock Hamiltonian requires knowledge of the molecular orbitals (MOs), whereas solution of the HFR equation is needed to determine the MOs, one must employ a **self-consistent-field** (SCF) procedure in which estimated or approximate MOs are used as input to construction of the Hamiltonian, and improved MOs are output from the solution of the HFR equation. In practice, this procedure does not always converge smoothly or quickly, and modern SCF programs usually contain numerous schemes for enhancement and acceleration of the SCF convergence.

The most important determinant of the quality of a HFR calculation is the size and flexibility of the expansion basis set. If the results of calculations using the HFR approach as a function of basis set size are examined, going from a minimum basis set with one function representing each occupied atomic orbital (e.g. the STO-3G basis, with three coupled Gaussian functions representing each Slater type orbital), through a polarized split valence basis with two functions representing each occupied valence orbital and additional functions of higher one quantum number on the non-H atoms (such as the 6-31G*), to a **near-HF** or **HF limit** basis with a very large number of functions, systematic trends in the values calculated for properties are generally found. For example, calculated bond lengths usually decrease as the flexibility of the basis set increases, and at the HF limit the calculated bond distance is a few hundredths of an Ångström shorter than that measured experimentally (Szabo and Ostlund, 1989). A good discussion of the terminology used to describe the different types of basis sets can be found in Hehre *et al.* (1986). Partial inclusion of electron correlation through many-body perturbation theory (MBPT) to low order makes the bond distances too long, while inclusion of higher order terms in the many-body perturbation expansion gives better agreement with experiment. Representative results are given for N_2 and CO in Table 3.1, taken from Szabo and Ostlund (1989).

Some general features of basis set truncation and correlation errors on molecular properties are:

- increasing the flexibility of the basis set decreases bond distances, increases vibrational frequencies, increases bond electron densities, and increases electron density anisotropies;
- at the HF limit, bond distances are generally too short by about 0.02–0.05 Å, stretching frequencies are too high by about 10%, bond energies are underestimated by as much as 50%, and measures of the anisotropy of the electron distribution, such as electric field gradients at nuclei, are exaggerated by about 15%.

For many properties of many molecules, the effects of basis set truncation and neglect of correlation are in opposite directions, so if one perfoms a

Table 3.1 Calculated vs. experimental bond lengths for N_2 and CO (in Å) as a function of basis set size and with inclusion of correlation effects (From Szabo and Ostlund, 1989)

Basis set employed	N_2			CO		
	HF	HF + second-order MBPT	HF + third-order MBPT	HF	HF + second-order MBPT	HF + third-order MBPT
STO-3G	1.134	1.229	1.176	1.146	1.198	1.173
4-31G	1.085	1.149	1.110	1.128	1.173	1.148
6-31G*	1.079	1.29	1.116	1.114	1.151	1.134
Experimental value		1.097			1.128	

Table 3.2 Ionization potentials (in eV) calculated for the two highest energy occupied molecular orbitals of N_2 at the Koopmans' theorem level (i.e. as negatives of the orbital eigenvalues) and at the HF + CI energy difference level using a Hartree–Fock limit basis, compared with experiment (from Szabo and Ostlund, 1989)

Orbital	Koopmans' IP	HF + CI energy difference IP	Exp. IP
$3\sigma_g$	17.3	15.8	15.6
$1\pi_u$	16.7	16.6	17.0

HFR SCF calculation (without any correlation correction) and employs a medium size basis set, it is often possible to get a numerical result in quite good agreement with experiment through a cancellation of errors. Plots of calculated values of Si—O and Si—H distance in disiloxane against basis set size (Nicholas *et al.*, 1992) show this effect very clearly, with the calculated bond distances matching experiment very well at the polarized split valence basis set HFR SCF level. It is common nowadays to calculate equilibrium geometries of large molecules at the medium basis set (e.g. polarized split valence) HFR SCF level and then to add correlation corrections, e.g. with MBPT, to get improved values for energies.

Ionization energies for electrons in molecular orbitals are also often calculated using HF techniques. The simplest approximation is to set the negative of the orbital eigenvalues equal to the orbital ionization potential (IP), utilizing Koopmans' theorem. A more accurate approach is to calculate directly the energy difference of the neutral molecule and the ion at either the HF or HF + CI level. Even for a case as simple as N_2, energy differences at the HF + CI level are necessary to obtain the correct ordering of the two highest MO ionizations, as shown in Table 3.2, taken from Szabo and

Ostlund (1989). Even worse, smaller basis sets give the right ordering of orbital ionization energies at the Koopmans' theorem level, while a HF limit basis does not. The relative ordering of these two orbitals is often explained qualitatively in terms of destabilization of the $3\sigma_g$ by s,p hybridization, but it is clear that the actual ordering depends significantly on rather subtle electronic relaxation and electron correlation effects. In transition metal compounds one commonly observes that HF orbital eigenvalues do not give the right order for metal d orbital vs. ligand p orbital ionizations. There are also many cases, even for small simple molecules, in which one cannot make a simple correspondence between experimentally observed IPs and single electron ionizations – i.e. the independent electron model for ionization can break down completely.

Let us now consider some features of the LDF method (for a much deeper and more detailed review see Von Barth, 1986). The LDF method involves:

- an orbital approximation to the many electron wave-function;
- orbitals determined by a variational method;
- a self-consistent-field electron distribution;
- exchange and correlation, which are local, expressed as a function of the electron density, and involve the electron density and often its gradient (although a gradient correction scheme is not strictly local);
- expansion of the MOs in a basis set.

Deficiencies of the LDF method include:

- an electron energy, which is a function of the electron density, but for which the exact form of the function is known only for simple model systems;
- less thorough testing than the HF method for molecules (but more testing in solid state calculations).

Improvements over the basic LDF method can include corrections incorporating functions of the electron density that give the exact exchange-correlation for simple systems. Such improved exchange-correlation potentials, e.g. Perdew and Zunger (1981), based on Ceperly and Alder (1980), give substantially better results than the original exchange-correlation function of Slater. The main advantages of the LDF method are:

- calculation of the exchange contributions is very much easier in the local formalism;
- the method scales better with the size of the system, as n^3 rather than n^4 as in HF theory;
- for some properties, LDF results are quite close to the results obtained from HF plus second-order perturbation theory calculations.

The disadvantages of the LDF approach are:

- procedures for improving on orbital approximation have not yet been developed;
- the method is not as widely available or as thoroughly tested for molecules as is the HF method.

3.5 Other properties that can be obtained from quantum mechanical calculations

Thus far we have considered the quantum mechanical calculation of equilibrium geometries and ionization potentials. In addition to these properties, several other properties are routinely available from HF calculations, and generally from LDF calculations as well:

- Mulliken orbital populations, which describe the decomposition of the electron charge density over the basis functions, and have comparative but not absolute significance in describing electron distributions within the material;
- spatial electron density distributions, which can be compared with X-ray data (in the formalism of Tersoff and Hamann (1985) often used to describe scanning tunnelling microscopy and spectroscopy, the tunnelling current depending on $\rho(r_0,E)$, where ρ is an electron density, r_0 is the distance between the sample and the centre of curvature of the tip, and E is the energy of the state);
- additional ground electronic state properties involving extra calculation (e.g. vibrational spectra);
- additional excited state or response properties, e.g. visible–UV spectra (requiring HF-CI in most cases) or NMR properties (requiring HF perturbation theory, although an LDF approach is currently under development, Malkin *et al.*, 1993);
- bond and reaction energies (bond energies must be corrected for correlation in HF but are often in fair agreement with experiment in LDF, while HF reaction energies are good for **isodesmic** reactions, in which the number of electron pairs is held constant and formal chemical bond types are conserved).

However, some subtle differences between apparently similar quantities in HF and LDF calculations should be kept in mind when comparing results using the two approaches:

- HF orbitals are electron orbitals, while LDF orbitals are, technically, orbitals of non-interacting quasiparticles (though, in practice, the significance of this distinction for density distributions and related properties does not appear to be large);
- in HF theory, $E = \Sigma_i \varepsilon_i - \Sigma_i \Sigma_{j>i}(J_{ij} - K_{ij})$, where J_{ij} and K_{ij} are electron–electron repulsion integrals, and are often large, while in LDF theory,

$E \approx \Sigma_i \varepsilon_i$ (though more accurate expressions can be used), so that excitation energies are more readily described in terms of LDF eigenvalue differences;

- neither HF nor LDF band structures and densities of states give good predictions for band gaps; HF overestimates band gaps and LDF underestimates them (both by factors of as much as two or more).

Both HF and LDF can be corrected to give better results, but such corrections are difficult to apply for periodic solids (although they have been performed in the calculations on GaAs, to be discussed later). For molecules, correction methods for HF are well established, consisting of SCF energy differences, CI or many-body perturbation theory, while in LDF calculations accurate energy differences can be calculated from orbitals in a transition state electron configuration. One intriguing technique for correcting band gaps and excitation energies in solids is to use molecular models to evaluate the corrections. For example, CI calculations on $Zn(SH)_4^{2-}$ give a correction to the highest energy occupied MO (HOMO) – lowest energy unoccupied MO (LUMO) gap of about 4.6 eV, which when applied to the ZnS band structure reduces the calculated gap from 10.7 eV to 6.1 eV (Saghi-Szabo and Tossell, unpublished results), bringing it into significantly better (but still not particularly good) agreement with the experimental value of 3.8 eV.

3.6 Advances in quantum computational capabilities and associated problems

There have been many important advances since the first appearance of HF self-consistent-field computer programs in the 1960s, including:

- development of new algorithms, e.g. analytical gradients of the energy, Moller-Plessett many-body perturbation theory, direct SCF techniques which avoid integral storage problems, effective core potentials for heavy atoms, localized orbital NMR, etc.;
- enormous increases in computer speed and storage capacity;
- expansions of properties packages in HF SCF programs;
- great improvements in the 'user-friendliness' of the programs;
- the beginning of attempts to establish minimum acceptable standards for quantum mechanical calculations (e.g. by Streitweiser, editor of *Journal of Organic Chemistry*, who now requires at least polarized double zeta basis sets for geometry optimizations that are to be reported in that journal).

However, along with these advances have come some problems:

- occasional misuse of the programs;
- over-reliance on 'rules of thumb', e.g. failure to test reliability of canned

basis sets when applied to new problems, failure to test independence of basis set and correlation errors, etc.;

- the incipient demise of qualitative bonding theory and systematic structural theory, as embodied in the idea 'if you can compute it, why bother to understand it?' (which is a general consequence of 'Tossell's first law of computing', which says that the product of human intelligence and computer intelligence is a constant, or 'as computers get smarter, people get dumber');
- a degradation of the status of quantum mechanical calculation to that of a 'computer experiment', rather than a comprehensive description of the overall electronic structure of the material;
- increased demands on the time and energy of the investigator in the new distributed 'workstation' environment, as described by Counts (1992).

3.7 HF band theory: differences from molecular HF theory and LDF band theory and accuracy of results

Since HF band theory or periodic HF theory also differs from molecular HF theory in some important respects, it is important to examine these differences, basing our description of the theory on the approach of Pisani, Dovesi and Roetti (1988), which has been implemented in the CRYSTAL programs, the newest version being CRYSTAL92.

Crystal orbitals (CO) must obey Bloch's Theorem:

$$\psi(\mathbf{k}; r + \mathbf{g}) = \psi(\mathbf{k}; r)\exp(i\mathbf{k}\cdot\mathbf{g}) \tag{3.11}$$

where $\mathbf{k}$ is a reciprocal lattice vector, $\mathbf{g}$ is a (direct) lattice vector and r is the distance from the origin. COs are expanded as linear combinations of Bloch functions:

$$\varphi_\mu(\mathbf{k}; r) = \sum_{\mathbf{g}} \chi_\mu(r - \mathbf{g})\exp(i\mathbf{k}\cdot\mathbf{g}) \tag{3.12}$$

Great care must be exercised in assuring balanced truncation of the Coulomb and exchange series of integrals, symmetry must be fully exploited, and special techniques must be adopted for constructing the HF Hamiltonian matrix at an adequate number of points in k space. Basis set construction is somewhat different to that for molecules – diffuse functions are not usually needed and their inclusion can dramatically increase both the number of integrals and the computational time, and lead to problems of linear dependence. Going beyond the HF level with CI is probably impossible (since nearly all CI is not size consistent – its contribution to the energy scales with the square root rather than the first power of the number of electrons) and using many-body perturbation theory is possible but very difficult. Simple correlation energy formulae developed by Colle and Salvetti (1983) based on the HF density are currently being used to correct for correlation approximately in the CRYSTAL92 program.

The main differences between HF band theory and the LDF approaches are:

- use of HF non-local exchange;
- expansion of crystal orbitals in a basis of Gaussians.

The HF band results appear to be similar to molecular HF results in many respects. For example, basis set truncation errors on bond distances and angles and other properties are of the same sense and magnitude as those in molecules, and errors in bond energetics are much like those in molecules. In fact, in at least some cases, HF band and molecular results seem to be very similar. For example, the comparison of HF band results for SiO_2 (Dovesi *et al.*, 1987) and previous molecular HF studies on disiloxane $(SiH_3)_2O$, shows substantial similarity in minimum energy structural parameters and electron density distributions.

Besides the relatively new HF band theory described above, there are a number of more commonly used LDF band theories, such as linearized augmented plane wave, linear muffin tin orbital, *ab initio* pseudopotential method and others. The differences between these various LDF band methods involve mainly differences in the choice of the expansion basis sets and in the averaging (if any) of the potential (see Srivastava and Weaire, 1987, for details). A newer, very powerful method is the combined molecular dynamics-LDF approach of Car and Parrinello (1985), in which one simultaneously minimizes the energy with respect to both electronic and nuclear coordinates. This approach, or variants of it, were used by Ancilotto *et al.* (1990) and Brommer *et al.* (1992) in their studies of the DAS reconstruction of the (111) surface of Si, and we can expect this powerful, but computationally intensive, technique to be used widely in the future.

3.7.1 A brief and selective history of the relationship between HF and LDF theory

In deciding between the use of the HF or the LDF approach for a particular problem, it is important to be aware of certain aspects of the previous competition and co-operation between HF and local density functional theorists. We might first ask what the commonly used term *ab initio* means, as in '*ab initio* Hartree-Fock calculation'. This is a term coined in the early days of molecular quantum mechanical calculations to identify HF calculations in which all integrals were evaluated accurately (it was highly tempting in those early days to approximate some of the more difficult electron–electron repulsion integrals required for calculation of the total energy). There is continuing debate as to whether the term *ab initio* should also be applied to LDF calculations, and in this chapter they are referred to by the slightly more general term 'first principles'.

In the beginning of the period when (rather approximate) LDF methods were first applied to molecules and (rather approximate) HF methods were first applied to solids there was fierce competition between the two schools, with each side trying to shift the dispute on to more favourable 'turf' and with many approximations, misconceptions and mis-statements on each side. Some general conclusions from the controversy concerning the electronic structure and spectra of TiO, as described by Jennison and Kunz (1977), Henrich, Zeiger and Reed (1978) and Tossell (1979), were that HF energy bands for transition metal compounds required large corrections to match experiment, while LDF bands were usually in better agreement with experiment. Conclusions from the intense controversy regarding the bonding in Cr_2 and Mo_2, as described by Goodgame and Goddard (1982) and Delley, Freeman and Ellis (1983), were that both HF + correlation and LDF methods could, with some thought, care and considerable computational difficulty, give equally good descriptions of the bonding in Cr_2 and Mo_2.

In the middle period there was acceptance and accommodation between the two camps, with Messmer (1985) suggesting: 'This dilemma regarding the efficacy of these two approaches, HF and DFT, clearly suggests that the time may be ripe for considering some hybrid approaches. Such approaches could take advantage of the merits of the LDF and *ab initio* methods while avoiding their more serious shortcomings'. This time also saw the creation of the very powerful and efficient programs DGauss and DMon for Gaussian

Table 3.3 Comparison of LDF, HF and HF plus second-order perturbation theory results with experiment for structural parameters of some molecules (results from Delley, 1991)

Molecule	Parameter* (distances in Å, angles in degrees)	LDF	Experimental value	HF	HF + second-order MBPT
H_2O	R (O–H)	0.978	0.958	0.947	0.969
	<H–O–H	104.3	104.5	105.5	104.0
NH_3	R (N–H)	1.025	1.012	1.002	1.017
	<H–N–H	106.3	106.7	107.2	106.3
CO_2	R (C–O)	1.165	1.162	1.143	
PF_3	R (P–F)	1.588	1.551	1.561	
	<F–P–F	97.0	97.7	97.1	
$VOCl_3$	R (V–O)	1.570	1.570	1.465	
	R (V–Cl)	2.125	2.142	2.108	
	<Cl–V–Cl	110.6	111.3	110.1	
$Cr(CO)_6$	R (Cr–C)	1.874	1.92	1.789	
	R (C–O)	1.146	1.16	1.167	

* R (A–B) = interatomic distance between A and B in Å; <A–B–A = bond angle A–B–A in degrees.

orbital LDF calculations on molecules, programs competitive with the Gaussian orbital-based HF programs. Delley (1991) and Andzelm and Wimmer (1992) have demonstrated the accuracy of the results obtained using such programs. Some results from Delley's work are shown in Table 3.3. It is apparent that LDF and HF results agree with experiment to about the same accuracy. While bond distances are slightly underestimated at the HF level, they are often slightly overestimated by LDF and are often quite close to the results of HF plus second-order many-body perturbation theory.

At the present time, we are seeing the beginnings of full co-operation between the two different schools with prominent *ab initio* theorists using LDF functionals in which HF charge densities are used as input to calculate correlation energy corrections to molecular energies, as in Clementi and Chakravorty (1990), Pople, Gill and Johnson (1992) and Becke (1993). Some of these later approaches, along with an earlier, parameterized approach to the correlation energy problem developed by Colle and Salvetti (1983) have been implemented in CRYSTAL92.

3.8 Qualitative MO theory of surface reconstruction: the example of the (111) surface of ZnS, sphalerite

Before examining other computational studies of surface properties, it is interesting to examine early qualitative interpretations of the surface relaxation observed on simple semiconductor surfaces, e.g. the enhanced planarity at Zn, and enhanced pyramidalization at S, on the ZnS(111) surface. These changes in surface atom positions have been explained by Harrison (1976), using the bond orbital model (a form of valence bond theory), and by Vaughan, Tossell and Stanley (1987), using molecular orbital theory for an AH_3 system, as applied to minerals in Tossell (1984). The electropositive Zn site is basically like a 6 valence electron AH_3 system, which is stable in planar geometry, and the electronegative S site is like an 8 valence electron AH_3 system, which is distorted to pyramidal geometry to stabilize a sigma-bonding MO. This simple argument explains the sense of the distortion and why sulfide surfaces show larger distortions than analogous oxides, since the gap between the highest energy occupied MO (HOMO) and the lowest energy unoccupied MO (LUMO) is lower in the sulfide, so the orbital mixing giving the stabilization is greater for the sulfide.

To go beyond this qualitative approach we need to carry out accurate numerical calculations of properties. HF band theoretical results for bulk ZnS (sphalerite) using CRYSTAL92 are shown in Table 3.4, where they are compared with other calculations and with experiment. Properties considered are the lattice constant, a_0, the bulk modulus, B_0, the pressure derivative of the bulk modulus, B_0' and the cohesive energy, E_0. The LDF results considered are from plane-wave-pseudopotential calculations (Martins, Troullier and Wei, 1991) and linearized augmented plane wave calculations (Engel

Table 3.4 Comparison of HF band theory results, LDF results (using plane-wave-pseudopotential, PPP, and linearized augmented plane wave, LAPW, approaches) and experiment for the properties of bulk ZnS

Property	LDF, PPP	LDF, LAPW	HF	Experimental values
Lattice constant, a_0 (Å)	5.349	5.353	5.356	5.404
Bulk modulus, B_0 (GPa)	82	87	86	76.9
Pressure derivative of bulk modulus, B'_0	4.6	4.9	3.6	4.91
Cohesive energy, E_0 (eV)	7.22	7.25	7.45	6.33

and Needs, 1990). The basis set used in the HF calculations was of valence double-zeta type. Although more work remains to evaluate other properties of ZnS and to examine further the basis set dependence of the results, it is apparent that HF theory does yield a reasonable description of bulk ZnS.

3.8.1 Other species resembling the surfaces of ZnS

Let us now consider various procedures for describing the surfaces of ZnS. One straightforward but difficult procedure is to perform HF band calculations on slab models for the various ZnS surfaces and such calculations are underway (Tossell, unpublished results). Alternatively, we can perform calculations on cluster models for the ZnS surface, starting with the simplest possible model for Zn on the (111) surface, $Zn(SH)_3^-$, with the H employed to saturate dangling bonds, and then work our way up to larger clusters. Another related approach is to find a real molecule that serves as a model for the surface. This procedure is similar to that used by inorganic biochemists interested in enzymes and in designing and studying organometallic molecules, which have the same local geometry about the transition metal atom(s) as do the complicated enzyme molecules of interest.

An example is the recently synthesized $Zn_3S_3R_3R'_3$ compound (Olmstead, Poswer and Shoner, 1991), where R and R′ are alkyl groups. This has a ring structure, slightly distorted to a chair conformation, as shown in Figure 3.5. The calculated structure for a $Zn_3S_3H_6$ model of this molecule, at the split valence basis set HF level (Saghi-Szabo and Tossell, unpublished results), for C_s geometry is shown in Figure 3.6 and has much the same structure as the experimental model, with a slight distortion to a chair conformation and a modest amount of pyramidalization at S (sum of angles at S is 353.7° by calculation and approximately 348° by experiment). Models of even greater complexity have been used to describe Zn chalcogenide surfaces by Toth *et al.* (1992), but using an approximate minimum basis set HF approach with only partial geometry optimization. The general procedure in doing cluster calculations of this type is to do one set of calculations on larger and larger

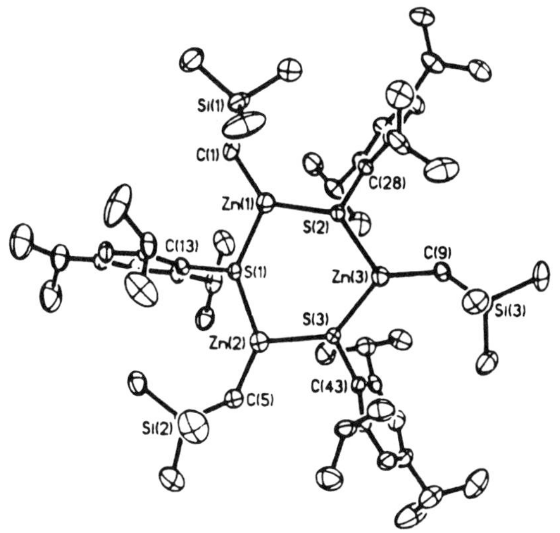

Figure 3.5 Structure of $Zn_3S_3R_3R'_3$, R = CH_2SiH_3, R′ = 2, 4, 6-i-$Pr_3C_6H_2$ from Olmstead, Poswer and Shoner (1991). H atoms are omitted for clarity.

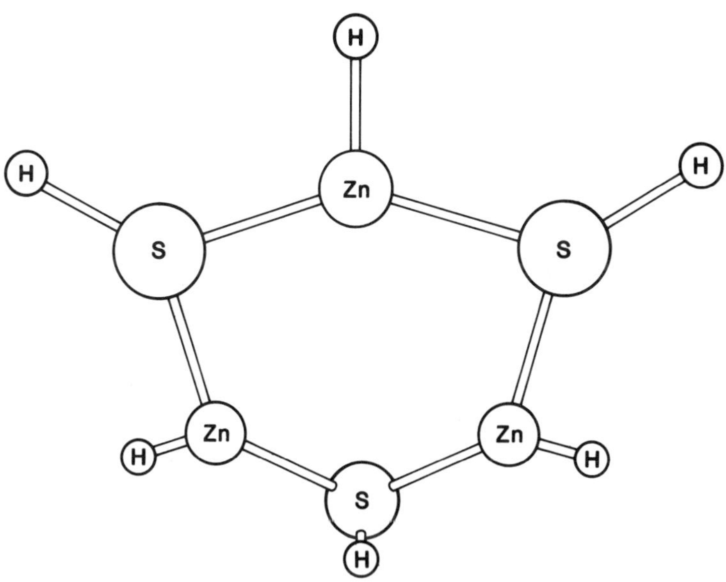

Figure 3.6 Optimized split valence Hartree–Fock C_s symmetry geometry of $Zn_3S_3H_6$ (Saghi-Szabo and Tossell, unpublished results). Atoms are labelled accordingly.

cluster models to establish convergence with cluster size, and a second set of calculations on a small cluster model at different levels of quantum mechanical rigour to establish convergence with respect to the rigour of the calculation. Focusing on a real molecule with a related structure can be useful because it provides a test for the accuracy of the results. Further studies on the properties and/or reactivity of this small molecule can also provide additional understanding of surface properties.

In this regard, it is useful to consider some new types of systems related to surfaces:

- nanocrystals, species so small they have a large fraction of atoms on the surface, e.g. nanocrystals of ZnS studied by Zhang *et al.* (1992);
- new inorganic materials, e.g. $Zn_3S_3R_3R'_3$ just discussed;
- solution species, e.g. $Zn(SH)_n^{2-n}$, studied by Tossell and Vaughan (1993).

It is important to recognize the parallels between species in different regimes – solution, surface, nanocrystal and bulk solid.

3.9 A case study comparing HF cluster and band theory and LDF cluster results for CO adsorbed on MgO

To get some idea of the degree of agreement between properties calculated by various cluster and band descriptions of adsorption phenomena, we now consider results for CO adsorbed on the MgO(100) surface from periodic HF calculations (Dovesi *et al.*, 1987), a perturbed cluster approach based on the periodic HF results (Pisani *et al.*, 1989), cluster HF calculations (Colburn and Mackrodt, 1984), and cluster LDF calculations (Neyman and Rosch, 1992), as shown in Table 3.5. This system is of substantial interest, since MgO is an effective catalyst for the oxidative conversion of methane.

The difficulty of the periodic HF calculation increases as the ratio of CO/MgO becomes smaller, since the surface unit cell must become larger. For the lowest ratio studied, the periodic HF results are similar to the

Table 3.5 Comparison of values calculated, using different methods, of interatomic distances (R in Å) and adsorption energy (ΔE in kcal mol^{-1}) for CO adsorbed on the MgO(100) surface

	Periodic HF 1:1 ratio of CO/MgO	Periodic HF 1:4 ratio of CO/MgO	Perturbed cluster based on periodic HF	Cluster HF	Cluster LDF
R (Mg–C) Å	2.408	2.358	2.401	2.49	2.22
R (C–O) Å	1.148	1.150	1.150	1.13 (assumed)	1.124
ΔE (kcal mol^{-1})	1.96	6.66	7.53	9.0	21.7, 8.5*

* Results for smaller and larger clusters, respectively.

perturbed cluster results based upon them (which in principle should be somewhat better). Given the differences in basis set between the periodic HF and cluster HF calculations, the results are about as close as one could expect. The cluster LDF results are somewhat different, with a larger adsorption energy of 21.7 kcal, although later results on larger clusters give the smaller value of 8.5 kcal. Neyman and Rosch (1992) quote experimental values of from 3.5 to 10.6 kcal for the adsorption energy.

3.10 Theory and experiment for the (111) and (110) surfaces of Li_2O, the (0001) surface of Al_2O_3 and the (110) surface of GaAs

The surface relaxation and electronic structure of Li_2O slabs have been studied by the periodic HF method (Lichanot *et al.*, 1991). Their three-layer slab geometry for the Li_2O surface is shown in Figure 3.4. Their relaxed geometry is only slightly modified from a truncation of the bulk – much less relaxed than that calculated by Mackrodt and Tasker (1985) using a pair-potential approach. In some cases, the percentage changes in lattice dimensions differed by a factor of four between the periodic HF and the pair-potential results. Lichanot *et al.* (1991) also report band structures, densities of states and electron density maps for the various Li_2O surfaces.

A similar comparison of LDF (Manassidis, De Vita and Gillan, 1993) and HF results for the (0001) surface of Al_2O_3 shows the same discrepancy – the LDF calculations show a large amount of surface reconstruction, while the HF results (albeit using a minimum basis set) show much smaller reconstruction. It would be extremely valuable to have reliable experimental data to help decide between these two results and to suggest which approach will most need improvement if it is to correctly describe surface reconstruction.

Srivastava and Jones (1992) have reviewed some calculations on the surface reconstruction and densities of states of GaAs. Calculated reconstructions from a number of methods are in reasonable agreement with each other and with experiment, as shown in Table 3.6, and the band structure calculated

Table 3.6 Calculated and experimental displacements of layer anions and cations on the GaAs (110) surface

Method	Perpendicular displacement of cation (Å)	Perpendicular displacement of cation (Å)	Parallel displacement of anion (Å)	Parallel displacement of cation (Å)
LDF band method	0.235	−0.510	−0.325	−0.530
Semi-empirical tight binding method	0.190	−0.460	−0.190	−0.350
Experimental values from LEED and ion scattering	0.159, 0.200	−0.527, −0.510	−0.235, −0.340	−0.512, −0.530

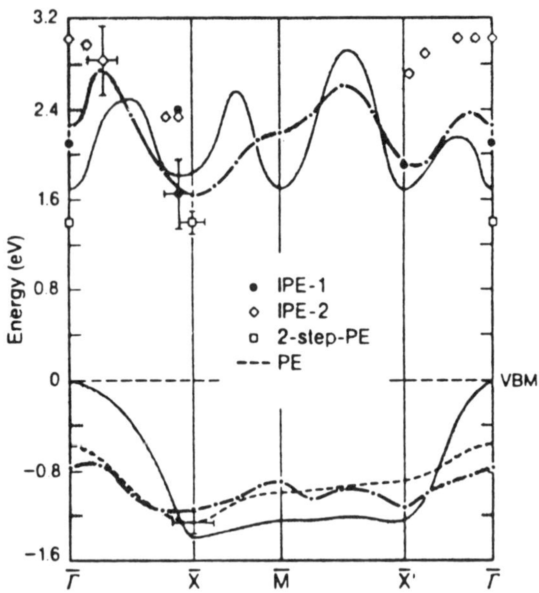

Figure 3.7 State-of-the-art calculation of the photoemission energies (PE) and inverse photoemission energies (IPE) on the GaAs(110) surface and comparison with experimental results. (From Srivastava and Jones, 1992.)

with a state-of-the-art quasiparticle approach (i.e. correcting the LDF bands for the electron self-energy) gives good agreement with the photoemission (PE) and inverse photoemission (IPE) data, as shown in Figure 3.7. At this high level of theory, agreement of calculated and experimental PE and IPE spectra is quite good, even though the self-energy correction is still only approximate.

3.11 Semi-empirical calculations and atomistic simulations in surface science

So far we have discussed only first principles quantum mechanical calculations and have not considered the role of approximate or semi-empirical quantum mechanical calculations and semi-classical pair-potential simulations in surface science. Semi-empirical calculations at the level of extended Huckel (EH) theory have been used for some time to understand qualitative, symmetry-related features of surfaces. For example, Messmer and Bennett (1972) successfully analysed the reactions of small molecules with graphite at the EH level. These methods continue to be used to address surface science questions and can be effective when proper account is taken of their limitations, and when they are closely meshed with experiment, e.g. as in

Parkinson, Ren and Whangbo (1991), who studied theoretically and experimentally the STM and AFM images of ReS_2.

More accurate semi-empirical methods are also still of use, mainly because of the n^4 dependence of HF calculations previously noted. Two particularly good methods are the INDO method as formulated by Zerner and co-workers (Anderson, Cundari and Zerner, 1991) for transition metal compounds, and the numerous variants of the MNDO procedure developed by Dewar *et al.* (1985), currently designated AM1, PM3, MOPAC, etc. A good example is a recent calculation by Rodriguez (1992) using INDO to study S compounds such as thiophene adsorbed on Mo surfaces. Such carefully parameterized semi-empirical methods will be valuable for some time to come.

Pair-potential methods have not been applied much to sulfides because of the lack of parameters. Many pair-potential studies of bulk properties of oxides and silicates exist and Garofalini (1990) has applied such methods to silicate surfaces. It was noted earlier that approximate tight binding studies have been performed by Qian and Chadi (1987) on semiconductor surfaces. These studies are intermediate between atomistic pair-potential studies and approximate quantum mechanical studies. A general problem with such methods applied to unusual bonding situations (such as surfaces) is that their parameters are usually obtained from normal bonding situations in bulk crystalline solids, and it is not certain that these parameters are transferable to lowered coordination sites on surfaces.

3.12 Conclusions

- The accuracy of present-day quantum mechanical methods is great enough to yield information of use to mineral surface scientists in interpreting properties of surfaces.
- Both infinite lattice and local models and both HF and LDF methods are valuable and will continue to be used for some time, with continued competition but enhanced co-operation between adherents to the different approaches.
- Changes in the computational environment will increase our ability to study theoretically mineral surfaces, but semi-empirical and qualitative approaches will be important for some time to come.
- Calculations at the level of rigour needed to describe definitively surface phenomena will continue to be very difficult.

Acknowledgements

This work was supported by grants from NATO (CRG 90015) and NSF (EAR 9000654).

References

Ancilotto, F., Andreoni, W., Selloni, A., Car, R. and Parrinello, M. (1990) Structural, electronic and vibrational properties of Si(111)-2×1 from ab initio molecular dynamics. *Phys. Rev. Lett.*, **65**, 3148–51.

Anderson, W.P., Cundari, T.R. and Zerner, M.C. (1991) An intermediate neglect of differential overlap model for second-row transition metal species. *Int. J. Quantum Chem.*, **39**, 31–45.

Andzelm, J. and Wimmer, E. (1992) Density functional Gaussian-type-orbital approach to molecular geometries, vibrations, and reaction energies. *J. Chem. Phys.*, **96**, 1280–303.

Bagus, P.S. and Pacchioni, G. (1992) The contribution of metal sp electrons to the chemisorption of CO: theoretical studies of CO on Li, Na and Cu. *Surf. Sci.*, **278**, 427–36.

Becke, A. (1993) A new mizing of Hartree–Fock and local density-functional theories. *J. Chem. Phys.*, **98**, 1372–7.

Brommer, K.D., Needels, M., Larson, B.E. and Joannopoulos, J.D. (1992) Ab initio theory of the Si(111)-7×7 surface reconstruction: a challenge for massively parallel computation. *Phys. Rev. Lett.*, **68**, 1355–8.

Car, R. and Parrinello, M. (1985) Unified approach for molecular dynamics and density functional theory. *Phys. Rev. Lett.*, **55**, 2471–4.

Ceperly, D.M. and Alder, B.J. (1980) Ground state of the electron gas by a stochastic method. *Phys. Rev. Lett.*, **45**, 566–9.

Clementi, E. and Chakravorty, S.J. (1990) A comparative study of density functional models to estimate molecular atomization energies. *J. Chem. Phys.*, **93**, 2591–602.

Colburn, E.A. and Mackrodt, W.C. (1984) A theoretical study of CO chemisorption at (001) surfaces of non-defective and doped MgO. *Surf. Sci.*, **143**, 391–410.

Coley, T.R., Goddard, W.A., III and Baldeschwieler, J.D. (1991) Theoretical interpretation of scanning tunneling microscopy images: application to the molybdenum disulfide family of transition metal dichalcogenides. *J. Vac. Sci. Tech.*, **B9**, 470–4.

Colle, R. and Salvetti, O. (1983) A general method for approximating the electronic correlation energy in molecules and solids. *J. Chem. Phys.*, **79**, 1404–7.

Counts, R.W. (1992) The plight of the simple worker. *J. Comp.-Aided Molec. Des.*, **6**, 427–8.

Delley, B. (1991) Analytic energy derivatives in the numerical local-density-functional approach. *J. Chem. Phys.*, **94**, 7245–50.

Delley, B., Freeman, A.J. and Ellis, D.E. (1983) Metal–metal bonding in Cr–Cr and Mo–Mo dimers: another success of local spin-density theory. *Phys. Rev. Lett.*, **50**, 488–91.

Dewar, M.J.S., Zoebisch, E.G., Healy, E.F. and Stewart, J.P. (1985) Development and use of quantum mechanical models. 76. AM1: a new general purpose quantum mechanical molecular model. *J. Am. Chem. Soc.*, **107**, 3902–9.

Dovesi, R., Pisani, C., Roetti, C. and Silvi, B. (1987) The electronic structure of α quartz; a periodic Hartree–Fock calculation. *J. Chem. Phys.*, **86**, 6967–70.

Engel, G.E. and Needs, R.J. (1990) Total energy calculations on zinc sulfide polytypes. *J. Phys. C., Condensed Matter*, **2**, 367–76.

Garofalini, S.H. (1990) Molecular dynamics computer simulations of silica surface structure and adsorption of water molecules. *J. Non-Cryst. Solids*, **120**, 1–12.

Goodgame, M.M. and Goddard, W.A., III (1982) Nature of Mo–Mo and Cr–Cr multiple bonds: a challenge for the local-density approximation. *Phys. Rev. Lett.*, **48**, 135–8.

Harrison, W.A. (1976) Surface reconstruction on semiconductors. *Surf. Sci.*, **55**, 1–19.

Hehre, W.J., Radom, L., Schleyer, P.V.R. and Pople, J.A. (1986) *Ab Initio Molecular Orbital Theory*. Wiley, New York.

Henrich, V.E., Zeiger, H.J. and Reed, T.B. (1978) Ultraviolet photoemission measurements of the band structure of TiO_x ($0.93 \leqslant x \leqslant 1.15$). *Phys. Rev.*, **B17**, 4121–3.

Jennison, D.R. and Kunz, A.B. (1977) Significant differences between Hartree–Fock and local-exchange energy bands for TiO. *Phys. Rev. Lett.*, **39**, 418–21.

Levine, I.N. (1991) *Quantum Chemistry*, 4th edn, Prentice-Hall, Englewood Cliffs, New Jersey.

Lichanot, A., Gelize, M., Larrieu, C. and Pisani, C. (1991) Hartree–Fock ab initio study of relaxation and electronic structure of lithium oxide slabs. *J. Phys. Chem. Solids*, **52**, 1155–64.

Mackrodt, W.C. and Tasker, P.W. (1985) Surfaces and interfaces in oxides. *Chem. Br.*, **21**, 366–9.

Malkin, V.G., Malkina, O.L. and Salahub, D.R. (1993) Calculations of NMR shielding constants beyond uncoupled density functional theory, IGLO approach. *Chem. Phys. Lett.*, **204**, 87–95.

Manassidis, I., De Vita, A. and Gillan, M.J. (1993) Structure of the (0001) surface of α-Al_2O_3 from first principles calculations. *Surf. Sci. Lett.*, **285**, L517–21.

Martins, J.L., Troullier, N. and Wei, S.-H. (1991) Pseudopotential plane-wave calculations for ZnS. *Phys. Rev.*, **B43**, 2213–17.

McCarthy, M. and Hess, A.C. (1992) Periodic Hartree–Fock studies of the properties of the Cl_2/MgO(001) interface. *J. Chem. Phys.*, **96**, 6010–17.

McCarthy, M., Hess, A.C., Harrison, N.M. and Saunders, V.R. (1993) A study of the energetics of the Cl_2/MgO(001) interface using correlation corrected periodic Hartree–Fock theory. *J. Chem. Phys.*, **98**, 6387–91.

Messmer, R.P. (1985) Theoretical overview of electronic structure and bonding of adsorbates. *Surf. Sci.*, **158**, 40–57.

Messmer, R.P. and Bennett, A.J. (1972) Orbital-symmetry rules for chemisorption and catalysis. *Phys. Rev.*, **B6**, 633–8.

Neyman, K.M. and Rosch, N. (1992) CO bonding and vibrational modes on a perfect MgO(001) surface: LCGTO-LDF model cluster investigation. *Chem. Phys.*, **168**, 267–80.

Nicholas, J.B., Winans, R.E., Harrison, R.J. (1992) An ab initio investigation of disiloxane using extended basis sets and electron correlation. *J. Phys. Chem.*, **96**, 7958–65.

Olmstead, M.M., Poswer, P.P. and Shoner, S.C. (1991) Synthesis and characterization of novel quasiaromatic zinc–sulfur aggregates and related zinc–oxygen complexes. *J. Amer. Chem. Soc.*, **113**, 3379–85.

Parkinson, B.A., Ren, J. and Whangbo, M.-H. (1991) Relationship of STM and AFM images to the local density of states in the valence and conduction bands of $ReSe_2$. *J. Amer. Chem. Soc.*, **113**, 7833–7.

Perdew, J.P. and Zunger, A. (1981) Self-interaction correction to density-functional approximations for many-electron systems. *Phys. Rev.*, **B23**, 5048–79.

Pisani, C., Dovesi, R. and Roetti, C. (1988) *Hartree–Fock Ab Initio Treatment of Crystalline Systems*. Springer-Verlag, Berlin.

Pisani, C., Dovesi, R., Nada, R. and Tamiro, S. (1989) Ab initio Hartree–Fock perturbed-cluster treatment of local chemisorption: isolated carbon monoxide on a periodic MgO(100) substrate. *Surf. Sci.*, **216**, 489–504.

Pople, J.A., Gill, P.M.W. and Johnson, B.G. (1992) Kohn-Sham density-functional theory within a finite basis set. *Chem. Phys. Lett.*, **199**, 557–60.

Qian, G.-X. and Chadi, D.J. (1987) Si(111)-7×7 surface: energy-minimization calculation for the dimer-adatom-stacking-fault model. *Phys. Rev.*, **B35**, 1288–93.

Rodriguez, J.A. (1992) Bonding and decomposition of thiophene, sulhydryl, thiomethoxy and phenyl thiolate on Mo surfaces. *Surf. Sci.*, **278**, 326–38.

Roothaan, C.C.J. (1951) New developments in molecular orbital theory. *Rev. Mod. Phys.*, **23**, 69–89.

Simons, J. (1991) An experimental chemist's guide to *ab initio* quantum chemistry. *J. Phys. Chem.*, **95**, 1017–29.

Srivastava, G.P. and Jones, B. (1992) *Ab-initio* computer modeling of semiconductor surfaces. *Advanc. Mater.*, **4**, 482–9.

Srivistava, G.P. and Weaire, D. (1987) The theory of the cohesive energies of solids. *Adv. Phys.*, **36**, 463–517.

Stich, I., Payne, M.C., King-Smith, R.D., Lin, J.-S. and Clarke, L.J. (1992) Ab initio total-energy calculations for extremely large systems: application to the Takayanagi reconstruction of Si(111). *Phys. Rev. Lett.*, **68**, 1351–4.

Szabo, A. and Ostlund, N.S. (1989) *Modern Quantum Chemistry: Introduction to Advanced Electronic Structure Theory*, 1st edn, revised. McGraw-Hill, New York.

Tersoff, J. and Hamann, D.R. (1985) Theory of the scanning tunneling microscope. *Phys. Rev.*, **B31**, 805–13.

Tossell, J.A. (1979) Comment on 'significant differences between Hartree–Fock and local-exchange energy bands for TiO'. *Phys. Rev.*, **B19**, 5443–6.

Tossell, J.A. (1984) A qualitative MO model for bridging bond angle variations in minerals. *Phys. Chem. Mineral.*, **11**, 81–4.

Tossell, J.A. and Vaughan, D.J. (1992) *Theoretical Geochemistry: Applications of Quantum Mechanics in the Earth and Mineral Sciences*. Oxford University Press, New York.

Tossell, J.A. and Vaughan, D.J. (1993) Bisulfide complexes of Zn and Cd in aqueous solution: calculation of structure, stability, vibrational and NMR spectra and of speciation on sulfide mineral surfaces. *Geochim. Cosmochim. Acta*, **57**, 1935–45.

Toth, K., Pakkanen, T.A., Hirva, P. and Muilu, J. (1992) Model calculations for surfaces and chemisorption processes of zinc–chalcogenide thin films by an ab initio cluster approach. *Surf. Sci.*, **277**, 395–406.

Vaughan, D.J., Tossell, J.A. and Stanley, C.J. (1987) The surface properties of bornite. *Mineral. Mag.*, **51**, 285–93.

Von Barth, U. (1986) Applications of density functional theory to atoms, molecules and solids. *Chem. Scripta*, **26**, 449–61.

Zhang, V., Raman, N., Bailey, J.K., Brinker, C.J. and Crooks, R.M. (1992) A new sol-gel route for the preparation of nanometer-scale semiconductor particles that exhibit quantum optical behavior. *J. Phys. Chem.*, **96**, 9098–100.

CHAPTER FOUR

New X-ray techniques and approaches to surface mineralogy

G.N. Greaves

4.1 Introduction

This chapter concentrates on the recent use of bulk X-ray methods for studying the internal atomic structure of microporous solids and on the application of grazing incidence X-ray techniques to record the atomic structure at the external surfaces of materials. In both cases, emphasis is given to ways of highlighting migrant species both on the internal and the external surfaces of solids. These considerations are especially relevant to surface mineralogy, as most 'corrosion' processes result from an imbalance between internal and external transport mechanisms.

Many of the so-called 'new' X-ray techniques (scattering, absorption and reflectometry) were discovered 60 or more years ago. In most cases, however, it is only the advent of dedicated synchrotron radiation sources in the past decade that has drawn such phenomena into the arena of practical measurement and it is therefore only in this sense that they are novel. Among the special properties of synchrotron radiation that distinguish it from radiation from laboratory X-ray sources are intensity, collimation and tunability (Greaves and Catlow, 1990), all of which have brought immediate benefits to X-ray methods. The high intensity and collimation of synchrotron radiation have greatly reduced data collection times and improved the signal-to-background ratio for the familiar techniques of single crystal diffraction and powder diffraction. These advances have enabled the structures of microcrystals (e.g. zeolites and aluminium phosphates, which are often difficult to crystallize) to be solved. Both single crystal methods (Harding, 1990) and powder methods (Cernik, Clark and Pattison, 1991) have been refined for *ab initio* structure solving. Less familiar are the X-ray techniques that exploit the tunability of synchrotron radiation. Given the availability of multiple wavelengths, anomalous diffraction effects can be utilized to phase enhance contributions from particular elements (e.g. Attfield, 1990; Nelmes *et al.*, 1992). X-ray absorption fine structure (XAFS) spectroscopy has become an important and versatile probe of local atomic environments, a technique that can be applied to crystalline and non-crystalline systems alike (see Gurman, 1990 for an introduction). For example, XAFS has been

Mineral Surfaces. Edited by D.J. Vaughan and R.A.D. Pattrick. Published in 1995 by Chapman & Hall, London. ISBN 0 412 56340 1

successful recently in locating ion-exchanged species in crystalline aluminosilicates (Dooryhee *et al.*, 1991) and glasses (Houde-Walter *et al.*, 1993). With fluorescence detection, XAFS can be made sensitive to the environments of impurities (Jaklevic *et al.*, 1977), with obvious application to the structural characterization of trace impurities in minerals (Henderson, Greaves and Smith, 1993). All these phenomena, whether diffraction or spectroscopic in character, become surface sensitive when X-rays are incident at glancing angles (Parratt, 1954). Indeed, by altering the angle of incidence, X-rays will probe into the material to a greater or lesser extent. As there is no requirement for a vacuum environment when hard X-rays are employed, the realistic 'weathering' conditions of air and water contact can be readily simulated (Greaves, 1991).

Introductions to all of these synchrotron radiation methods can be found elsewhere (e.g. Catlow, 1990; Gurman, 1990). The following sections offer brief descriptions of X-ray powder diffraction and XAFS techniques using synchrotron radiation, taking examples from problems in mineralogy. This forms the basis for reviewing several applications of direct relevance to surface mineralogy. In particular, recent work on the structural chemistry of zeolitic catalysts, on the synthesis of minerals from the amorphous state, and on the elucidation of the species resulting from corrosion at glass surfaces, will be highlighted.

4.2 X-ray powder diffraction

4.2.1 X-ray diffraction

X-rays are diffracted by crystalline solids at specific angles, θ, dependent on the geometry of the atomic planes. In particular, if d is the distance between atomic planes, this is related to θ by the familiar Bragg equation:

$$n\lambda = 2d \sin\theta \tag{4.1}$$

where λ is the wavelength of the X-rays used and n is the order of the reflection. As X-rays are scattered primarily by electrons (rather than by nuclei), so the atoms in a crystal are located by their respective electron clouds. In particular, the electron density for a crystalline material at a given point $\mathbf{r}$ in the unit cell is given by

$$\rho(r) = 1/V \sum_G S(G) \exp(i\mathbf{G}\cdot\mathbf{r}) \tag{4.2}$$

where V is the unit cell volume, $\mathbf{G}$ is the reciprocal lattice vector of the atomic planes and $S(G)$ is the structure factor,

$$S(G) = \sum_i f_i \exp(-i\mathbf{G}\cdot\mathbf{r_i}) \tag{4.3}$$

In the above expression, the f_i are the atomic form factors of the scattering centres at the various $\mathbf{r_i}$, and, as these are electron clouds, the f_i exhibit a strong dependence on the scattering angle, decreasing with increasingly θ. Now the intensity of the X-rays diffracted from a crystalline solid, I, is given by

$$I = S(G)\,S*(G) \tag{4.4}$$

where $S*(G)$ is the complex conjugate of the structure factor. Accordingly, as only the intensities (I) and scattering angles (θ) of the diffracted beams are measured in X-ray crystallography, the fundamental problem is to extract the phase associated with the above product (equation 4.4) so that the structure factor (equation 4.3) can be extracted, the electron density (equation 4.2) computed, and the positions of the atoms determined.

Account has to be taken of any inherent disorder in the atomic positions. Thermal vibrations at a particular temperature (T) result in a reduction in the measured intensities

$$I(T) = I(0)\exp - \frac{(\sigma^2 k^2)}{2} \tag{4.5}$$

where σ^2 is the mean square displacement of atoms about their mean positions and k the X-ray wave vector ($4\pi \sin\theta/\lambda$). Static disorder will also make a contribution to the height of diffraction peaks, I. In the extreme case of a glass of liquid, Bragg diffraction is replaced by diffuse scattering from interatomic correlations, r_{ij}. This is characterized by a weakly sinusoidal background given by

$$k\,I(k) = \sum_i \sum_j N_{ij} f_i(k) f_j(k)\,[(\sin kr_{ij})/r_{ij}]\exp - (k^2\sigma_{ij}^2/2) \tag{4.6}$$

where N_{ij} are the number of type j atoms surrounding atoms of type i at distance r_{ij} and σ_{ij}^2 is the variance in r_{ij} in the interatomic distance (Warren, 1969). The double sum in equation (4.6) includes all atoms, but in practice diffuse scattering is dominated by nearest neighbours. Nearest neighbour values of r_{ij} are governed by chemical forces and are usually extremely precise, so σ_{ij}^2 values are small, whereas correlations with more distant neighbours are progressively less well defined with increasingly large σ_{ij}^2 values. Equation (4.6) is often Fourier transformed into real space to give the radial distribution function (RDF) of the average atom. For all but the simplest compositions, the structural content of diffuse scattering and its associated RDF is ambiguous.

In single-crystal X-ray crystallography the background is generally easy to assess, reflections are usually discrete, and the number of X-rays whose intensity can be unambiguously measured typically amounts to several thousand. Direct methods (Sheldrick, 1986) assign the phase of $S(G)$

(equation 4.3) on a probability basis and, from a knowledge of the positions and intensities of the diffracted X-rays, the symmetry and dimensions of the unit cell and the positions of the atoms within it can be obtained in real space.

4.2.2 Powder diffraction

Many natural materials, however, are not available as single crystals but as microcrystalline powders. If the individual particles are randomly orientated, then the Bragg 'spots' are smeared into Debye–Scherrer 'rings' and $|\mathbf{G}|$ is measured rather than $\mathbf{G}$. A powder pattern can be measured quite simply by sweeping a detector around the specimen in a 2θ circle. While this is simple enough in principle, in practice Debye–Scherrer rings overlie one another increasingly as 2θ advances and the number of resolvable peaks is extremely limited compared with single-crystal methods. Added complications relate to preferred orientation and background. As hard X-rays typically penetrate only a few microns into materials (the X-ray attenuation length being given by $1/\mu$, where μ is the X-ray absorption coefficient of the material at wavelength, λ), the microcrystals involved in X-ray scattering may not be randomly orientated, in which case the Debye–Scherrer rings will not be uniform and the profile recorded by the 2θ detector will be atypical. This problem can be removed, in part, by rotating the specimen. The background underlying the diffraction peaks in an X-ray powder pattern may be difficult to determine precisely, and although largely instrumental, it may also contain an indeterminate diffuse scattering component. As a result of all these uncertainties, *ab initio* structure solving by X-ray powder methods has generally been considered impracticable for all but the simplest structures and analysis of X-ray Debye–Scherrer patterns has been traditionally limited to the determination of unit cell parameters.

4.2.3 High resolution method crystal structure refinement

This situation has changed radically with the development of synchrotron radiation techniques. The experimental arrangements commonly used for powder diffraction are illustrated in Figure 4.1. The angle-dispersed geometry shown in Figure 4.1(a) was developed by Hart and Parrish (1989). Compared with a laboratory X-ray source, where the intensity and collimation of the radiation is limited, a synchrotron source can now be obtained with very high resolution powder patterns and very low instrumental background, as illustrated in Figure 4.2. Peak widths (FWHM) are better than 0.03° in 2θ, compared with ~0.15° for a typical laboratory source. Moreover, the peak shapes obtained from synchrotron powder diffraction are much simpler (readily simulated by pseudo-Voigt functions), all of which reduces the fitting parameters in any least squares analysis. Recent devel-

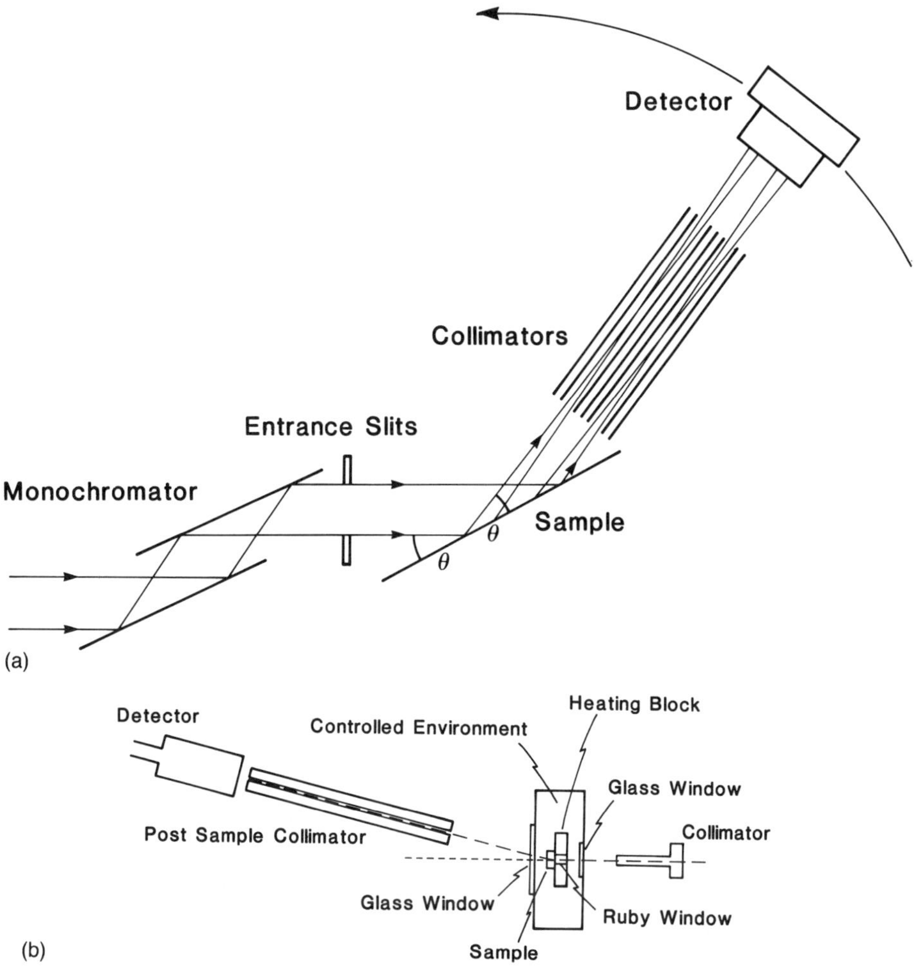

Figure 4.1 Experimental arrangements for powder diffraction using synchrotron radiation. (a) Angle-dispersed powder diffraction (Hart and Parrish, 1989): the white radiation from the storage ring is monochromatized and the Debye–Scherrer pattern intercepted using a $\theta/2\theta$ diffractometer; the long multiple collimators, which are matched to the source/monochromator angular resolution, enable large flat plate (rotating) samples to be employed with much improved signal. (b) Energy-dispersed powder diffraction (Clark, 1989): white radiation is highly collimated and diffracted X-rays at a fixed angle, θ, are collected using a solid state detector; the diffracted volume is indicated by the intersection of the incident and diffracted beams.

opments of image plate X-ray detector systems are also particularly exciting (Nelmes *et al.*, 1992). These have been specifically devised for high pressure experiments using diamond anvil cells, but have more general application. By imaging and integrating the intensities from complete Debye–Scherrer rings, orientational effects can be all but eliminated, and by increasing the

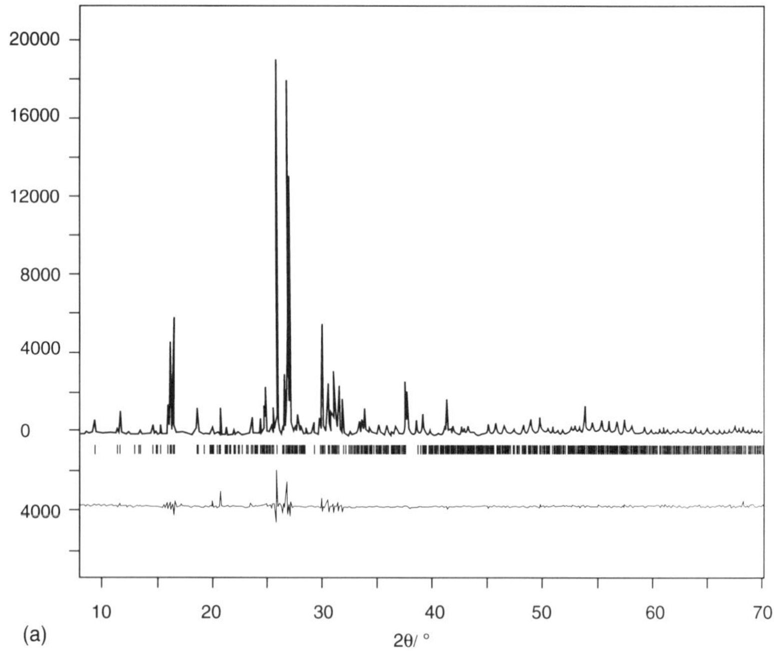

(a)

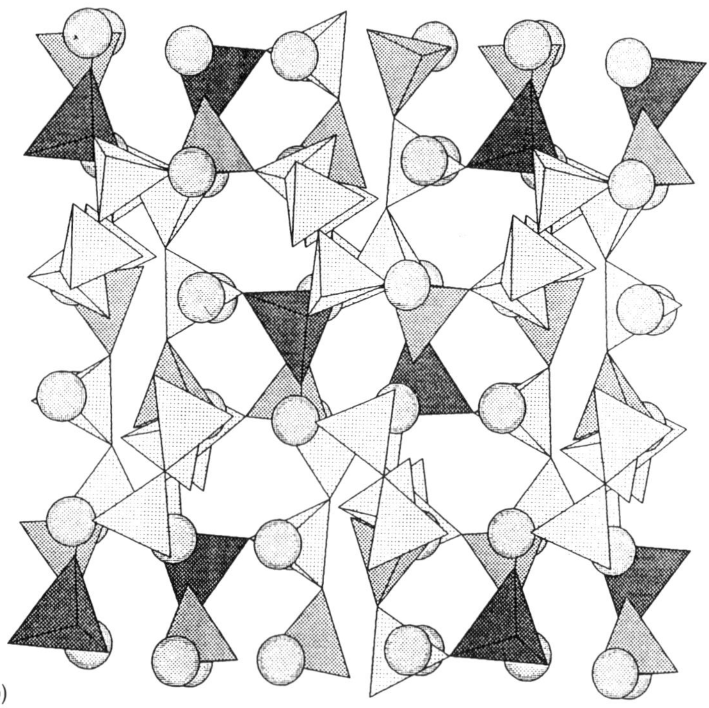

(b)

Figure 4.2 High resolution powder diffraction and crystallographic refinement. (a) The pattern of a monoclinic leucite analogue $K_2MgSi_5O_{12}$, including the raw data (top), the indexed reflections (middle) and the residuals (bottom) from refinement (Bell *et al.*, 1993). (b) The chemically ordered atomic structure obtained for $K_2MgSi_5O_{12}$ ((100) projection). Mg sites – dark shaded tetrahedra; Si sites contiguous to four Si atoms – medium shaded tetrahedra; Si sites contiguous to three Si atoms and one Mg atom – light shaded tetrahedra, K – spheres.

scattering solid angle, the signal-to-background for a given exposure time is greatly enhanced.

As a result of these improvements, many more diffraction lines (typically several hundred) in an X-ray powder pattern can now be deconvoluted, to the point where a combination of direct methods and refinement procedures (Rietveld, 1969) can be successfully applied to solve structures directly (Cernik *et al.*, 1991). Accordingly, once the crystal symmetry has been determined and the X-ray powder pattern indexed, the intensity of selected lines can be used in conjunction with direct methods to obtain possible phases for the $S(G)$. This furnishes various likely Fourier maps of the electron density (equation 4.2), which can then be used as possible starting model structures for refinement. Atomic coordinates and Debye–Waller (or temperature) factors are iterated to obtain the best agreement between calculated diffracted intensities, I_{calc}, and the experimental values, I_{obs}. The fitting parameter

$$R = \sum (I_{obs} - I_{calc})/I_{obs}$$

records the quality of fit. Because R is summed over all the measured diffraction peaks (including those that are superimposed), the information content available from the powder pattern is maximized.

4.2.4 *Monoclinic leucite*

As an example, Figure 4.2 shows synchrotron radiation results for a hitherto unknown monoclinic form of a synthetic analogue of leucite, $K_2MgSi_5O_{12}$, formed by hydrothermal synthesis (Bell *et al.*, 1993). The high quality of the raw data (Figure 4.2(a)) can be readily appreciated. The positions of indexed reflections are shown as are the differences between calculation and experiment ($I_{obs} - I_{calc}$), demonstrating the close agreement. The refined atomic structure is shown in Figure 4.2(b). The monoclinic symmetry was determined from electron diffraction, enabling the powder pattern to be indexed. The starting atomic positions used for refinement were based on those of the structure of $K_2MgSi_5O_{12}$ produced by 'dry' synthesis. This is cubic, with Mg and Si occupying tetrahedral sites but chemically disordered, the considerable difference between Si—O and Mg—O separations (1.6 Å compared to 1.9 Å) resulting in large temperature factors for tetrahedral sites. In contrast, magic angle spinning ^{29}Si NMR distinguished ten Si sites for the monoclinic hydrothermally synthesized form, pointing to the ordering of tetrahedral cation sites. Therefore, a model based on the cubic form but in which Mg and Si atoms were ordered was used as a starting point from which the fit shown in Figure 4.2 was eventually obtained. The ordering resulted in a reduction of tetrahedral site temperature factors, and this appears to compensate for the lower symmetry. This example illustrates the importance of shoring up the information explicit in the high resolution X-ray powder pattern with complementary clues from other probes of atomic structure. Another example of the interplay between spectroscopy and diffraction will be given in the account of ion-exchanged zeolites (see Figures 4.11 and 4.12).

4.2.5 Energy dispersive powder diffraction

The experimental arrangement shown in Figure 4.1(a), however, is often impracticable for *in situ* experiments on samples in extreme environments, such as the conditions of high pressure and temperature relevant to Earth processes. Under these circumstances, it is usually desirable to collect the whole diffraction pattern at one time rather than sequentially with a scanning detector, when the conditions may vary with k. This can be done by fixing the scattering angle, θ, and measuring the diffracted intensities, $I(E)$, as a function of X-ray energy (Catlow, 1990). By fixing the diffraction angle, θ, and replacing λ by ch/E (where c is the speed of light and h is Planck's constant) in equation (4.1), the Bragg equation can be reformulated as

$$E = ch/(2d \sin \theta)$$

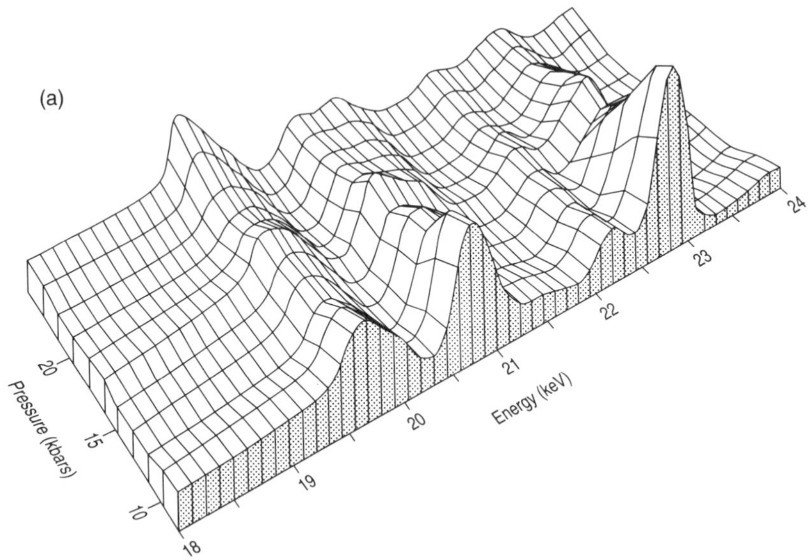

Figure 4.3 High pressure energy dispersive diffraction. The study of a ferroelastic phase transition in gillespite (Redfern *et al.*, 1993). (a) Pressure-dependent powder patterns; (b) pressure dependence of the unit cell volume (top) and variation of parameters (bottom) – the a parameter splits into a + b. (10 kbar = 1 GPa.)

(b)

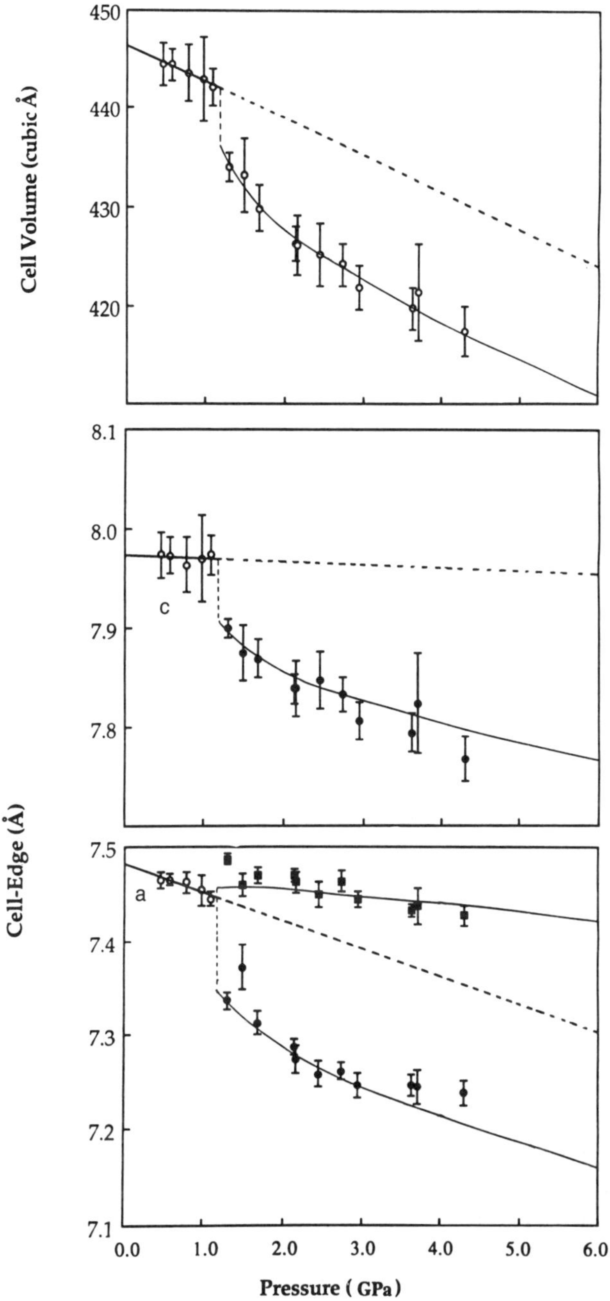

Figure 4.3 *Continued*

The **white** character of synchrotron radiation makes this an ideal source for measuring powder patterns in an energy-dispersive mode. A suitable experimental geometry is presented in Figure 4.1(b), which shows the white beam collimated directly from the storage ring and incident on to a specimen held in a controlled environment (Clark, 1989). The fixed angle is defined by long collimator slits and the angular resolution, $\Delta\theta$, matched to the angular divergence of the incident beam. Routine values for θ and $\Delta\theta$ are 10° and 0.01°, respectively. For this arrangement, the incoming beam is typically ~0.1 mm in diameter, which clearly confers huge advantages on measuring diffraction from very small specimens in restricted environment cells.

4.2.6 First-order phase transition in gillespite

Figure 4.3 provides a striking example of energy dispersive powder diffraction used to follow the details of a high pressure phase transition. The example illustrated reveals the sharp, first-order tetragonal–orthorhombic transition in gillespite ($BaFeSi_4O_{10}$). This can be seen in the abrupt splitting of peaks in the powder patterns as the pressure is changed. Raising and lowering the pressure reveals surprisingly little hysteresis, the measurements locating the transition very precisely at 1.2 ± 0.1 GPa. The changes in the lattice parameters analysed from these patterns reveal, however, an anomalous 1.5% volume change at the transition (also shown in Figure 4.3). Consideration of the volume mismatch and the switch in unit cell dimensions across the tetragonal–orthorhombic transition indicate that this symmetry transformation is more than a simple ferroelastic modification and involves two coupled transition processes which most likely embody some displacement character (Redfern *et al.*, 1993). Another example of the utility of energy dispersive diffraction for characterizing materials synthesis will be given later (Figure 4.9).

4.3 X-ray absorption fine structure

4.3.1 Fundamentals

Examples of X-ray absorption thresholds can be seen in Figures 4.6, 4.8, 4.11, 4.14 and 4.19. These occur whenever the X-ray energy is sufficient to excite a core level photoelectron. The fine structure or XAFS occurring above the threshold results from the scattering of the photoelectron by atoms neighbouring the excited atom. The simplest expression to describe normalized fine structure, χ, the plane wave approximation, bears a marked resemblance to the Debye formula for diffuse scattering (equation 4.6) except that the double sum is replaced by a single sum

$$k\chi_i(k) = \sum_j N_{ij} f_j(k,\pi)\,[(\sin(2kr_{ij} + 2\delta_i + \varphi_j))/r_{ij}^2]\exp - (2k^2\sigma_{ij}^2)\exp - (2r_{ij}/\lambda) \tag{4.7}$$

This simplification derives from the fact that the ith atom is the excited atom generating the photoelectron probe of its immediate environment. As a result, most of the ambiguities of determining the local structure of individual elements from diffuse scattering are removed. The wave-vector k in equation (4.7) is that of a free electron, that is

$$h^2k^2/8\pi^2m = h\nu - E_b + E_0$$

where E_b is the photoelectron binding energy and E_0 the free electron energy minimum. Then $f_j(k,\pi)$ and φ_j in equation (4.7) are the photoelectron backscattering amplitude and phase shift for the neighbouring atom j, respectively, δ_i is the change in phase of the photoelectron interacting with the excited or central atom, and λ is its mean free path. These replace the atomic form factors in crystallography. All of these photoelectron quantities are calculable (Lee and Pendry, 1975) and can be refined using model systems (Gurman, 1990). The remaining quantities in equation (4.7), N_{ij}, r_{ij} and σ_{ij}^2, are the structural parameters of interest and have the same meaning as given above for equation (4.6). Although equation (4.7) can be Fourier transformed to yield a radial distribution function specific to atom-type i, the plane wave expression is not exact (the photoelectron wave is far from plane in character except when k is large (typically $>\sim 5\,\text{Å}^{-1}$)). As a result, XAFS analysis is generally performed in k-space using the spherical wave formalism (Lee and Pendry, 1975) rather than in r-space using the Fourier transform of equation (4.7).

4.3.2 *Local structure by least squares analysis*

Because of the structurally underdetermined character of XAFS (in most cases it is a radial average of the local three-dimensional structure), analysis is a least squares refinement procedure. A model for the structural environment of the element in question is chosen and then N_{ij}, r_{ij} and σ_{ij}^2 are adjusted until the best match with experiment is achieved (Stephenson, Greaves and Gurman, 1991). The goodness of fit is generally given by an index of the form

$$f = \sum_n [(x_{\text{obs}} - x_{\text{calc}})k^3]^2/n$$

where n is the number of experimental points, which might typically be around 300 for a spectral range extending to a $k_{\max}$ of $12\,\text{Å}^{-1}$.

Experimentally, χ is obtained from the measured absorption, μ, by normalizing the fine structure with respect to the unstructured absorption, μ_a, of a free atom, that is

$$\chi = (\mu - \mu_a)/(\mu_a - \mu_b)$$

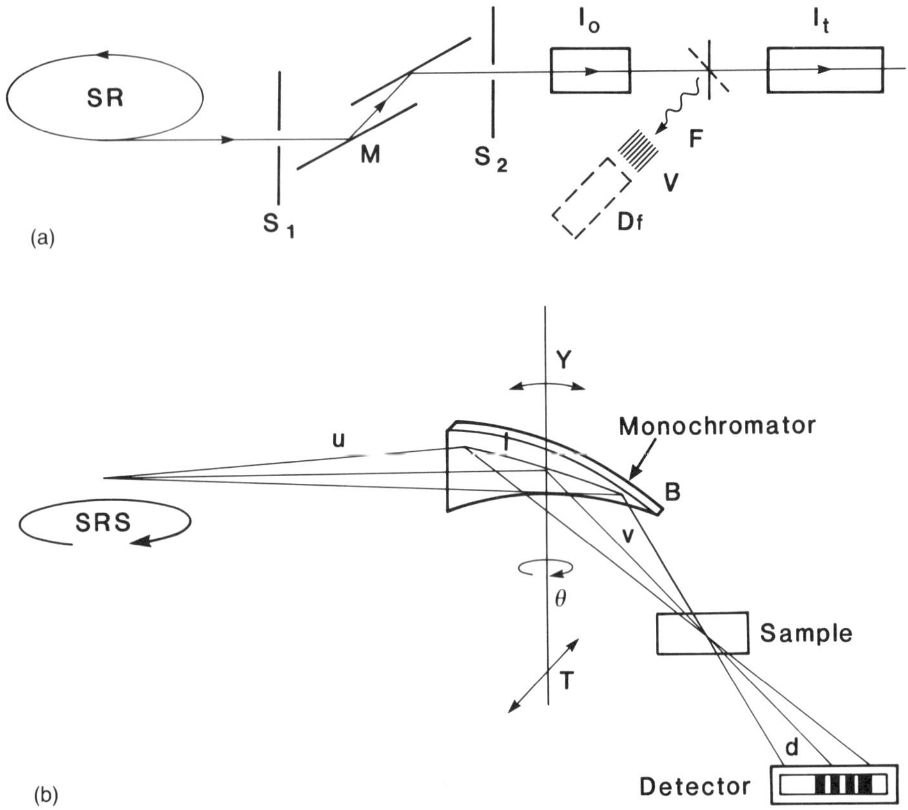

Figure 4.4 Experimental arrangements for X-ray absorption fine structure (XAFS). (a) Conventional scanning mode: the slits, S_1 and S_2, define the energy resolution of the vertically diffracting monochromator, M; the detectors I_0 and I_t record the incident and transmitted intensities, respectively, and D_f, the X-ray fluorescence. (b) Energy dispersive mode: the bent crystal monochromator, radius R_M, selects the geometrically dispersed band pass, ΔE (= cot Θ/R_M); the transmitted spectrum is measured by a position-sensitive detector (Allinson *et al.*, 1988).

where μ_b is the background X-ray absorption coefficient due to the remaining absorption processes. However, $\mu - \mu_a$ is typically only a few per cent, or less, of $(\mu_a - \mu_b)$ and the demands on maximizing signal-to-noise in the measured μ are critical. Generally, the number of absorbed photons required for each experimental point should exceed 10^8, which, for all practical purposes, is beyond the bremsstrahlung intensity capability of even a rotating anode generator. On the other hand, synchrotron radiation sources deliver in excess of $10^{11}\,ps^{-1}$ of monochromatic X-rays and enable high quality XAFS spectra to be obtained routinely.

The standard experimental arrangement is illustrated in Figure 4.4(a).

Monochromatic radiation is incident on a specimen and the absorbed X-rays are detected either in transmission (generally using ion chambers) or from the X-ray fluorescence emitted, often detected with solid state detectors (Bilsborrow, 1992; Derbyshire *et al.*, 1992). Details of the different experimental geometries and various trade-offs in terms of signal-to-noise can be found elsewhere (Greaves, 1990a). The following illustrations taken from recent work give an indication of the versatility of the XAFS technique.

4.3.3 Modifying cations in glasses

XAFS has proved particularly powerful in revealing the atomic environments of modifying cations in oxide glasses, which have led to new microscopic mechanisms for ionic transport (Greaves *et al.*, 1991). Monovalent cations readily ion exchange for the alkalis in glasses and presumably share similar sites and diffusion pathways to the cations they replace. In Figure 4.5, the K-edge XAFS of Na in an aluminosilicate glass ($Na_4Al_2Si_5O_{15}$) is contrasted with that of Ag ion exchanged into the same material (Houde-Walter *et al.*, 1993). The results of least squares fitting to experiment are shown by the dashed curves, which indicate significant changes in the atomic environment between host (a) and dopant (b) cations. Compared with Na, where the number of nearest neighbour oxygens in the aluminosilicate glass is found to be close to 3 (Figure 4.5(a)), the co-ordination number for ion-exchanged Ag (Figure 4.5(b)) is approximately 2. The interatomic distance is also sharply reduced from 2.30 Å (r_{Na-O}) to 2.08 Å (r_{Ag-O}) and with it the Debye–Waller factor ($2\sigma^2_{Na-O} = 0.015$ Å^2, $2\sigma^2_{Ag-O} = 0.008$ Å^2). Clearly Ag locates more tightly to the aluminosilicate framework than the Na it replaces. Moreover, the different nearest neighbour oxygen distances point to different electrostatic binding energies for Na^+ and Ag^+ cations, which accord well with the different self-diffusion coefficients of the two cations. We will return to the theme of ion exchange later, using the benefits of XAFS to identify the environments of ion-exchanged cations in zeolites (see Figure 4.11).

4.3.4 Cu^{2+} in tetrahedrites

The second example relates to the fine structure at the X-ray absorption threshold or near-edge structure. The scattering of the photoelectron for energies $<\sim 50$ eV is much stronger than for XAFS at higher energies. Multiple scattering paths prevail for low energy photoelectrons, rather than the simple back scattering mechanism referred to above, which applies when kinetic energies are greater. Accordingly, near-edge structure carries the imprint of local symmetry in addition to the radially averaged N_{ij}, r_{ij}, σ^2_{ij} quantities (Durham, Pendry and Hodges, 1981). Of particular relevance to mixed valence cations, the absolute position of the absorption threshold

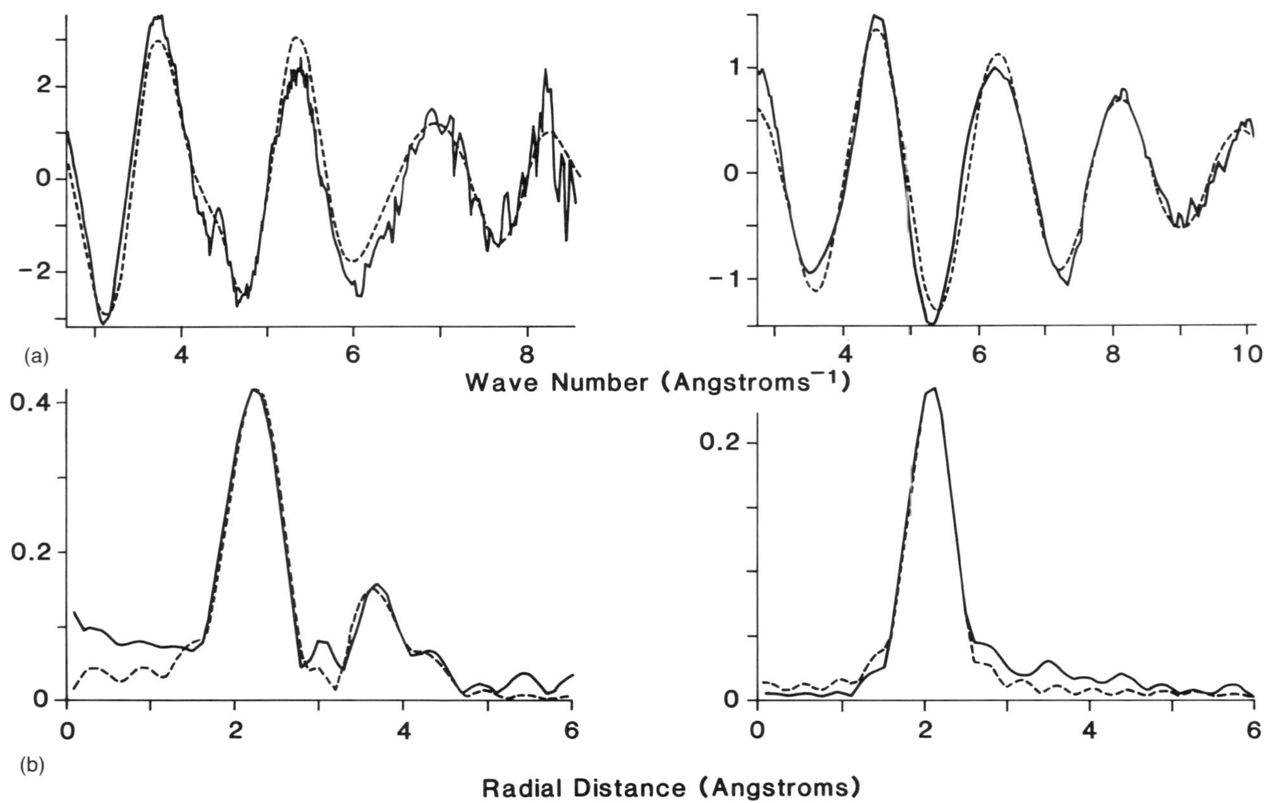

Figure 4.5 Host cation environments for an aluminosilicate glass and dopant cation environments following ion exchange (Houde-Walter *et al.*, 1993). Normalized weighted XAFS is shown above, with the corresponding Fourier transform below. Raw data are represented by the solid line and the results of least squares fitting by the dashed line. (a) Na K-edge XAFS reveals host cations are threefold co-ordinated, with an Na—O radius of 2.30 Å. (b) Dopant cation environments are obtained from Ag K-edge XAFS. Ag

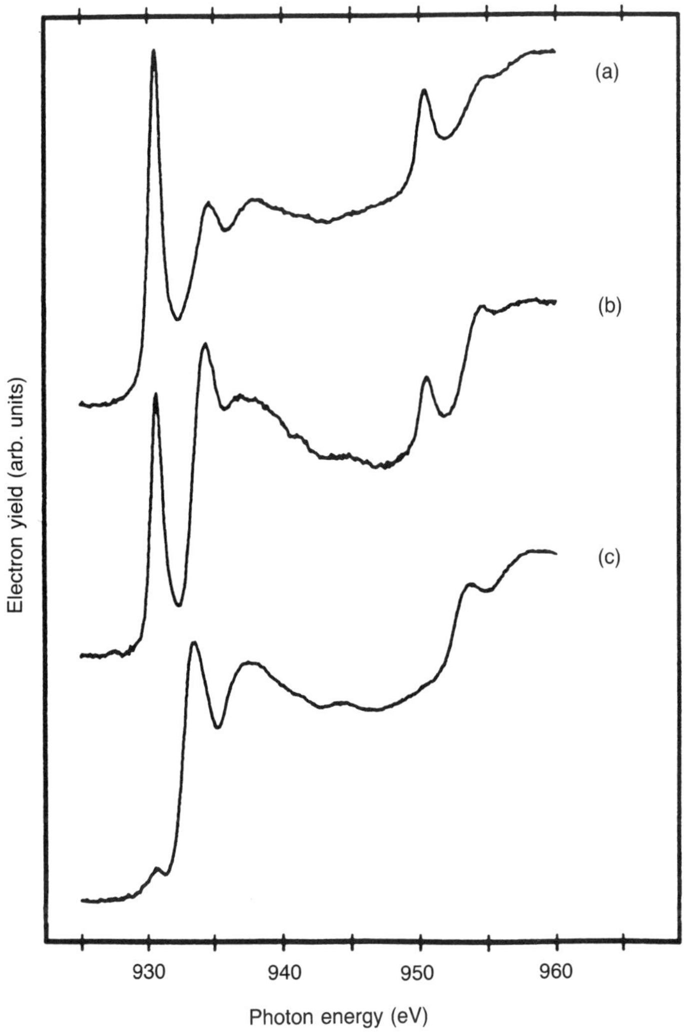

Figure 4.6 Oxidation states of Cu in tetrahedrites obtained from near-edge structure for Cu L_{III}/L_{II} (van der Laan *et al.*, 1992). The peak at 931 eV corresponds to 2p absorption from Cu^{2+} and the peak at 934 eV to 2p absorption from Cu^{+}. The relative sizes of these features reflects the proportions of each oxidation state. The tetrahedrites are: (a) $Cu_{11}FeSb_4S_{13}$; (b) $Cu_{10}FeZnSb_4S_{13}$; (c) $Cu_{10}Te_4S_{13}$.

varies with oxidation state. These advantages are demonstrated in Figure 4.6, which shows the Cu 2p absorption spectra of various complex tetrahedrites exhibiting Cu in different oxidation states (van der Laan *et al.*, 1992). Although Cu is generally assumed to be present in monovalent form for most tetrahedrite compositions, in those tetrahedrites containing an excess of copper (displacing the zinc and iron in the structure) the presence of divalent copper is indicated. The near-edge structure results shown in Figure 4.6 reveal just this. In particular, L_{III} thresholds for Cu^{2+} and Cu^{+} are well separated at 931 and 934 eV. From charge balance considerations in $Cu_{11}FeSb_4S_{13}$, the amount of Cu^{2+} expected is one formula unit, which is close to that observed (Figure 4.6(a)). In addition, the presence of Cu^{2+} was detected in compositions such as $Cu_{10}(FeZn)_2Sb_4S_{13}$, where none was predicted (Figure 4.6(b)). For the tetrahedrite-group mineral goldfieldite $Cu_{10}Te_4S_{13}$, the substitution mechanism proposed in current models is Te^{4+} + void $\rightleftharpoons$ $Cu^{+}Sb^{3+}$, suggesting the presence of no Cu^{2+}. This is confirmed by the Cu 2p spectra, in which the 931 eV peak is all but absent (Figure 4.6(c)). Thus the first direct evidence for Cu^{2+} in these minerals has been provided. (Note that the same changes are also mirrored in the L_{II} peaks at 950 and 953 eV.)

4.3.5 Chemical state imaging

The chemical contrast in near-edge structure can be harnessed to image different oxidation states of the same element when these are present in the same material. Figure 4.7 illustrates this with images of hematite and magnetite inclusions in a specimen of sintered iron ore (Iida *et al.*, 1993). The different near-edge structure at the iron K-edge for Fe_2O_3 compared with Fe_3O_4 enables two energies to be chosen where $\mu(Fe_2O_3) > \mu(Fe_3O_4)$ for one, and $\mu(Fe_2O_3) < \mu(Fe_3O_4)$ for the other. By comparing the X-ray fluorescence from the specimen at these two energies, and by normalizing this to the emission from reference samples of Fe_2O_3 and Fe_3O_4, the relative concentrations of hematite and magnetite in the specimen can be obtained pixel by pixel as shown. A tunable microprobe is required, and one possible optical arrangement for a hard X-ray source (Hayakawa *et al.*, 1992) is included in Figure 4.7, the chemically sensitive image being created by rastering the specimen at the different threshold energies. The diameter of the focal spot obtained by Hayakawa and co-workers is 3.6 μm FWHM and is responsible for the 4 μm × 4 μm pixel dimension in the images included in Figure 4.7. The source image size is a function of optical aberrations and the size of the synchrotron source, both of which can be expected to reduce in the future with improved optics and increased source brilliance. (See Rivers, Sutton and Jones (1991) for a review of synchrotron X-ray fluorescence microscopy).

4.3.6 Impurity sites in staurolite

Detecting characteristic X-ray fluorescence rather than X-ray transmission brings increased signal-to-noise for dilute concentrations of the excited element (Jaklevic *et al.*, 1977). This is because the X-ray absorption signal

(a)

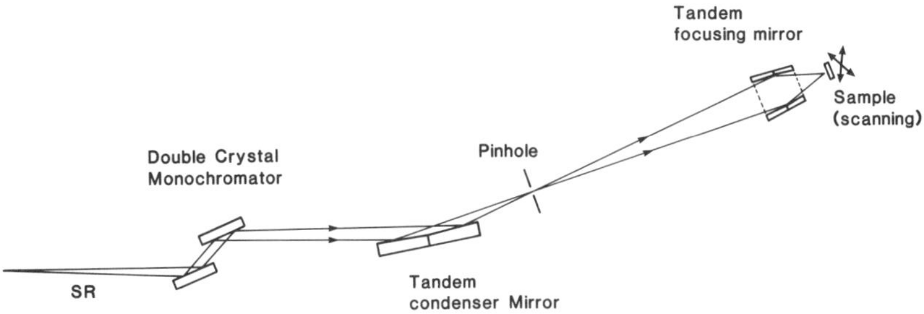

(b)

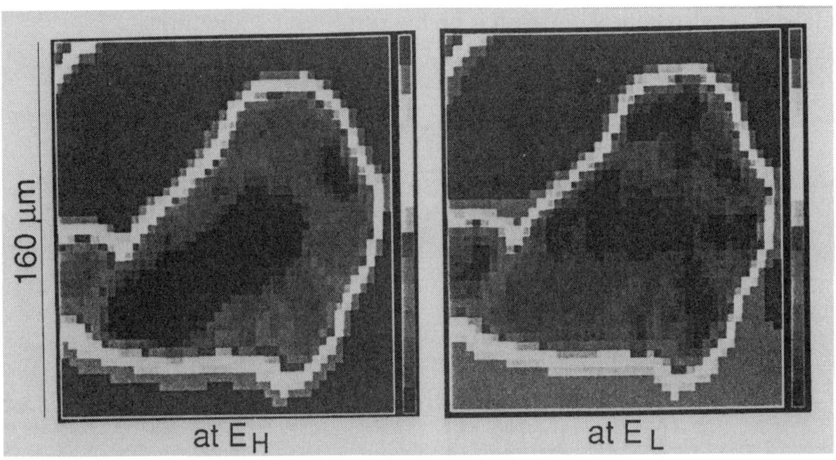

Figure 4.7 Chemical state imaging using microfocus X-ray fluorescence. (a) Optical arrangement using Kirkpatrick-Baez mirror pair (Hayakawa *et al.*, 1992); (b) images for hematite (left) and magnetite (right) obtained from analysing the spatially resolved X-ray fluorescence above and below the Fe K-edge (Iida *et al.*, 1993).

for an impurity measured in fluorescence is proportional to $(\mu - \mu_b)$ rather than μ. The background due to X-ray fluorescence and scatter from the remaining material can usually be reduced, and sometimes eliminated, by narrowing the energy acceptance of the detector to that of the characteristic line of the impurity (see Derbyshire *et al.*, 1992). An example is given in Figure 4.8, where the K-edge near-edge structure of Ti and Mn impurities in staurolite $[(Fe,Mg)_4Al_{17.3}[Si_{7.6}Al_{0.4}]O_{48}H_3]$ are compared with various model compounds (Henderson, Greaves and Smith, 1993). The atomic structure of staurolite obtained from X-ray diffraction was refined on the basis of seven octahedral sites occupied by Al, and two tetrahedral sites – one occupied by Si and the other by Fe (Smith, 1968). Much controversy has been associated with the assignment of impurity cations sites in this structure, most authors placing Ti, Co and Zn in the tetrahedral Fe^{2+} site, with Mn in the Al octahedral sites. Figure 4.8 demonstrates the exact opposite to be the case. The similarity in the near-edge structure of Mn (present at 0.4 wt%) and Fe K-edges in staurolite is striking, and points firmly to a tetrahedral assignment for Mn^{2+}. This is corroborated by the Mn—O distance of 2.01 Å analysed from the extended XAFS, which is typical of a tetrahedral Mn^{2+}—O distance. On the other hand, the K-edge near-edge structure of Ti, which is also included in Figure 4.8, shows no similarity to that of tetrahedral Ti^{4+} in Ba_2TiO_4, but is far closer to that of octahedral Ti^{4+} in rutile, TiO_2. The best agreement between the K-edges is, in fact, with Ti in titanite, where these sites are known to be distorted octahedra. Moreover, the Ti—O distance of 1.86 Å, obtained from XAFS analysis, is also consistent with a highly distorted octahedral site.

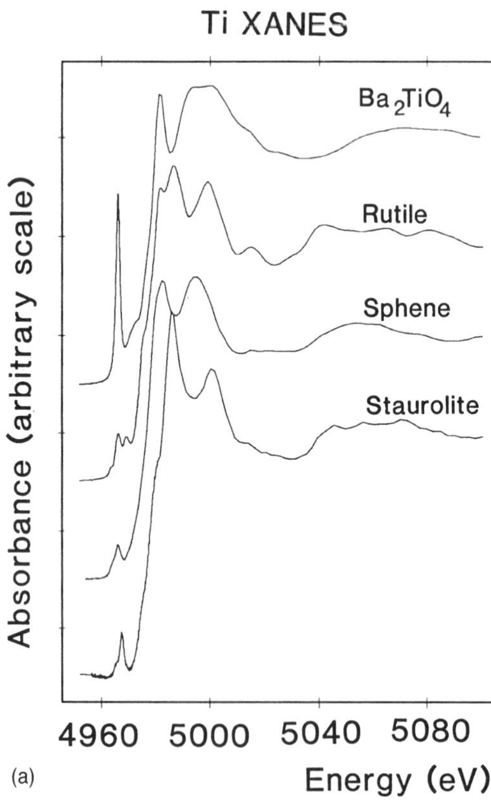

Figure 4.8 Impurity crystallography using fluorescence XAFS. The impurity sites for Ti and Mn in staurolite $[(Fe,Mg)_4Al_{17.3}[Si_{7.6}Al_{0.4}]O_{48}H_3]$ can be differentiated and rationalized (Henderson, Greaves and Smith, 1993). (a) K-edge of Ti in staurolite compared with model tetrahedral and octahedral compounds indicating distorted octahedral sites for this impurity; (b) K-edge of Mn in staurolite directly compared with that of tetrahedrally co-ordinated Fe in the same mineral.

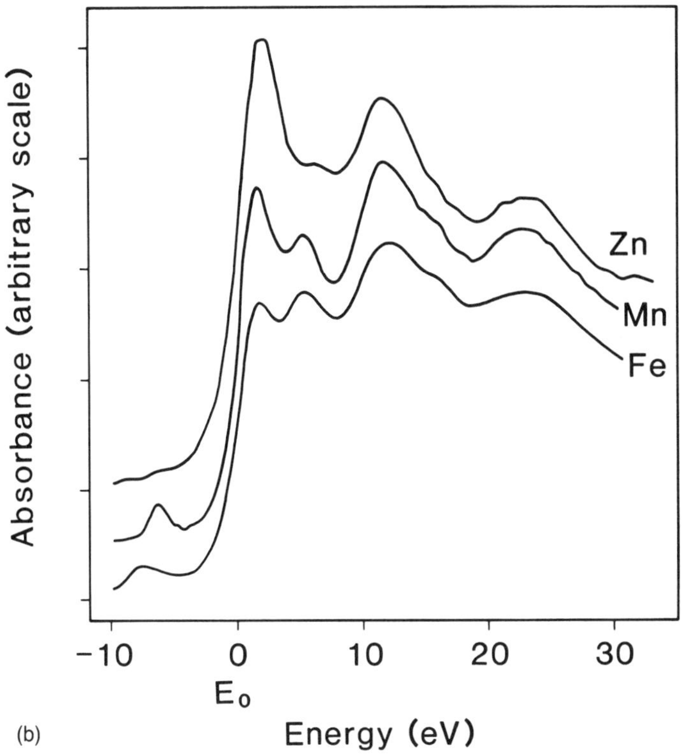

(b)

Figure 4.8 *Continued*

Fluorescence detection is often used for glancing angle experiments (see Figure 4.16(c)) and has been employed to identify the atomic geometries of cations released at glass surfaces undergoing aqueous corrosion, to be described later (Figures 4.19 and 4.20).

4.3.7 Energy dispersive XAFS

Finally, returning to Figure 4.4, arrangement (b) illustrates an energy-dispersed geometry suitable for time resolved XAFS studies. The bent crystal selects a suitable bandpass, straddling the absorption edge of interest, and the transmitted spectrum is recorded on a position-sensitive detector (Allinson *et al.*, 1988). An example of energy dispersive XAFS is given in Figure 4.11(d), where the dehydration of Ni-exchanged zeolite Y (to be discussed in the next section) is followed as a function of time. The obvious change in XAFS marked by the arrow pinpoints the transfer of Ni from the supercage into the hexagonal prisms illustrated in Figure 4.10 (Couves *et al.*, 1990). In the remaining sections, we will show how the XAFS and powder

X-ray diffraction techniques can be applied to tackle this and other problems associated with the physical chemistry of mineral surfaces – both internal and external.

4.4 Structure of microporous materials

4.4.1 Zeolite synthesis

The highly porous nature of zeolitic structures is well known. With large internal surfaces, these materials have shape-selective catalytic properties directly related to the specific labyrinths of their cage-like structures. Zeolite synthesis can be readily followed *in situ* using energy-dispersive diffraction (Figure 4.1(a)). Figure 4.9 illustrates the application of this technique to follow the formation of sodalite from a silica-alumina gel at 95 °C in an autoclave (Munn *et al.*, 1992). Interestingly, zeolite A is an intermediate phase, identified by the ephemeral peaks at 28 and 34 keV. The peaks that are common to both zeolite A and sodalite are clearly established at an early stage, as the diffuse scattering from the gel decreases, and persist throughout the synthesis as sodalite is gradually established. The application of energy dispersive diffraction to quantify the reaction kinetics of phase transitions is obvious and the outcome elegant as Clark and Dooryhee (1992) have recently shown.

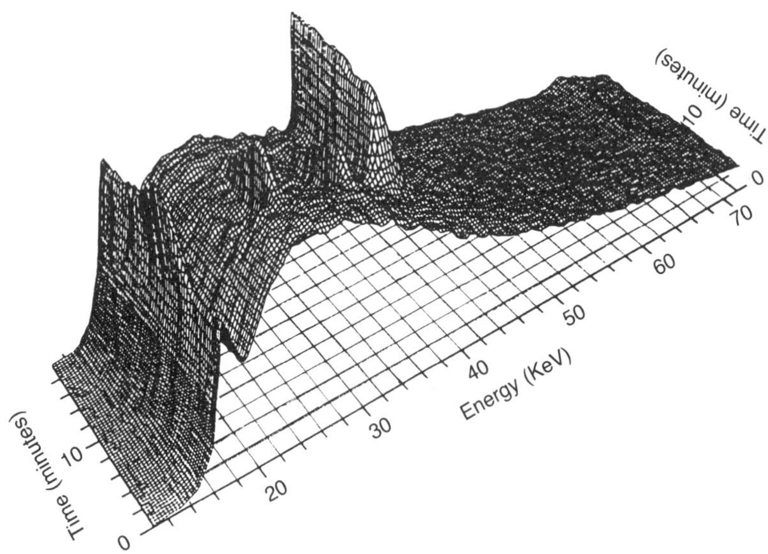

Figure 4.9 Structural changes occurring during the synthesis of sodalite from a silica-alumina gel using energy dispersive diffraction (Munn *et al.*, 1992). Note the transitory peaks close to 25 and 33 keV, indicating the formation of the intermediate zeolite A phase.

4.4.2 Ni-exchanged zeolite Y: X-ray diffraction and XAFS

The bulk X-ray diffraction from zeolites reveals the network geometry responsible for the large internal alumino-silicate surfaces; these can have effective areas as great as $10^3\,m^2\,g^{-1}$. On the other hand, XAFS can be exploited to pinpoint the geometry of the ion-exchanged metal, enabling this to be followed as the catalyst is activated. Accordingly, a combination of X-ray diffraction and XAFS techniques can provide unique insights into the surface and catalytic properties of these high-area solids (Dooryhee *et al.*, 1991).

The structure of zeolite-Y ($\sim Na_{59}Al_{59}Si_{133}O_{384}xH_2O$) is illustrated in Figure 4.10, showing the usual polyhedral structure, including a large supercage. Each vertex refers to an Si or Al site, and the full lines to the bridging oxygens. The sites available to Na atoms and ion-exchanged species are also marked, S_I sites being situated within the hexagonal prisms

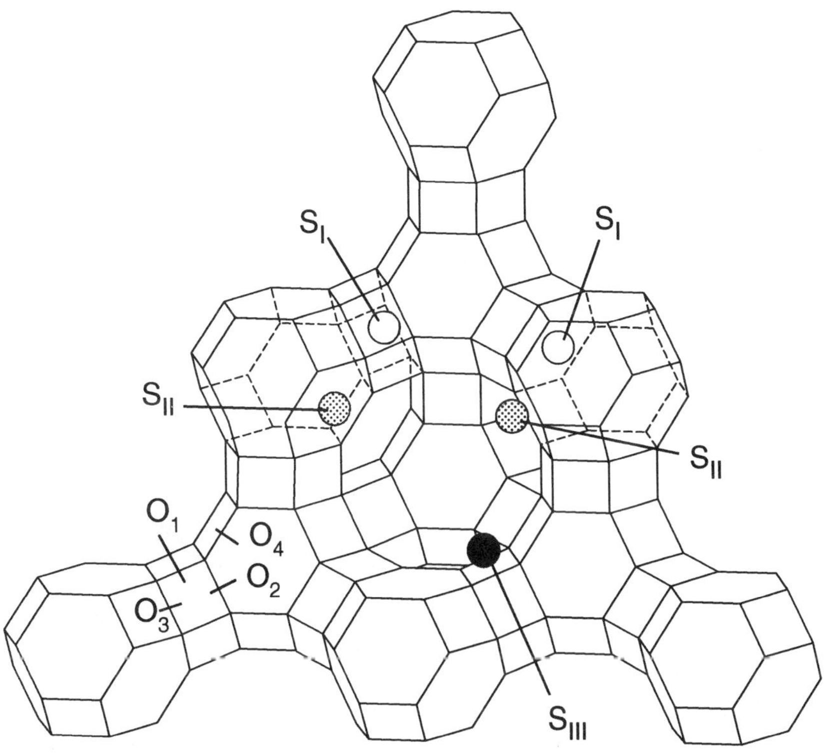

Figure 4.10 Structure of zeolite Y. The straight lines indicate the network bridging oxygens, with aluminiums and silicons at each apex. The S sites for extra-framework cations are also shown, as well as the neighbouring oxygens. S_I sites occur within the hexagonal prisms between sodalite cages, while S_{II} and S_{III} sites decorate the supercage walls. The $S_{I'}$ and $S_{II'}$ sites, which are mirror images of the S_I and S_{II} sites, are located on the walls of the sodalite cages.

connecting pairs of sodalite cages, with S_{II} and S_{III} sites hugging the walls of the supercage. The $S_{I'}$ and $S_{II'}$ sites, the mirror images of the S_I and S_{II} sites, are located inside the sodalite cages. By part ion exchanging with Ni, zeolite Y can be converted into a model catalyst, facilitating the oligimerization of acetylene into benzene. Figure 4.11 collates the Fourier transforms of the Ni K-edge XAFS and charts the environment of Ni from the hydrated ionexchanged structure, through calcination, to reduction. The powder diffraction pattern of the calcined material is displayed in Figure 4.12. The dotted lines in Figures 4.11 and the residuals in 4.12 are the results of least squares fitting to model structures that are self-consistent with respect to both XAFS and X-ray diffraction, using the Daresbury programs EXCURVE and PDPL, respectively.

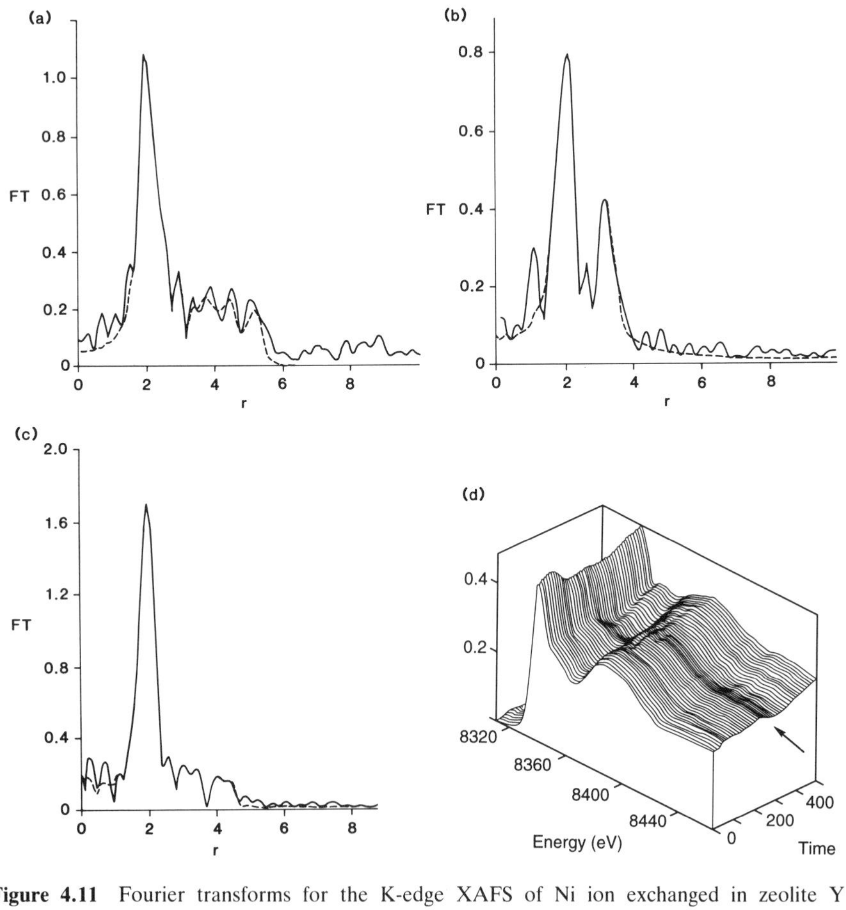

Figure 4.11 Fourier transforms for the K-edge XAFS of Ni ion exchanged in zeolite Y (Dooryhee *et al.*, 1991), (a) hydrated, (b) calcined and (c) reduced. The dehydration process has been followed dynamically at the K-edge of Ni by Couves *et al.* (1990), as shown in (d). The arrow indicates the point at which the Ni ions transfer to the S_I sites.

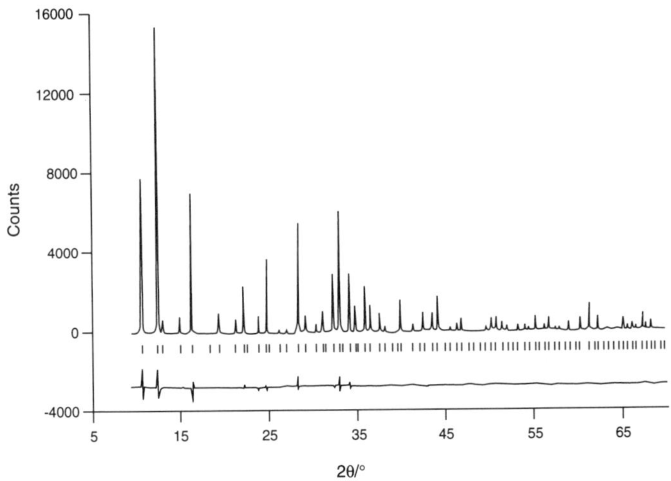

Figure 4.12 High resolution powder diffraction pattern for calcined Ni zeolite Y. Experiment and theory are shown above, with indexed reflections and residuals plotted below (Dooryhee *et al.*, 1991).

Taking the Ni XAFS first, the simple structure of the ion-exchanged cation is dominated by a single shell of oxygens at 2.06 Å. This is readily modelled by a hydrated Ni^{2+} cation, whose O co-ordination is very close to that of Ni in $Ni(NO_3)_2 \cdot 6H_2O$. The only location of sufficient girth to accommodate this unit in the framework structure, shown in Figure 4.10, is the supercage. The Ni environment, however, becomes more complex following dehydration, as Figure 4.11 demonstrates. Modelling the XAFS reveals new Ni sites. The principal member of these sites is octahedral, but with a shorter Ni—O distance of 2.02 Å. Other features in the Fourier transform confirm that Ni occupies a variety of other sites in the dehydrated network, as detailed below. Moreover, in addition to the principal O peak in Figure 4.11(b), there is a strong peak around 3.5 Å signifying strong correlations between these new Ni sites in the calcined material and the framework Al and Si atoms.

Turning now to the powder diffraction pattern of the calcined material (Figure 4.12), Rietveld analysis confirms that Ni^{2+} atoms are located in the hexagonal S_I sites – the smallest interstices within the framework (Figure 4.10). The first O shell radius for the S_I site obtained from diffraction, however, is 2.25 Å, apparently contradicting the 2.02 Å value obtained from XAFS. At the same time, the occupancy of the S_I site analysed from the X-ray powder pattern (Figure 4.12) is only 75%, and the difference in oxygen shell radii between X-ray diffraction and XAFS is, in fact, attributable to the different geometries of unoccupied and occupied sites. When the S_I site is empty, the radius of the O shell is 2.6 Å. On occupancy, however, the hexagonal oxygens clearly relax inwards to a radius of 2.02 Å. As X-ray diffraction records the average radius, allowing for partial occupancy, the mean O radius should be 2.17 Å in reasonable agreement with the measured value of 2.25 Å. Without complementary measurements of high resolution X-ray powder diffraction and XAFS, this sizable distortion caused by the principal ion-exchanged site in the calcined zeolite would have been missed.

Further analysis of the powder pattern shown in Figure 4.12 indicates partial occupancy at the 10% level of Ni^{2+} at other metal sites within the internal framework. Approximately half of these lie on the walls of the supercage (S_{II} sites), with the rest located on the interior of the sodalite cages ($S_{I'}$ sites). These are responsible for the extra features in the Ni XAFS Fourier transform mentioned above. In particular, Ni^{2+} in the $S_{I'}$ site is bound to three oxygens from the zeolitic network and three oxygens from residually bound water molecules – the peaks at 1.9 Å and 2.6 Å in Figure 4.11. Interestingly, referring to the final Fourier transform in this figure, it is these features that disappear when the calcined material is reduced (Figure 4.11(c)). Their place is taken by Ni–Ni correlations at 2.5 Å, 4.2 Å and 5.1 Å – the first, second and third shells of the fcc structure of Ni metal. Evidently, Ni at the $S_{I'}$ sites of the dehydrated material may well be the source from which metal particles are formed when the catalyst is activated.

Analysis at this level of detail would be speculative for either the Ni XAFS or the powder pattern, each taken on its own. A combined approach, however, removes many of the uncertainties and demonstrates the degree of detail with which the micro-environments of Ni can be ascertained in this heterogeneous catalyst.

4.5 Ceramic formation and mineral synthesis

4.5.1 Combined detection of XAFS and XRD for in situ *experiments*

Minerals are often synthesized by reheating glasses of appropriate composition above the glass transition temperature. This is also the traditional way of making glass ceramics (MacMillan, 1979). More commonly, ceramics are formed by sintering powdered mixtures of the crystalline constituents. In any event, for many compositions heat treatment temperatures are high, which may not always be desirable for materials' fabrication or from the standpoint of geological relevance. In the past decade, novel low temperature routes for the formation of ceramics have been devised, exploiting sol gel techniques (Rao and Golapakrishnan, 1987; Zarzycki, 1990). Most recently, and perhaps most fascinating of all, is the route afforded by the collapse and subsequent recrystallization of zeolitic structures (Chowdhry, Corbin and Subramanian, 1989; Bedard and Flanigen, 1993). The processes of nucleation and growth, whereby crystalline phases are formed from the amorphous state, customarily require a range of techniques to differentiate

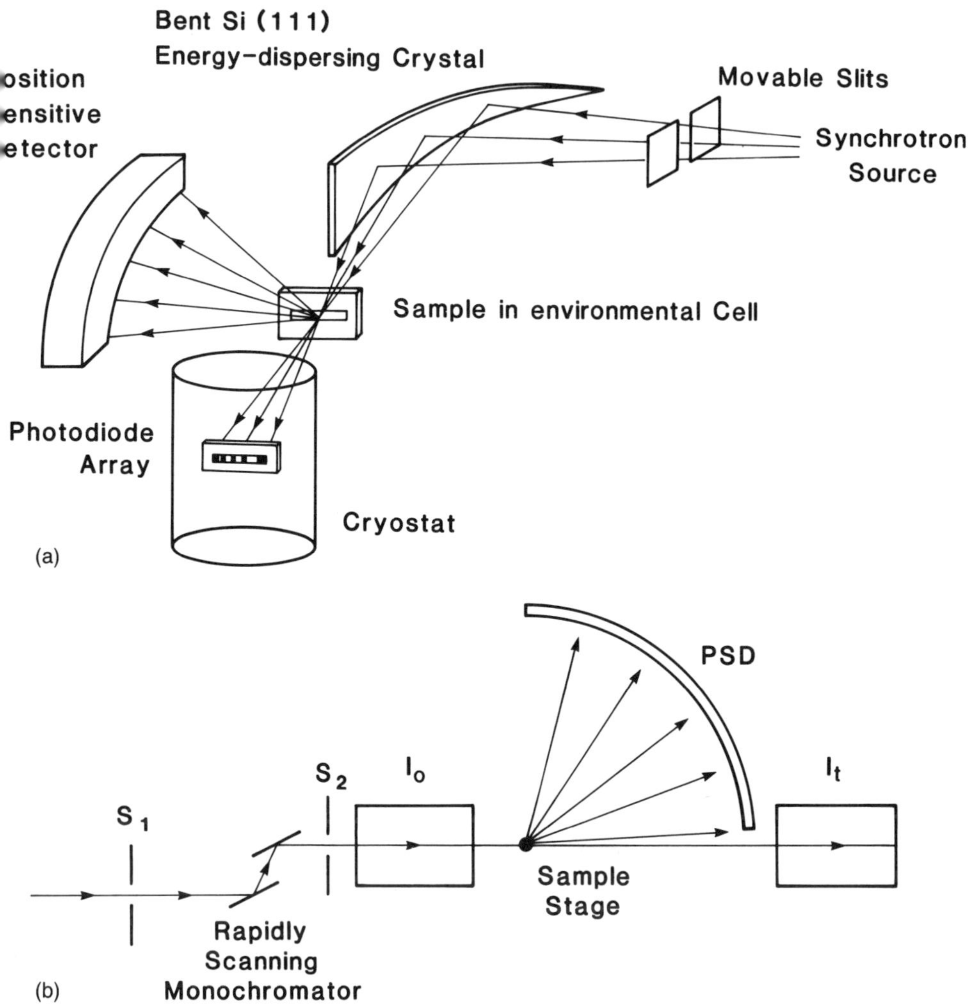

Figure 4.13 Combined X-ray diffraction/XAFS experimental arrangements. (a) Dispersive XAFS mode; XAFS is recorded in transmission with the photodiode array, and XRD with the curved position-sensitive detector (psd); entrance slits enable a wide bandpass to be selected for XAFS and a narrow bandpass for diffraction (Couves *et al.*, 1991). (b) Rapidly scanning XAFS mode; monochromator scans quickly through the relevant edge(s) and then to a suitable fixed wavelength for XRD before repeating (Thomas and Greaves, 1993).

and fully to characterize them. The X-ray methods outlined so far, promise to play a powerful role in elucidating the structural mechanisms that lead to the ceramic state by whatever route. This appears to be particularly true for *in situ* experiments when X-ray powder diffraction and XAFS can be recorded in tandem. Two such combined detector arrangements are illustrated in Figure 4.13. Each of these employs a curved position-sensitive detector (PSD) to collect the Debye–Scherrer pattern from the specimen. On the other hand, XAFS is measured using either the scanning monochromator or the energy dispersive geometry, as shown earlier in Figure 4.4.

4.5.2 Catalyst formation from aurichalcite

We begin by describing the formation of a Cu catalyst from the mineral precursor, aurichalcite, $(Cu_{1-x}Zn_x)(CO_3)_2.(OH)_6$ (Couves *et al.*, 1991). With Cu and Zn in the proportions 1:4, this generates a highly dispersed catalyst, promoting methanol synthesis. Aurichalcite is a clay mineral with a layered structure (Cernik, 1993; Harding *et al.*, 1994) not dissimilar to that of hydrozincite $(Zn_5(CO_3)_2.(OH)_6$. In order to follow both the overall changes in structure and the establishment of the active Cu environment, Couves and co-workers used the dispersive XAFS geometry shown in Figure 4.13(a). Throughout the calcination stage, which was complete by 450 °C, the Cu K-edge XAFS hardly altered, indicating that the immediate environment of the metal remained intact. By contrast, radical changes occurred in the powder patterns as the hydroxycarbonate decomposed into an intimate mixture of Zn and Cu oxides. Following calcination, the mixed oxide was returned to room temperature and reduced. Figure 4.14(a) shows the combined diffraction/XAFS data. Close to 270 °C, the threshold resonance in the XAFS typical of CuO disappears, to be replaced by the weaker near-edge structure of Cu metal. This is accompanied by commensurate changes in the powder pattern where the peaks associated with CuO are lost in favour of Cu metal peaks. At the same time, the ZnO structure stays intact throughout. Estimates from the line broadening point to Cu particles in the region of 100 Å in size.

4.5.3 Synthesis of cordierite

The unusual route to ceramic formation afforded by heat-treating a suitably ion-exchanged zeolite (Chowdhry, Corbin and Subramanian, 1989) can be tracked for these relatively slow transformations using the second combined diffraction/XAFS method illustrated in Figure 4.13(b). In the following example, zeolite B ion exchanged with Mg to yield the composition $Mg_2Al_4Si_5O_{18}$ was heat treated *in situ* to elevated temperatures to produce synthetic cordierite, a ceramic with new applications in the packaging of integrated circuits and as a support for exhaust catalysts. It has been found that this transformation from zeolite to ceramic is aided by ZnO (~1 wt %). Figure 4.14(b) shows dramatic changes occurring in the X-ray powder pattern and Zn K-edge XAFS measured *in situ* as the temperature is progressively increased (Thomas and Greaves, 1993). Subtle transformations in zeolite B at low temperatures (<150 °C) are detected first. These processes

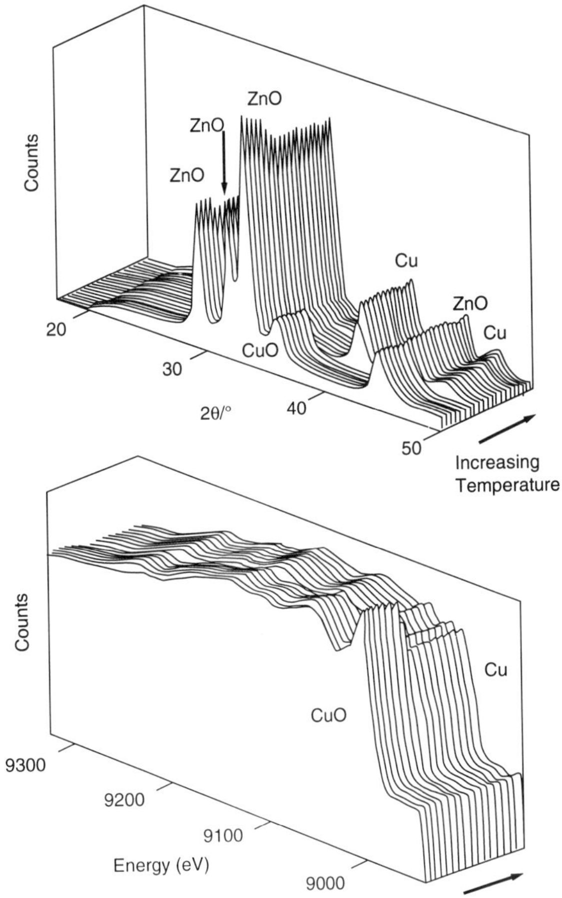

(a)

Figure 4.14 Experimental results of combined X-ray diffraction/XAFS. (a) Cu K-edge XAFS (bottom) and powder diffraction (top) following the formation of a Cu catalyst by thermal reduction of calcined aurichalcite between 100 °C and 500 °C (Couves *et al.*, 1991); (b) Zn K-edge XAFS (bottom) and XRD (top) following the collapse of Mg ion-exchanged zeolite B doped with ZnO and the subsequent formation of cordierite at temperatures up to 1050 °C (Thomas and Greaves, 1993).

are overtaken when the zeolite collapses and an amorphous phase is formed at around 650 °C. The formation of an intermediate stuffed quartz phase can just be detected in the powder pattern at the point where the Zn XAFS indicates dissolution of ZnO. The final conversion to cordierite close to 1000 °C is obvious in the powder diffraction. There is an accompanying modification recorded in the Zn XAFS. Zinc, which is tetrahedrally co-ordinated in ZnO, retains this local configuration when it is assimilated into

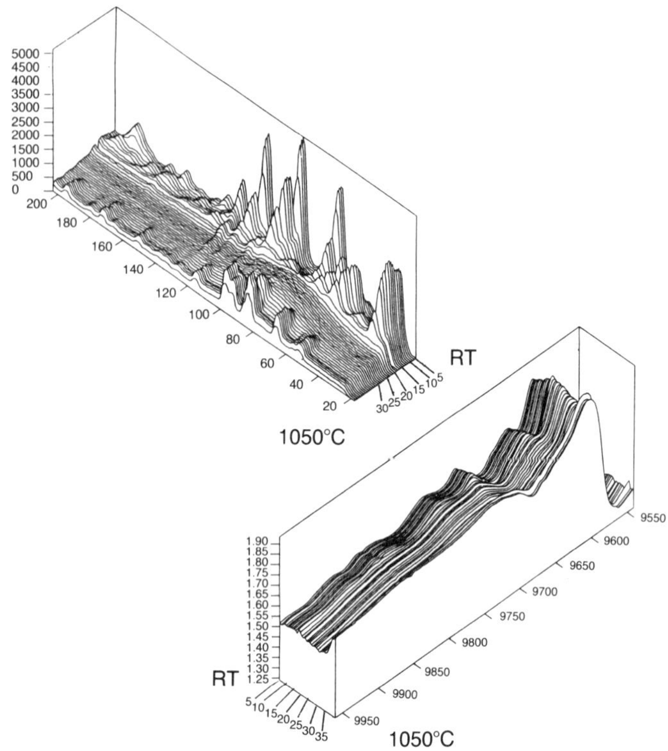

(b)

Figure 4.14 *Continued*

the collapsed material, and it is this ingredient which promotes the rapid conversion of the intermediate stuffed quartz phase into the final cordierite product. If Zn is not present, the whole process is inhibited until much higher temperatures are reached.

By contrast, when Cr is used as a nucleating agent, the stuffed quartz phase is much more persistent. This behaviour has been examined in cordierite glass made conventionally by quenching from the melt (Oversluizen, Clark and Greaves, 1993). Incorporating Cr_2O_3 (~0.3 wt %) into the glass provides a source of nucleation sites which develop at around 850 °C, close to the glass transition temperature. Subsequent heat treatment at 950 °C results in a fine precipitate of the stuffed quartz phase. This precipitate and the associated heterogeneity of the glass ceramic can be seen in Figure 4.15(a), where the small angle X-ray scattering (SAXS) for the glass heat treated at 950 °C (A) is compared with the homogeneous quenched glass (B). The SAXS intensity at low q increases dramatically as a result of scattering from the microcrystallites. Figure 4.15(a) also includes the complementary wide angle X-ray scattering (WAXS) measured simultaneously

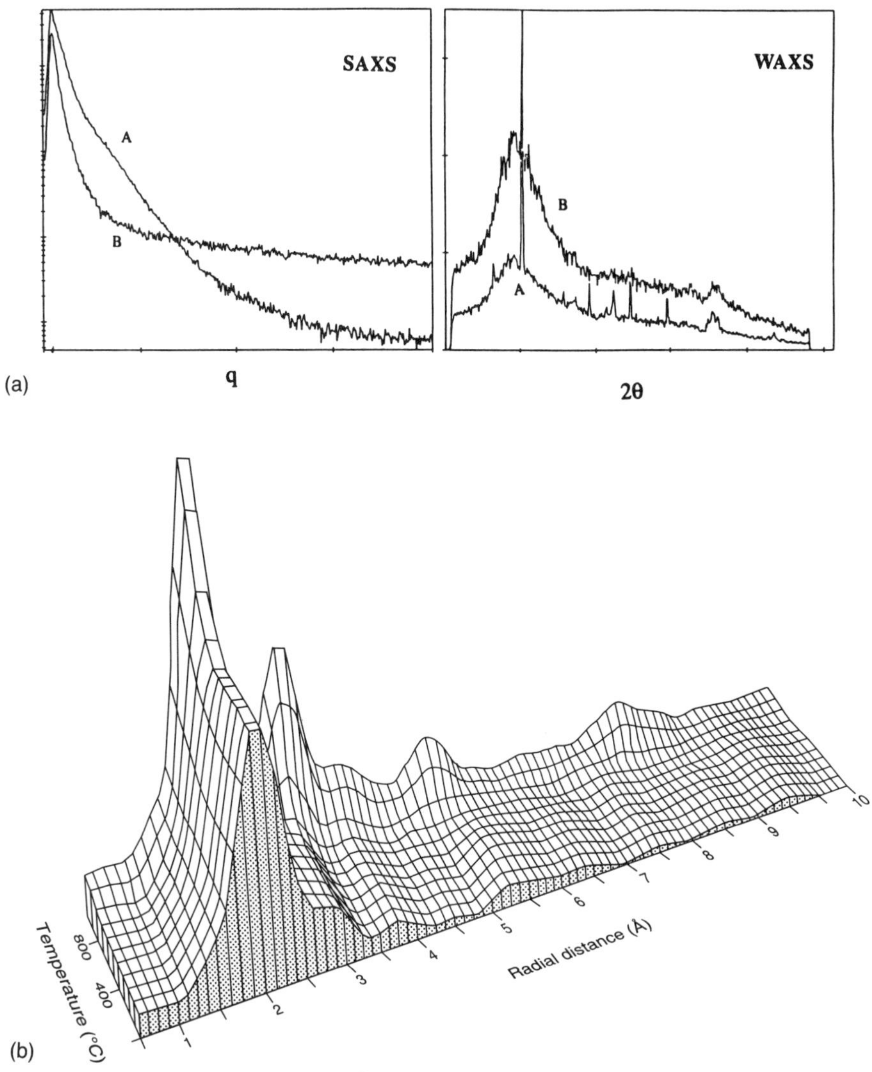

Figure 4.15 Formation of a stuffed quartz phase from a Cr-doped cordierite glass. (a) Simultaneous small and wide angle X-ray scattering (SAXS/WAXS) from the glass ceramic (A) compared with the quenched glass (B); (b) The changing environment of Cr used as a nucleating agent obtained from K-edge fluorescence XAFS (Oversluizen, Clark and Greaves, 1993).

with the SAXS (Bras *et al.*, 1993). Here, the diffuse scattering can be clearly seen, and the way in which this reduces once the coherent scattering from the quartz phase is established. The corresponding change in the Cr K-edge XAFS (collected in fluorescence) is illustrated in Figure 4.15(b). Initially,

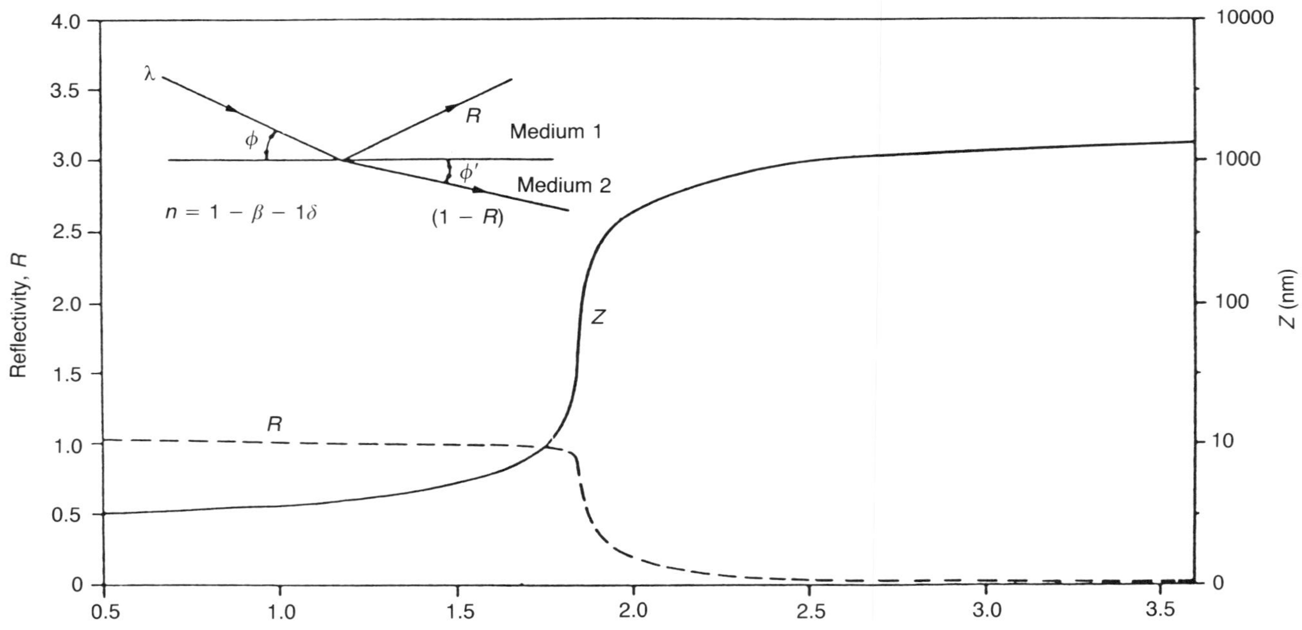

(a)

Figure 4.16 Total external reflection of X-rays. (a) Depth of penetration and reflectivity as a function of angle of incidence for 0.7 Å X-rays (Greaves *et al.*, 1991); (b) experimental arrangement for glancing angle powder diffraction (Huang, 1991); (c) geometry for glancing angle XAFS (Thornley *et al.*, 1986).

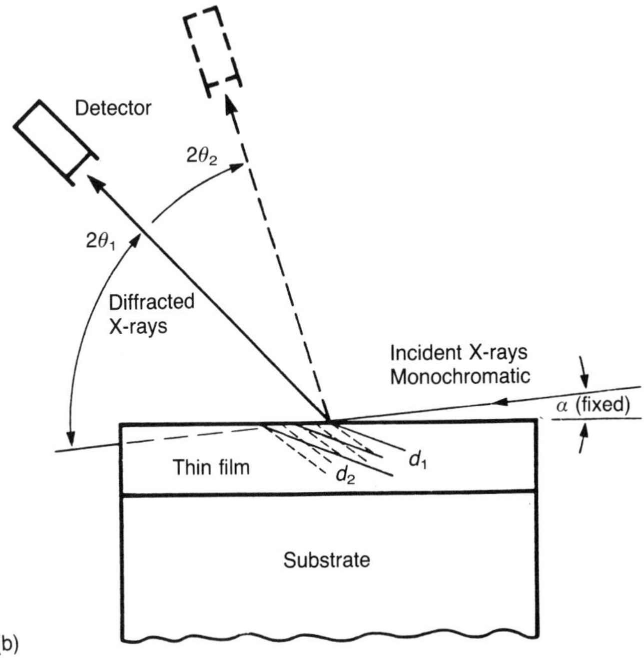
Detector
2θ2
2θ1
Diffracted X-rays
Incident X-rays Monochromatic
α (fixed)
Thin film
d1
d2
Substrate

(b)

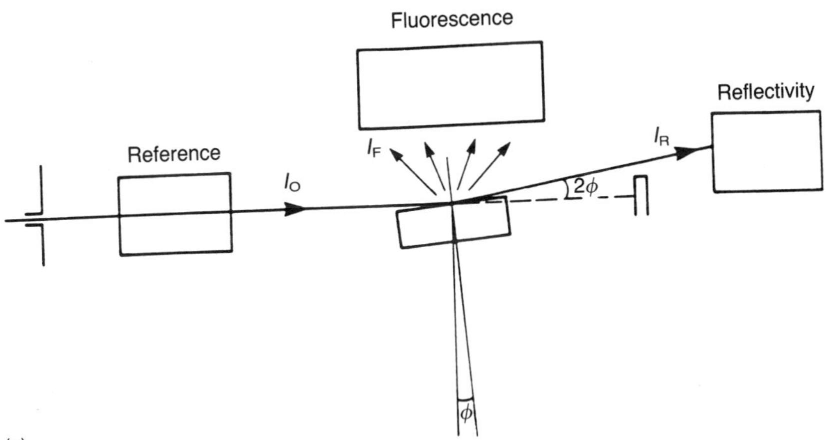
Fluorescence
Reflectivity
Reference
IF
IR
IO
2φ
φ

(c)

the Cr is located within the glassy structure, it is octahedrally co-ordinated and the Debye–Waller factor associated with nearest neighbours reflects significant static disorder. The latter dramatically reduces on heat treatment, signifying the onset of local crystallization. In addition, the composition of the shell of next nearest neighbours alters. The EXCURVE analysis reveals that the Si and Al atoms present in the as-quenched glass may well be supplemented by Mg atoms on heat treatment – pointing to the possible formation of a Cr-doped $MgAl_2O_4$ spinel. This microphase appears to be the precursor of the principal stuffed quartz phase in these Cr-doped glasses. Because the quartz phase is silica-rich compared with cordierite, the silica level in the remaining glass will be correspondingly low. This spatial redistribution of composition associated with the crystalline and glassy phases would appear to be instrumental in 'locking out' the transition to cordierite in this Cr-doped glass until temperatures in excess of 1250 °C are reached.

The results presented in Figures 4.14 and 4.15 illustrate the wealth of information concerning structural transformations in minerals that can be assembled by combining X-ray powder diffraction with XAFS (and also with SAXS) measurements on the same specimen.

4.6 Surface structure

4.6.1 Grazing incidence X-rays

Finally, we turn to the potential offered by employing glancing angles of incidence to surface sensitize X-ray diffraction and XAFS. The geometry for the total external reflection of X-rays is illustrated by the inset of Figure 4.16(a), where medium 1 is less dense than medium 2. As the refractive index of materials for X-rays is slightly less than unity, the refracted beam is bent towards the surface ($\phi' < \phi$) and, when the angle of incidence falls below a critical value, ϕ_c, an evanescent wave is generated at the surface or interface and X-rays are totally externally reflected. The depth of penetration of X-rays, z, is typically a few nanometres falling to a minimum value, z_{min} when $\phi = 0$. Both ϕ_c and z_{min} are dependent on the surface electron density, which is approximately proportional to the physical density, ρ ($g\,cm^{-3}$), according to the following expressions:

$$\phi_\chi \sim (20/h\nu)\,\rho^{1/2}\,\text{mrad} \qquad z_{min} \sim 48/\rho^{1/2}\,\text{Å}$$

The dependence of the X-ray reflectivity, R and the penetration depth, z, on ϕ is illustrated in Figure 4.16(a) for 0.7 Å X-rays. At ϕ_c, R falls rapidly towards 0 and z increases, approaching $\sin\phi/\mu$, the penetration depth of the refracted beam. By suitable choice of the angular position of the incident X-rays, therefore, depth-specific structural information can be obtained for any X-ray diffraction or XAFS technique. Experimental arrangements are given in Figure 4.16(b) and (c). Huang (1991) has developed the asymmetrical angle dispersed geometry shown in Figure 4.16(b). Diffraction from different atomic planes of crystallites at the surface can be detected by scanning in 2θ, keeping the angle of incidence, α, very small (i.e. $\alpha \ll \theta$) to limit diffraction to the surface region. By fixing θ and removing the monochromator, white X-rays can be used to obtain surface energy dispersive powder patterns, as Cernik, Clark and Pattison (1991) have shown by using a flat plate version of Figure 4.1(b). The geometry for glancing angle XAFS is shown Figure 4.16(c), where spectra can be detected in either the reflected or the fluorescence signal (Thornley *et al.*, 1986).

There have been only a few applications of these techniques to problems allied to geology, but the following examples of grazing incidence X-ray powder diffraction and XAFS are intended to give an indication of the prospects in this area for surface mineralogy. It is worth emphasizing that the use of X-rays in these surface structure experiments removes the necessity for vacuum environments. Indeed, experiments can be performed in air on exposed surfaces, or at liquid–solid interfaces, provided the liquid is less dense than the solid. The chief requirement is for a polished surface, which is often 'stock in trade' for the mineralogist.

4.6.2 *α-Fe_2O_3 at surface of γ-Fe_2O_3*

Powder patterns from thin films have been obtained using the asymmetrical arrangement shown in Figure 4.16(b), which has been exploited to identify the antiferromagnetic α-Fe_2O_3, which forms on the surface of thin magnetic films of γ-Fe_2O_3 after oxidation (Huang, 1991). Debye–Scherrer plots for an oxidized film measured at various angles of incidence (λ = 1.83 Å; ϕ_c = 0.39°) are shown in Figure 4.17(a). The long wavelength used avoided the Fe fluorescence (inevitable with a laboratory Cu Kα source), and the grazing angles of incidence restricted the X-ray penetration to around 100 Å or less. For the 2θ range, presented reflections from both α-Fe_2O_3 and γ-Fe_2O_3 can be identified, and the way these vary in intensity with the angle of incidence is quite evident. In particular, the α(104) line decreases with increasing incidence angle, while the γ(313) line increases. As a result, α-Fe_2O_3 can clearly be seen to be associated with the surface. Indeed, from these changes of line intensities with grazing angle, the thickness of the α-Fe_2O_3 overlayer has been measured to be 90 Å ± 15 Å. The results of modelling the variation in the peak intensities of the α(104) and γ(313) lines to this geometry are shown in Figure 4.17(b).

4.6.3 *X-ray reflectivity*

If the surface structure has a lower density than the substrate, then X-ray reflectometry (Figure 4.16(c)) can be used to estimate the thickness. Inter-

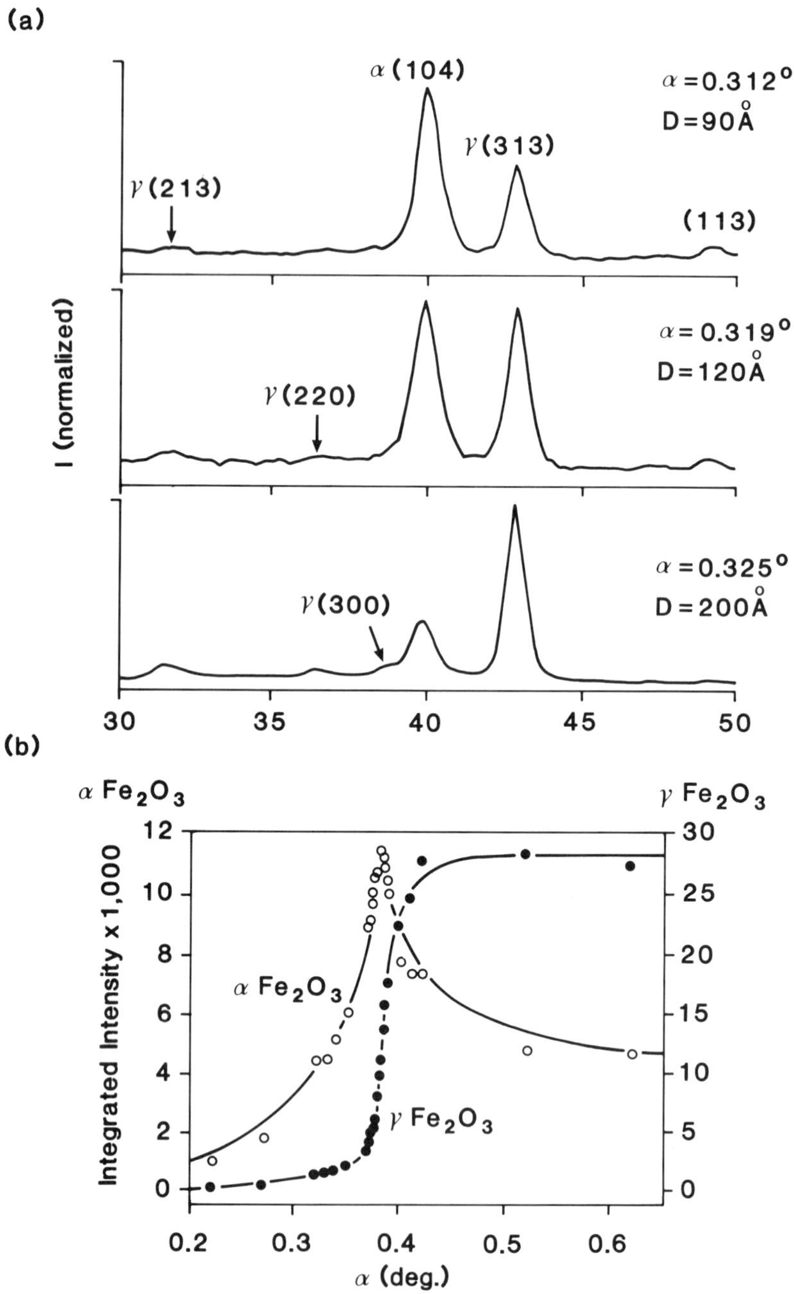

Figure 4.17 Grazing incidence powder diffraction from a magnetite film, revealing lines from an α-Fe_2O_3 overlayer as well as the underlying γ-Fe_2O_3 (Huang, 1991). (a) As the angle of incidence, α, decreases and with it the depth of penetration, d, the (104) reflection of α-Fe_2O_3 can be seen to increase and the (313) reflection of γ-Fe_2O_3 decrease; (b) the variation of the α(104) and the γ(313) intensities enable the step profile of the α-Fe_2O_3 overlayer to be estimated.

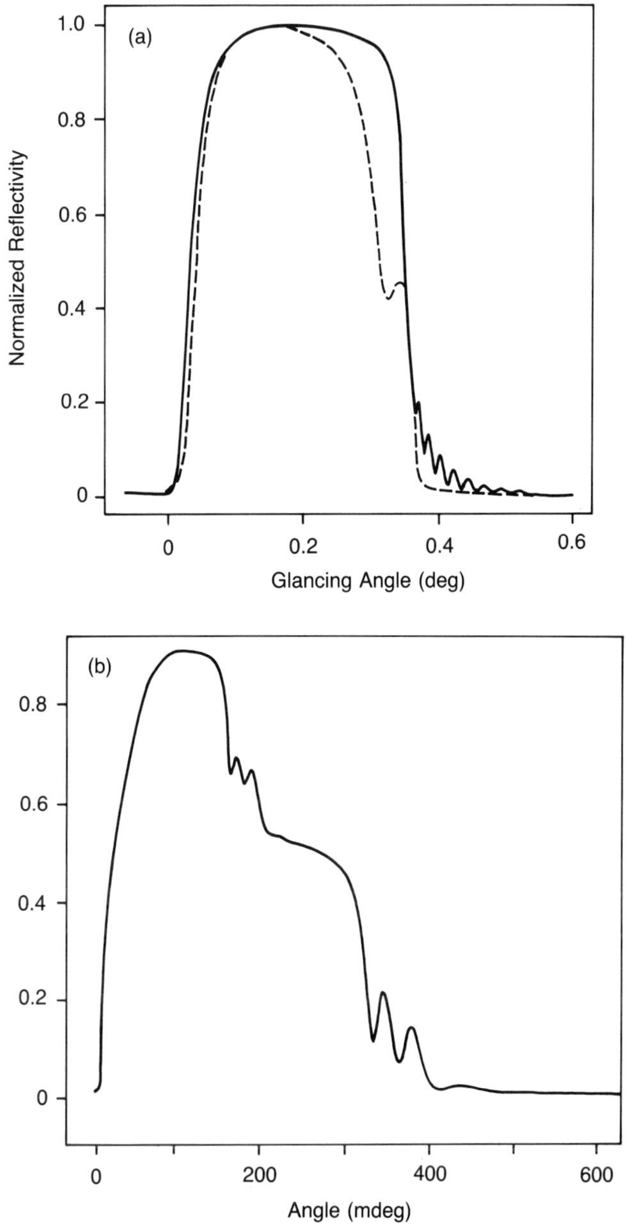

Figure 4.18 X-ray reflectivity from surface coatings. (a) Cu film on glass before oxidation, solid; and after oxidation, dashed (Pizzini *et al.*, 1991). (b) Leached glass surface (Sprenger, personal communication).

ference between X-rays reflected from the top surface and the interface with the substrate results in a fringe pattern – Kiessig fringes – whose spacing relates to the thickness of the surface coating (Parratt, 1954). Figure 4.18 shows two examples of different coatings but of comparable thickness (~1000 Å). The first is a film of Cu evaporated on to a glass slide (Pizzini *et al.*, 1991). The X-ray reflectivity, R, from the as-deposited film, exhibits many fringes (solid line). Note how, after oxidation, the single reflectivity threshold is replaced by two reflectivity thresholds: one relating to the oxide on the surface and the other to the interface between this and the underlying metal (dashed line). The second case relates to a leached glass surface (Sprenger, personal communication). Again two critical angles for total external reflection can be distinguished, the smaller relating to the leached surface and the other to the bulk glass. The retention of the fringes in this reflectivity profile indicates that the corroded layer is extremely well defined, both in thickness and in limited surface roughness.

4.6.4 Corrosion at glass surfaces

When XAFS is conducted at glancing angles of incidence, it enables the structural environment of particular elements at the surface to be determined and compared with their environment in the bulk. This technique has been applied to a wide variety of problems (Greaves, 1991). We use the corrosion of borosilicate glasses to illustrate the technique (Greaves *et al.*, 1989; Greaves, 1990b). Glasses loaded with small quantities of Fe and U were examined after leaching in water at 100 °C for progressively longer periods. The XAFS at the Fe K-edge and the U L_{III}-edge were obtained using X-ray fluorescence detection, as shown in Figure 4.16(c). Specimens were measured immediately after treatment, while still in a 'wet' state. Dramatic changes were observed in the edge region and in the extended XAFS, both as a function of depth and as a function of the extent of aqueous corrosion. Figure 4.19 demonstrates the way the K-edge of Fe altered with advancing corrosion (Greaves, 1990b). From the position of the threshold (which remained fixed), Fe is present throughout as Fe^{3+}. The edge itself is characterized by a pre-edge feature (B) relating core level transitions to the incomplete 3d band, with the initial resonance (A) marking the onset of core level–conduction band excitation. Notably, the resonance A sharpens as the pre-edge feature (B) reduces (see inset). Initially, the pre-edge is pronounced, the threshold diminished, and together with a shorter O distance (obtained from XAFS), this reflects a tetrahedral environment for Fe^{3+} ions in the glass prior to treatment. As corrosion advances, the size of the threshold increases, signifying the migration of Fe to the surface. At the same time, the resonance (A) sharpens and the pre-edge feature (B) is reduced, pointing to a change in co-ordination for Fe^{3+} at the surface from tetrahedral to octahedral. Similar distinctions can be found between

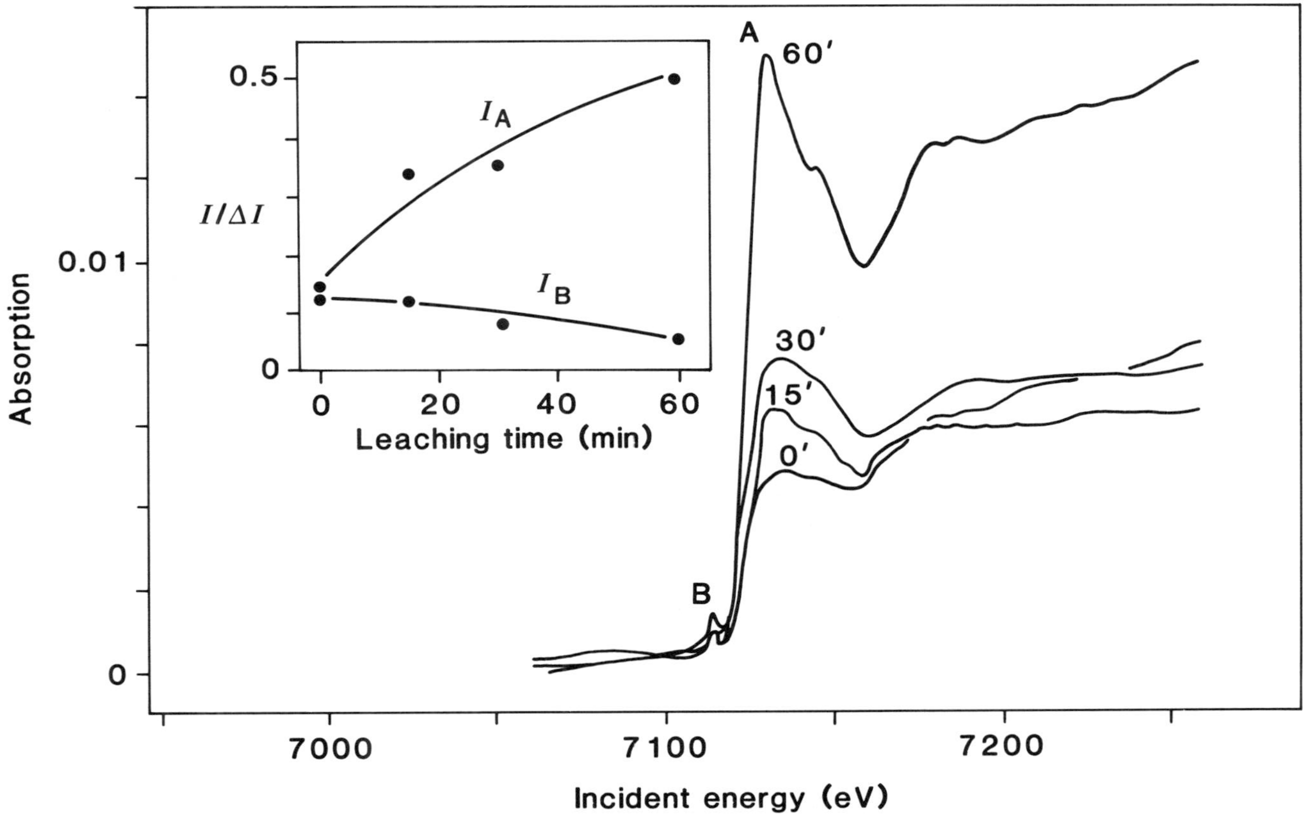

Figure 4.19 Changing near-edge structure at the K-edge of Fe in borosilicate glass at $0.8\,\phi_c$ as a function of leaching. The inset shows the counterwise behaviour of the white line resonance (A) and the pre-edge feature (B) (Greaves, 1990b).

the environment of Fe^{3+} in glasses and minerals (Binsted, Greaves and Henderson, 1986). The upshot is that Fe^{3+}, present in the bulk glass as a tetrahedral network former, migrates to the surface with aqueous corrosion, where it precipitates, possibly as a ferric silicate.

As far as the near-edge structure of U is concerned, the chemical shift in the L_{III} edge indicates the oxidation state is U^{6+}. Compared to the behaviour of Fe^{3+}, however, the same degree of corrosion brings less of this element to the surface. The U^{6+} environment, on the other hand, is particularly sensitive to the presence of water, and marked changes can be detected in the extended XAFS, as Figure 4.20 demonstrates (Greaves *et al.*, 1989). This collates the various partial radial distribution functions (PRDFs) for U within the first 40 Å as a function of aqueous attack (b–c). The PRDFs of U at the glass surface are compared with the PRDF obtained for the bulk structure (a), where the nearest neighbour O shell comprises two subshells at 1.9 Å and 2.3 Å. This splitting, which characterizes all the PRDFs, is the hallmark of the layered uranyl configuration found in crystalline uranates (Wiegel, 1986). The precise length of the axial and equatorial bonds are influenced by the types of cation neighbouring U^{6+} and the degree of interlayer bonding. Noticeably, we find that for U at the surface, these subshells are often further divided into subsidiary shells at ~1.8 Å and ~2.4 Å. Where the 1.9 Å and 2.3 Å distances found in the bulk glass structure mimic that U environment in alkali uranates such as Na_2UO_4, the 1.8 Å and 2.4 Å distances found at the corroded surface closely match the U

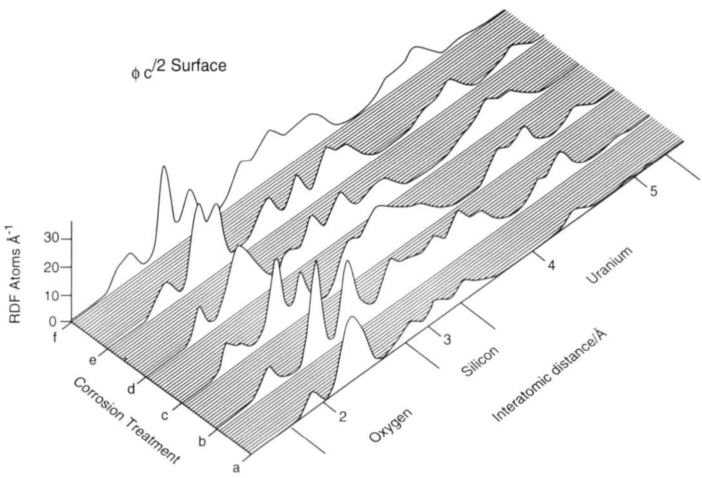

Figure 4.20 Uranium environments in borosilicate glasses exposed to different degrees of corrosion obtained from L_{III}-edge XAFS measured at $0.5\,\phi_c$ (z = 40 Å). (a) Bulk structure measured at $0.8\,\phi_c = \pi/4$; (b) polished surface; (c) 15 min leaching; (d) 15 min leaching followed by 12 h drying in vacuum; (e) 30 min leaching; (f) 90 min leaching (Greaves *et al.*, 1989).

configuration found in the layered hydroxide $UO_2(OH)_2$. This subdivision of axial and equatorial bond lengths for U atoms at the surface strongly suggests that water is intruding into the environment of these atoms as a result of aqueous corrosion. Further confirmation comes from comparing (c) and (d) in Figure 4.20 which both correspond to glasses leached for 15 min. Glass (c), however, was measured in the 'wet' state, while glass (d) was first dried under vacuum for 12 h. The $UO_2(OH)_2$-like configuration evident for U in (c) disappears for (d), where the Na_2UO_4-like environment found in the bulk glass (a) is restored. Compared with tetrahedral Fe^{3+} then, octahedral U^{6+} enters the bulk glass as a glass modifier rather than a network former. Accordingly, we would expect that U might be juxtaposed initially with Na in the bulk glass. As corrosion progresses, however, this neighbouring cation should be replaced by protonated water. The surface PRDFs for U obtained from glancing angle U L_{III}-edge XAFS demonstrate that this process is indeed taking place.

4.6.5 X-ray fluorescence analysis with ng/g sensitivity

Finally, attention is drawn to the important use of the same grazing incidence geometry presented in Figure 4.16(c) to improve the sensitivity of X-ray fluorescence (XRF) for trace element analysis. By recording the X-ray emission intensity as a function of ϕ it is possible to map out the depth profiles of impurities at material surfaces for very low concentrations, with sensitivities down to 10^9 atoms cm^{-2} (Iida, 1990). Perhaps more importantly, by utilizing a suitable reflector surface as a substrate (e.g. a Si wafer), the detection limits for trace element analysis of powders have been greatly improved (Wobrauschek *et al.*, 1991). The evanescent wave at the surface acts as a very efficient 'hot plate' for exciting fluorescence, where the material is scarce and the grazing incidence geometry greatly reduces the background scatter endemic in XRF. Detection limits in the pg range are claimed with ultimate sensitivities of ng/g.

Acknowledgements

The Science and Engineering Research Council (SERC) is thanked for the provision of synchrotron radiation facilities at the Synchrotron Radiation Source. The co-operation is also acknowledged of many colleagues from Materials Science Division at Daresbury Laboratory, from the Geoscience Research Institute at the University of Manchester, and from the Royal Institution of Great Britain.

References

Allinson, N.M., Baker, G., Greaves, G.N. and Nicoll, J.K. (1988) PDA system for energy dispersive EXAFS. *Nucl. Instrum. Methods*, **A266**, 592–7.

Attfield, J.P. (1990) Determination of valence and cation distributions by resonant powder X-ray diffraction. *Nature*, **343**, 46–9.

Bedard, B.D. and Flanigen, E.M. (1993) *Proc. of 9th Int. Zeolite Assoc. Meet., Montreal* (in press).

Bell, A.M.T., Henderson, C.M.B., Redfern, S.A.T. *et al.* (1993) Structure of synthetic $K_2MgSi_5O_{12}$-leucites by integrated X-ray powder diffraction, electron diffraction and 29Si MAS NMR methods. Submitted to *Acta Cryst.*

Bilsborrow, R.L. (1992) An investigation into the use of filters as a means of enhancing the performance of solid state detectors of extended X-ray absorption fine structure. *Rev. Sci. Instrum.*, **63**, 890–2.

Binsted, N., Greaves, G.N. and Henderson, C.M.B. (1986) Fe K-edge X-ray absorption spectroscopy of silicate minerals and glasses. *J. de Physique*, **C8**, 837–40.

Bras, W., Derbyshire, G.E.D., Mant, G. *et al.* (1993) Simultaneous time resolved SAXS and WAXS experiments using synchrotron radiation. *Nucl. Instrum. Methods*, **A326**, 587–91.

Catlow, C.R.A. (1990) X-ray crystallography from powders and crystallites, in *Applications of Synchrotron Radiation* (eds C.R.A. Catlow and G.N. Greaves), Blackie, London, pp. 39–64.

Cernik, R.J. (1993) Crystal structure of minerals. *Acta Cryst.*, **A49**, 246.

Cernick, R.J., Clark, S.M. and Pattison, P. (1991) Study of thin films and multilayers using energy dispersive diffraction of synchrotron radiation. *Adv. X-ray Anal.*, **33**, 101–7.

Cernik, R.J., Cheetham, A.K., Prout, D.J., Warkins, D.J. and Willis, B.T.M. (1991) The structure of cimetidine ($C_{10}H_{16}N_6S$) solved from synchrotron radiation X-ray powder diffraction data. *J. Appl. Cryst.*, **24**, 222–6.

Chowdhry, U., Corbin, D.R. and Subramanian M.A. (1989) Anorthite–cordierite based ceramics from zeolites. US Patent 4,814,303.

Clark, S.M. (1989) Energy dispersive powder diffraction at the SRS. *Nucl. Instrum. Methods*, **A276**, 381–7.

Clark, S.M. and Dooryhee, E. (1992) A quantitative kinetic study of the I $\rightleftharpoons$ II phase transition of ammonium chloride. *J. Phys. Cond. Mat.*, **4**, 8969–74.

Couves, J.W., Thomas, J.M., Catlow, C.R.A. *et al.* (1990) In situ studies of the dehydration of zeolite catalysts by time-resolved energy-dispersive X-ray absorption spectroscopy. *J. Phys. Chem.*, **94**, 6517–19.

Couves, J.W., Thomas, J.M., Waller, D. *et al.* (1991) Simultaneous time resolved SAXS and WAXS experiments using synchrotron radiation. *Nature*, **354**, 465–8.

Derbyshire, G.E., Dent, A.J., Dobson, B.R. *et al.* (1992) A multielement solid-state detector system for use on the SRS materials science station 9.3. *Rev. Sci. Instrum.*, **63**, 814–15.

Dooryhee, E., Steel, A.T., Maddox, P.J. *et al.* (1991) A study of cation environment and movement during dehydration and reduction of nickel-exchanged zeolite-Y by X-ray absorption and diffraction. *J. Phys. Chem.*, **95**, 1229–36.

Durham, P.J., Pendry, J.B. and Hodges, C.H. (1981) XANES: determination of bond angles and multi-atom correlations in ordered and disordered systems. *Solid State Commun.*, **38**, 159–62.

Greaves, G.N. (1990a) X-ray absorption spectroscopy, in *Glass Science and Technology*, Vol. 4A (eds D.R. Uhlmann and N.J. Kreidl), Academic Press, New York, pp. 1–76.

Greaves, G.N. (1990b) EXAFS for studying corrosion of glass surfaces. *J. Non-Cryst. Solids*, **120**, 108.

Greaves, G.N. (1991) Glancing angle X-ray absorption spectroscopy. *Adv. X-ray Anal.*, **34**, 13–22.

Greaves, G.N. and Catlow, C.R.A. (1990) Synchrotron radiation characteristics and techniques in *Applications of Synchrotron Radiation* (eds C.R.A. Catlow and G.N. Greaves), Blackie, London, pp. 1–38.

Greaves, G.N., Barratt, N.T., Antonini, G.M. *et al.* (1989) Glancing angle X-ray absorption spectroscopy of corroded borosilicate glass surfaces containing uranium. *J. Am. Chem. Soc.*, **111**, 4313–24.

Greaves, G.N., Gurman, S.J., Catlow, C.R.A. *et al.* (1991) A structural basis for ionic diffusion in oxide glasses. *Phil. Mag.*, **A65**, 1059–72.

Gurman, S.J. (1990) EXAFS and structural studies of glasses, in *Applications of Synchrotron Radiation* (eds C.R.A. Catlow and G.N. Greaves), Blackie, London, pp. 140–70.

Harding, M.M. (1990) Crystal structure studies with synchrotron radiation. *Chem. Brit.*, **26**, 956–8.

Harding, M.M., Kariuki, B.M., Cernik, R.J. and Cressey, G. (1994) The structure of aurichalcite determined from a microcrystal. *Acta Cryst.*, **B** (in press).

Hart, M. and Parrish, W. (1989) Polycrystalline diffraction and synchrotron radiation. *Mat. Res. Soc. Symp. Proc.*, **143**, 185–95.

Hayakawa, S., Iida, A., Aoki, S. and Gohshi, Y. (1992) Development of a scanning X-ray microprobe with synchrotron radiation. *Rev. Sci. Instr.*, **60**, 2452–5.

Henderson, C.M.B., Greaves, G.N. and Smith, J.V. (1993) X-ray absorption spectroscopy of Fe, Mn, Zn and Ti structural environments in staurolite. *Amer. Mineral.*, **78**, 477–85.

Houde-Walter, S., Inman, J.M., Dent, A.J. and Greaves, G.N. (1993) Sodium and silver environments and ion exchange processes in silicate and aluminosilicate glasses. *J. Phys. Chem.*, **97**, 9330–6.

Huang, T.C. (1991) Surface and ultra-thin film characterisation by grazing-incidence asymmetric Bragg reflection. *Adv. X-ray Anal.*, **33**, 91–100.

Iida, A. (1990) Semiconductor surface characterisation by synchrotron X-ray fluorescence analysis. *Adv. X-ray Anal.*, **34**, 23–33.

Iida, A., Noma, T., Hayakawa, S., Takahashi, M. and Gohshi, Y. (1993) High spatial resolution XAFS and its imaging applications. *Jpn. J. Appl. Phys.*, **32-2**, 160–4.

Jaklevic, J., Kirby, J.A., Klein, M.P. *et al.* (1977) Fluorescence detection of EXAFS: sensitivity enhancement for dilute species and thin films. *Solid State Commun.*, **23**, 679–82.

Laan, G. van der, Pattrick, R.A.D., Henderson, C.M.B. and Vaughan, D.J. (1992) Oxidation state variations in copper minerals studied with Cu 2p X-ray absorption spectroscopy. *J. Phys. Chem. Solids*, **53**, 1185–90.

Lee, P.A. and Pendry, J.B. (1975) Theory of the extended X-ray absorption fine structure. *Phys. Rev.*, **B11**, 2795–811.

MacMillan, P. (1979) in *Glass Ceramics* (eds J.T. Roberts and P. Popper), Academic Press, New York.

Munn, J., Barnes, P., Haussermann, D., Axon, S.A. and Klinowski, J. (1992) In-situ studies of hydrothermal synthesis of zeolites using synchrotron energy-dispersive X-ray diffraction. *Phase Transitions*, **13**, 129–34.

Nelmes, R.J., Hatton, P.D., McMahon, M.I. *et al.* (1992) Angle-dispersive powder diffraction techniques for crystal structure refinement at high pressure. *Rev. Sci. Instrum.*, **63**, 1039–42.

Oversluizen, M., Clark, S.M. and Greaves, G.N. (1993) Elucidating structural mechanisms for cordierite ceramic formation using synchrotron radiation. *Mat. Res. Symp. Proc.*, **307**, 39–44.

Parratt, L.G. (1954) Surface studies of solids by total reflection of X-rays. *Phys. Rev.*, **95**, 359–69.

Pizzini, S., Roberts, K.J., Dring, I,S., Oldman, R.J. and Greaves, G.N. (1991) Glancing angle X-ray absorption spectroscopy: applications to the structural characterisation of condensed interfaces, in *SR Light: Applications and Related Instrumentation II*. World Scientific, Singapore and London, pp. 67–107.

Rao, C.R.N. and Golapakrishnan, J. (1987) Synthesis of complex metal oxides by novel routes. *Accounts Chem. Res.*, **20**, 228–35.

Redfern, S.A.T., Henderson, C.M.B., McGuinn M.D. and Clark, S.M. (1993) High-pressure orthorhombic-tetragonal transition in gillespite – order parameter behaviour for synchrotron powder diffraction. *Terra Abstracts, Suppl. 1 Terra Nova*, **5**, 364–5.

Rietveld, H.M. (1969) A profile refinement method for nuclear and magnetic structures. *J. Appl. Cryst.*, **2**, 65–71.

Rivers, M.L., Sutton, S.R. and Jones, K.W. (1991) Synchrotron radiation X-ray fluorescence microscopy. *Synch. Rad. News*, **4**, 23–6.

Sheldrick, G. (1986) *SHELXS-86: Programs for Crystal Structure Solution*. University of Göttingen.

Smith, J.V. (1968) The crystal structure of staurolite. *Amer. Mineral.*, **53**, 1139–55.

Stephenson, P., Greaves, G.N. and Gurman, S.J. (1991) XAFS analysis using EXCURV90. *Synch. Rad. News*, **4**, 29–31.

Thomas, J.M. and Greaves, G.N. (1993) Combined EXAFS and XRD for the *in-situ* structural elucidation of solid catalysts under operating conditions. Submitted to *Cat. Lett.*, **20**, 337–43.

Thornley, F.R., Barrett, N.T., Greaves, G.N. and Antonini, G.M. (1986) EXAFS with grazing incidence: application of leached nuclear waste glasses. *J. Phys.*, **C19**, L563.

Warren, B.E. (1969) *X-ray Diffraction*, Addison-Wesley, Palo Alto, p. 116.

Wiegel, F. (1986) *Chemistry of the Actinide Elements*, Vol. 1, Chapman & Hall, London, p. 169.

Wobrauschek, P., Kregamer, P., Streli, C. and Aiginger, H. (1991) Recent developments and results in total reflection X-ray fluorescence analysis. *Adv. X-ray Anal.*, **34**, 1–12.

Zarzycki, J. (1990) Special methods of obtaining glasses and amorphous materials. *Mat. Sci. Technol.*, **9** (eds R.W. Cahn, P. Haasen and F.J. Krammer), VCH, New York, pp. 91–113.

CHAPTER FIVE

Sorption at mineral–water interfaces: macroscopic and microscopic perspectives

G.E. Brown, Jr, G.A. Parks and P.A. O'Day

5.1 Introduction

The processes by which chemical species accumulate on the surfaces of solids are referred to as **sorption**, and may involve solid–solid, solid–gas, or solid–liquid interactions. Sorption processes of the last type, in which chemical species partition from aqueous solutions to solid surfaces, are important in many contexts, including mineral growth from aqueous solution, mineral dissolution and weathering, formation of certain types of mineral and precious metal deposits, dispersal of contaminants in ground waters, processing of ores, treatment of chemical waste in sewage and industrial effluents, design of waste containments, electrochemical reactions, corrosion, and semiconductor processing. Therefore they have received much attention since Irving Langmuir (1916) first pointed out the important role that surfaces play in chemical reactions involving solids. This chapter deals with sorption reactions of hydrated metal ions at solid–water interfaces and discusses bulk (macroscopic) and molecular-level (microscopic) measurements that provide different levels of information about the products of these reactions. Such reactions control many terrestrial processes and have become the focus of a growing area of geochemistry referred to as mineral–water interface geochemistry.

Oxide, oxyhydroxide and silicate mineral surfaces in contact with water typically are electrically charged because of ionization of $\underline{S}OH$ functional groups. ($\underline{S}$ represents a metal ion at the solid surface.) The magnitude and sign of this surface charge vary with solution pH (Parks, 1990). Thus positively- or negatively-charged ions in solution accumulate in a diffuse swarm at or near the charged interface. Sorption at mineral–water interfaces may involve further reaction of some of these ions (referred to as adsorbates) with $\underline{S}OH$ sites on the mineral surface (referred to as the adsorbent), resulting in the loss of one or more waters of hydration from the adsorbate ion (or adion) and the formation of a relatively strong chemical bond(s) between adsorbate and adsorbent. This process is known as **specific**

Mineral Surfaces. Edited by D.J. Vaughan and R.A.D. Pattrick. Published in 1995 by Chapman & Hall, London. ISBN 0 412 56340 1

adsorption or **chemisorption**, and the adsorbed species is referred to as an **inner-sphere adsorption complex**. A weaker interaction may occur between the hydrated ion and the mineral surface in which waters of hydration are retained. In this case the adsorbate is not bonded directly to the surface, and the attraction is due mainly to longer-range coulomb forces and hydrogen bonding. Adsorption of this type is termed **non-specific**, and the adsorbed species is referred to as an **outer-sphere adsorption complex**. Sorption may also involve formation of a **precipitate** on the mineral surface, which is characterized by a three-dimensional structure that can be crystalline or non-crystalline. When the precipitate consists of chemical species derived

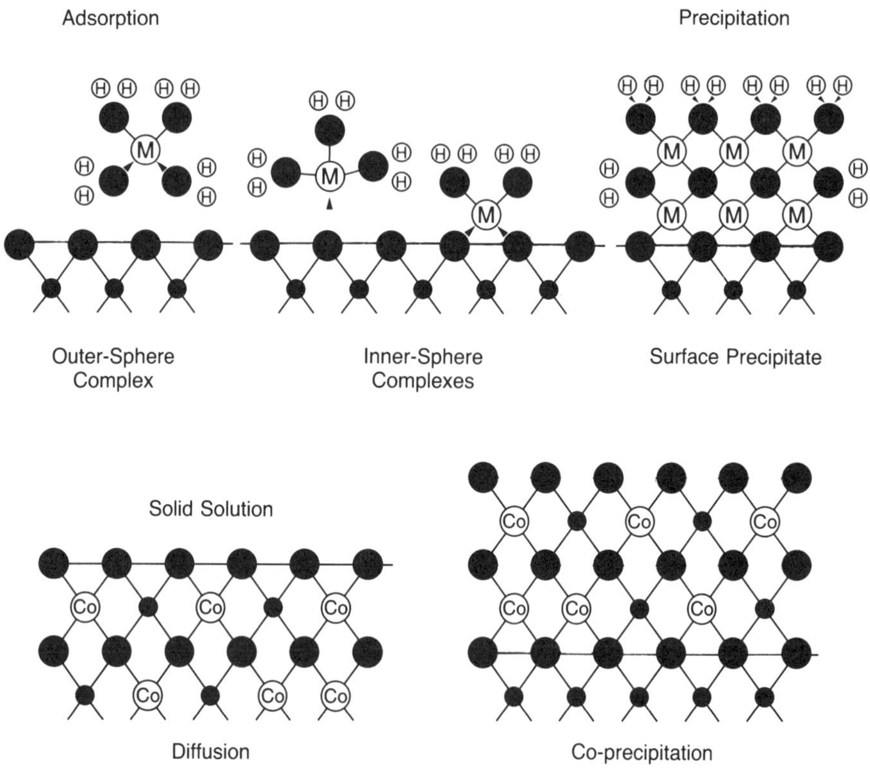

Figure 5.1 Possible sorption complexes at the mineral–water interface (represented by horizontal line – aqueous solution above, mineral substrate below). The large shaded circles represent oxygens and the small shaded circles represent metal ions in the mineral. The large circles labelled M represent aqueous metal ions, and the small circles labelled H represent protons. Different sorption modes involving adsorption, precipitation, and absorption (diffusion and co-precipitation) are shown. Outer-sphere and inner-sphere adsorption complexes are shown, including two different modes of inner-sphere binding (monodentate and bidentate). The figures representing diffusion and co-precipitation show two possible modes of Co absorption in $CaCO_3$ (calcite).

from both the aqueous solution and dissolution of the mineral, it is referred to as a **co-precipitate**. If the adions from solution are similar in size and of the same valency as ions in the mineral and the co-precipitate structure is the same as that of the adsorbent, the co-precipitation process may result in a true **solid solution**.

Absorption refers to any sorption process by which the adsorbate becomes incorporated in the crystal structure of the adsorbent, including the formation of a solid solution by co-precipitation or solid-state diffusion. Schematic illustrations of surface complexes resulting from these processes are shown in Figure 5.1. Further complications may arise in sorption processes where the adsorbent is capable of reducing or oxidizing the adsorbate. For example, under certain solution conditions aqueous Cr(VI) may be reduced to Cr(III) when it adsorbs on Fe_3O_4 (magnetite) or other minerals capable of donating electrons to the adsorbate (White, 1990). A surface reaction may be catalysed or a surface complex modified by heterogeneous photochemical processes (e.g. Waite, 1990). In addition to these sorption processes, particles may interact, resulting in their agglomeration into larger aggregates (e.g. O'Melia, 1987), or small particles may form by homogeneous precipitation and adhere to the surfaces of other solids.

Until relatively recently, sorption processes at solid–water interfaces were studied predominantly by macroscopic methods, such as measurement of adsorbate uptake and adsorbent surface charge as a function of solution pH, adsorbate concentration, complexing ligand concentration, ionic strength, and time (Kinniburgh and Jackson, 1981; Schindler and Stumm, 1987; Davis and Kent, 1990; Stumm, 1992). Such measurements have provided a wealth of information on the nature of sorption processes and have led to most of our conceptual equilibrium models of sorption at the solid–water interface, which are based on mass law equations. Some of these surface complexation models have been incorporated into computer codes which are used widely to predict sorption behaviour of aqueous metal ions on mineral surfaces (e.g. HYDRAQL, Papelis, Hayes and Leckie, 1988; see Davis and Kent, 1990, Dzombak and Morel, 1990, and Haworth, 1990, for reviews of surface complexation models). However, all such macroscopic models of sorption require a thermodynamic database and assumptions about the nature of reactive sites on hydrated solid surfaces, the nature of sorbed species (their composition and how they are attached or ligated to the solid surface) and the nature of the electrical double layer to predict the dependence of ion partitioning on solution pH, metal ion concentration and ionic strength. Even though such models have proven successful in simulating some laboratory and field observations of cation and oxyanion uptake, they are not uniquely sensitive to the composition and structure of sorption complexes or to the geometrical details of the oxide–water interface. Use of surface complexation models to predict sorption and partitioning behaviour of metal ions over wide ranges of solution conditions requires an accurate molecular-level description of the sorption process. However, the macroscopic measurements on which these models are based are fundamentally incapable of providing this molecular-level information (Sposito, 1986).

To illustrate this last point, consider the reactions below describing sorption of an aqueous oxyanion at an oxide–water interface (analogous reactions can be written for cation adsorption). One reaction sequence favoured by many soil chemists for inorganic oxyanion absorption on soil minerals involves the two-step ligand exchange

$$\underline{S}OH(s) + H^+(aq) = \underline{S}OH_2^+(s) \tag{5.1}$$

$$\underline{S}OH_2^+(s) + L^{q-}(aq) = \underline{S}L^{1-q}(s) + H_2O(l) \tag{5.2}$$

where L^{q-} is an oxyanion (e.g. SeO_3^{2-}) with charge $q-$ (Sposito, 1984). The protonation step in reaction (5.1) is thought to enhance the exchangeability of the surface hydroxyl group. Reaction (5.2) assumes that the oxyanion bonds to a single metal site on the oxide surface in an inner-sphere, monodentate fashion (Figure 5.1). Another equally plausible ligand-exchange reaction could be one that describes the formation of an inner-sphere, bidentate surface complex

$$2\underline{S}OH_2^+(s) + L^{q-}(aq) = \underline{S}_2L^{1-q}(s) + 2H_2O(l) \tag{5.3}$$

in which the oxyanion bonds to two adjacent metal sites on the oxide surface. In reactions (5.2) and (5.3), water molecules are displaced from the surface reaction site and from the hydration sheath of the oxyanion. Other oxyanions such as SO_4^{2-} and SeO_4^{2-} may not adsorb by a ligand-exchange mechanism on oxide surfaces. For example, SO_4^{2-} adsorption on α-FeOOH (goethite) and α-Cr_2O_3 (eskolaite) is similar in rate and extent on the two solids, which is consistent with the formation of an outer-sphere surface complex as described by

$$\underline{S}OH_2^+(s) + L^{q-}(aq) = \underline{S}OH_2^+L^{q-}(s) \tag{5.4}$$

in which no ligand exchange occurs (Yates and Healy, 1975). However, the point of zero salt effect (Sposito, 1984) for sulfate sorption on these two solids is shifted to higher pH in a sulfate background solution, which could reflect the formation of inner-sphere sorption complexes (Yates and Healy, 1975). Macroscopic measurements alone cannot distinguish between these possibilities.

A necessary complement to macroscopic studies of sorption are microscopic measurements capable of providing direct, molecular-level information on the nature of hydrated mineral surfaces, the mode or mechanism of sorption, the geometrical and electronic structures of the sorption complex, including oxidation state, and the electrical double layer. Within the past decade, relatively new microscopic methods have been developed and used for such measurements at the mineral–water interface, *in situ*, i.e. with bulk water present and in contact with the mineral surface (see Brown, 1990, and Hochella, 1990 for reviews). In particular, spectroscopic methods that utilize extremely intense synchrotron radiation provide some of the molecular-level information on sorption complexes needed to test macroscopic surface complexation models and to quantify our knowledge of chemical reactions at mineral–water interfaces.

Our objectives in this chapter are threefold:

1. to present examples of macroscopic sorption measurements commonly used to infer different modes of sorption of oxyanions and cations at mineral–water interfaces;
2. to discuss some of the microscopic methods that provide molecular-level information on adsorbed species, including water, with an emphasis on synchrotron-based X-ray absorption spectroscopy (XAS);
3. to present several examples of XAS studies of chemisorption of aqueous ions on oxide, oxyhydroxide and silicate surfaces.

A more detailed account of the structure and properties of these hydrated surfaces and the molecular details of metal ion sorption complexes at mineral–water interfaces can be found in Brown (1994) and references therein.

5.2 Macroscopic studies of sorption at mineral–water interfaces

Among the most common macroscopic methods used in studies of aqueous ion sorption on solids is measurement of the uptake of aqueous species as a function of solution pH and ionic strength at constant temperature and presssure (see, e.g. Davis and Leckie, 1978, 1980). Figure 5.2 shows representative uptake curves, or **adsorption isotherms**, for anion uptake on solids, specifically the sorption of selenate (SeO_4^{2-}) and selenite (SeO_3^{2-}) ions on powdered, high surface area ($\approx 55\,m^2\,g^{-1}$) goethite (α-FeOOH) using data from Hayes (1987). Figure 5.3 shows the uptake behaviour of Pb(II) (Hayes and Leckie, 1987) and Ba(II) (Hayes, 1987) on the same solid, illustrating typical behaviour of cations.

In general, anion sorption on oxide, oxyhydroxide and silicate surfaces is characterized by an increase in uptake as pH is lowered. Anion sorption on these solids also commonly results in an increase in solution pH. These observations suggest an exchange of the anion for a surface hydroxide. Examination of Figure 5.2 shows that sorption of selenite ions on goethite begins at a pH near 12 and that uptake is complete at a pH where this surface is still negatively charged, i.e. above the point of zero charge (pzc) of goethite (pH = 7–8), defined as the pH value where the total net particle charge is zero. In addition, selenite uptake shows little or no dependence

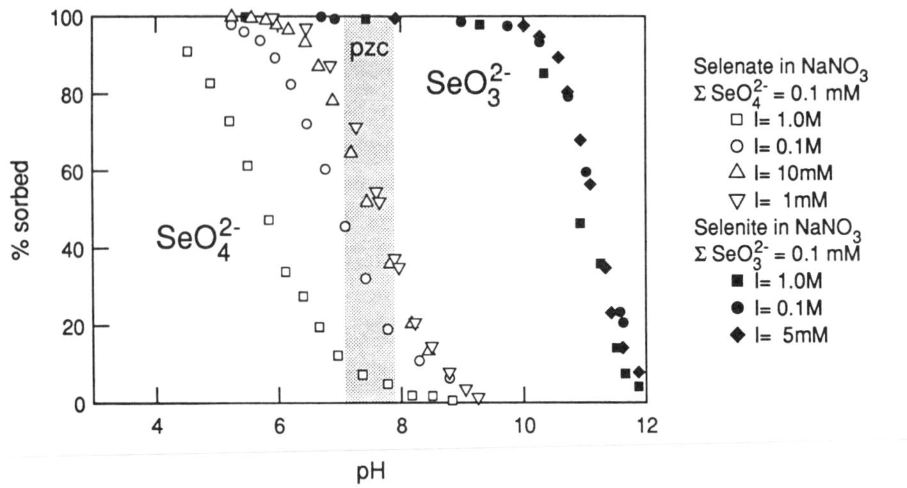

Figure 5.2 Uptake of selenate (SeO_4^{2-}) (open symbols) and selenite (SeO_3^{2-}) (closed symbols) as a function of pH and ionic strength of the background $NaNO_3$ solution. The shaded vertical area shows the point of zero charge (pzc) for goethite. Total selenate and selenite concentrations were fixed at 0.1 mM. At pH values below the pzc, the goethite surface in contact with water is positively charged, whereas at pH values above the pzc, the goethite surface is negatively charged. (After Hayes, 1987.)

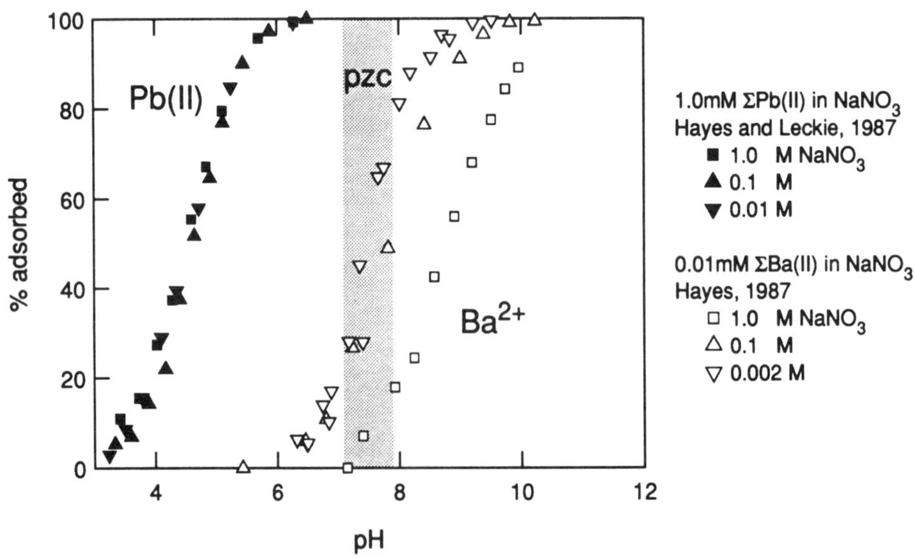

Figure 5.3 Uptake of Pb(II) (closed symbols) and Ba(II) (open symbols) on goethite as a function of pH and ionic strength of the background $NaNO_3$ solution. See caption of Figure 5.2 for explanation of pzc. (After Hayes, 1987; Hayes and Leckie, 1987.)

on the solution's ionic strength. These observations support the idea that selenite adsorbs strongly to the goethite surface, perhaps forming an inner-sphere complex and losing one or more waters of hydration (Figure 5.1 and reactions 5.1 and 5.2).

Selenate ions begin sorbing on goethite at pH $\approx$ 9, and uptake is not complete until the goethite surface is positively charged, i.e. at a pH below the pzc. Figure 5.2 shows that selenate uptake is dependent on the ionic strength of the $NaNO_3$ solution, with higher ionic strength inhibiting uptake. These observations suggest that NO_3^- ions interfere with selenate sorption and may be interpreted as indicating that selenate is weakly attracted to the goethite surface relative to selenite and that it forms an outer-sphere complex without significant loss of waters of hydration (reaction 5.4) (Hayes and Leckie, 1988).

Sorption of cations on oxide, oxyhydroxide and silicate surfaces is typically low at low pH and increases with increasing pH (Figure 5.3), which is the reverse of uptake behaviour observed for anions. Solution pH typically decreases when cations adsorb on these surfaces, suggesting exchange with one or more protons from a surface site. The two examples chosen to illustrate cation uptake follow these general trends but exhibit quite different uptake behaviour as a function of pH and solution ionic strength. Sorption of Pb(II) begins at much lower pH than Ba(II), well below the pzc of goethite, and is complete before the surface charge of goethite becomes negative. The Ba(II) uptake is complete only after the goethite surface is negatively charged. Pb(II) uptake shows little dependence on ionic strength, whereas Ba(II) uptake decreases with increasing ionic strength, suggesting that Na^+ from the background electrolyte may interfere with Ba(II) adsorption. These observations suggest that Pb(II) forms a strong chemical bond or bonds with the goethite surface and that Ba(II) forms a weaker bond. One possible microscopic interpretation of these macroscopic observations is that Pb(II) forms an inner-sphere adsorption complex at the goethite–water interface and that Ba(II) forms an outer-sphere complex (Figure 5.1) (Hayes and Leckie, 1987). However, little can be inferred about the exact nature of the sorption complex(es), including how the adsorbed cation is ligated to the surface (e.g. monodentate vs. bidentate attachment) or whether the cation adsorbs as a monomer, a dimer, or a more complex oligomer.

As the concentration of a cation in solution in contact with a solid increases, eventually the solution will become saturated with respect to a solid phase containing the cation, and the solid phase should precipitate. Various workers have suggested that the adsorption isotherm should exhibit a steep rise when precipitation occurs (Parks, 1990). Figure 5.4 shows a series of adsorption isotherms for Co(II) uptake on silica at different Co concentrations (James and Healy, 1972). 'Normal' uptake behaviour is observed at total Co concentrations ranging from 1.2×10^{-6} M to 1.2×10^{-4} M. However, at a concentration of 2×10^{-3} M, the slope of the uptake

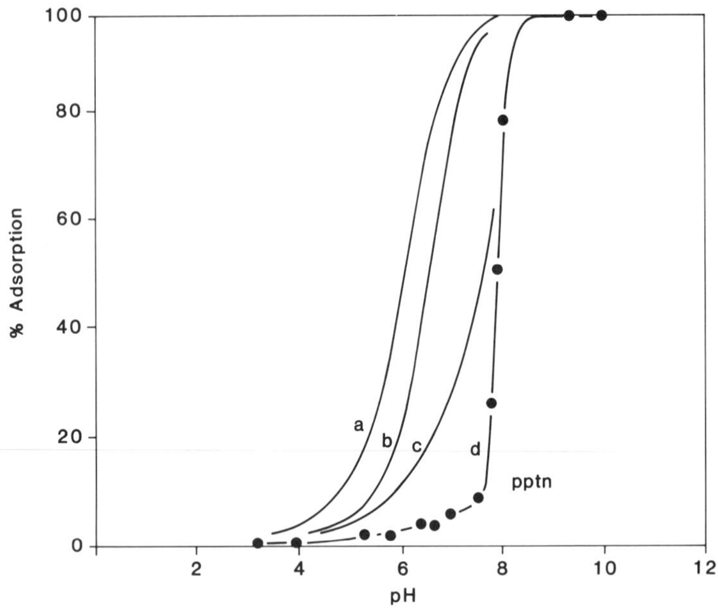

Figure 5.4 Uptake of Co(II) on silica at Co concentrations of (a) 1.2×10^{-6} M, (b) 1.2×10^{-5} M, (c) 1.2×10^{-4} M and (d) 2×10^{-3} M. Precipitation (pptn) was observed at the highest Co concentration (d), beginning at pH $\approx$ 7.5. No precipitation was observed for the other Co concentrations. (From James and Healy, 1972.)

curve rises abruptly at pH of 7.5. This type of uptake behaviour and the accompanying electrical charge reversal are often interpreted as evidence for precipitation of a Co-containing phase, even though the bulk solution is undersaturated with respect to any known hydroxide or oxide of Co in this case. However, this macroscopic behaviour alone is not sufficient evidence for precipitate formation (Sposito, 1986).

5.3 Macroscopic properties of water at the mineral–water interface

The structure and solvent properties of water near mineral surfaces may affect the type of sorption complex formed. The properties of water near oxide and clay mineral surfaces have been examined using measurements of the dielectric constant (ratio of static permittivity to the permittivity of a vacuum), dielectric relaxation (dipole response to an alternating external field) and dielectric loss (heat loss from relaxation) (Zettlemoyer, Micale and Klier, 1975; Sposito, 1984). As in heat-of-immersion studies, dielectric measurements are usually conducted on high surface area powders exposed to variable amounts of water vapour. Surface coverage is often expressed as a function of a monolayer, or as monolayers, of adsorbed water using the

BET method and assuming a cross-sectional area for the water molecule (usually in the range of 10–12 Å^2). The results of dielectric measurements show, in general, that water molecules near a mineral surface are less able to re-orient their dipoles in the presence of an electric field than they are in bulk water, similar to the dielectric response of ice. This behaviour has been attributed to stronger hydrogen bonding of the adsorbed water and an increase in molecular orientation or ordering relative to that of liquid water (McCafferty, Pravdic and Zettlemoyer, 1970). Despite a number of investigations (see summaries in Dukhin, 1971; Zettlemoyer, Micale and Klier, 1975; Texter, Klier and Zettlemoyer, 1978), there is little agreement on the values of the static dielectric constant of water (ε_s) near a mineral surface, owing in part, to the difficulty of separating out electrical effects from the surface and to differences in experimental technique (Sposito, 1984). For water near insulating mineral surfaces that have been studied, ε_s is depressed quite significantly from its value in liquid water (78.2 at 25 °C), perhaps to as low as 2–10 for the first monolayer of adsorbed water, and increases with adsorption of multiple layers of water. A reduction in solvent permittivity would favour the formation of metal adsorption complexes of low charge density (e.g. neutral or polar species) near a surface (e.g. Davies, 1962; Conway, 1981).

5.4 Microscopic studies of sorption on oxide, oxyhydroxide and silicate surfaces

5.4.1 Methods

Although many experimental methods are commonly used to characterize solid surfaces and adsorbed species (Brown, 1990; Table 5.1), only a small number are capable of providing information on their structure and composition in the presence of bulk water (*in situ*) (Goodman, 1986; Brown, 1990; Abruña, 1991a). Drying a sorption sample and placing it in an ultra-high vacuum (UHV) chamber for *ex situ* measurements poses the risk of altering the adsorption complex, particularly if it is weakly bonded to the surface (e.g. Abruña, 1991b). Thus *in situ* measurements are preferred.

In general, methods that utilize electron excitation and/or electron detection are not suitable for *in situ* measurements because they require that the sample be placed in UHV, owing to the relatively short mean free path of electrons in solids, water and air (escape depths in solids are of the order of 5–10 Å for electrons with kinetic energies of 30 to ≈400 eV; Lindau and Spicer, 1974). Thus surface-sensitive methods such as X-ray photoelectron spectroscopy (XPS) (e.g. Rhodin and Gadzuk, 1979; Sherwood, 1985), Auger electron spectroscopy (AES) (e.g. Hochella, 1988) and photoemission (e.g. Lindau and Spicer, 1980) that are routinely used to study solid surfaces in a UHV environment cannot be used for *in situ* characterization of adsorption complexes at mineral–bulk water interfaces. Among the exceptions to this statement is non-vacuum scanning tunneling microscopy (STM), which involves the tunnelling of electrons from a sharp metal tip over a very short distance (a few Å) to a conducting or semiconducting surface or vice versa (Binning *et al.*, 1982a, b; Hochella *et al.*, 1989; Hochella, 1990). Spectroscopic or scattering methods that employ photons as incident and excited (or detected) particles can be used to characterize adsorbates at mineral–water interfaces with varying limitations. They include vibrational spectroscopies [infrared (e.g. Stole, Popenoe and Porter, 1991; Kuhn, He and Goodman, 1992) and surface-enhanced Raman scattering (e.g. Pemberton, 1991)], resonance spectroscopies [electron paramagnetic resonance (e.g. Motschi, 1984; McBride, 1986), nuclear magnetic resonance (e.g. Grandjean, 1992; Haw, 1992), Mössbauer spectroscopy (e.g. Scherson, 1991)] and other methods reviewed in Hochella and White (1990), Abruña (1991a), Dines, Rochester and Thomson (1992), and Dwyer and Hoffman (1992).

5.4.2 *Overview of* in situ *studies of adsorption on mineral surfaces*

Among the more commonly used microscopic methods for *in situ* studies of adsorption processes on mineral surfaces are infrared (IR), nuclear magnetic resonance (NMR) and electron paramagnetic resonance (EPR) spectroscopies. Infrared spectroscopy and proton NMR spectroscopy can provide unique structural information on water molecules, hydroxyl groups and organic adsorbates on solid surfaces (Harrick, 1979; Fripiat *et al.*, 1982; Kiselev and Krylov, 1985; Tjedor-Tjedor and Anderson, 1986; Zeltner *et al.*, 1986). Examples of IR investigations of adsorbed water on SiO_2 and a few other silicates and oxides can be found in studies by Peri and Hannan (1960), Peri (1965), Little (1966), Hair (1967), Kiselev (1971), Gallei and Parks (1972), Poisignon, Cases and Fripiat (1978), and Kiselev and Krylov (1985), among others. These studies together with proton NMR studies (Kiselev and Krylov, 1985) show that both hydroxyl groups and water molecules populate the surfaces of these materials under ambient conditions and that the environment and properties of water molecules bonded to or located near (within ≈ 10 Å) an oxide or silicate surface are different from those of bulk water (see earlier discussion and Cases and François, 1982; Fripiat *et al.*, 1982; Giese and Costanzo, 1986). Estimates of the concentration of bound water on the surface of amorphous SiO_2 (non-porous aerosil) from proton NMR studies range from 10^{13}–10^{14} molecules cm^{-2} (Kiselev and Krylov, 1985, and references therein). Proton NMR studies of sorbed water on SiO_2 as a function of temperature suggest that adsorbed water molecules are lost at $T > 500$–600 K and that surface hydroxyl groups are not affected until $T > 1000$ K (Kiselev and Krylov, 1985). The strong adsorption of water on the SiO_2 surface has been explained by the formation of co-ordination bonds between water molecules and Si atoms at the SiO_2

surface (Kiselev and Krylov, 1985). Similar information has been provided by ^{29}Si NMR studies of hydrated SiO_2 surfaces (Kinney, Chuang and Maciel, 1993). Most of these studies involve gas–solid systems and elucidate the properties of water adsorbed from the vapour state. Cases and François (1982) and Fripiat *et al.* (1982), however, used measurements of the heat of immersion of solids with known surface densities of pre-sorbed water to demonstrate the important point – that water in the first 10 Å adjacent to the **immersed** surface is bound strongly and has properties different from bulk liquid water.

Infrared studies also have provided valuable information about the nature of reactive sites (e.g. Parfitt, Russell and Farmer, 1976), the coordination of anions, and the mode of attachment of anions on mineral surfaces (e.g. Parfitt, 1978, and references therein). Such information is usually obtained by exchanging OH with D_2O to give OD surface functional groups, then comparing the IR spectra of the OH and OD groups before and after exchange. One of the most studied systems is phosphate adsorption on goethite (α-FeOOH). Infrared spectroscopy has provided evidence for the formation of a bridging $(\underline{Fe}O)_2POOH$ surface complex (Russell *et al.*, 1975; Parfitt, Russell and Farmer, 1976). This bidentate mode of attachment is consistent with the slow rate of exchange of ^{32}P on phosphated goethite (Atkinson, Posner and Quirk, 1972), the strong adsorption of phosphate on goethite (Atkinson, 1969; Yates and Healy, 1975) and the observation that phosphate could not be desorbed from goethite by washing at low pH (Hingston, Posner and Quirk, 1974).

Electron paramagnetic resonance spectroscopy is an element-specific, *in situ* probe, but it is applicable only to paramagnetic ions such as Mn(II), Fe(III) and Cu(II) or paramagnetic molecules such as nitroxide radical cations (e.g. TEMPAMINE). When used to study such ions sorbed at solid–water interfaces, the solids must be diamagnetic or only weakly paramagnetic in order to avoid the spectral broadening caused by significant quantities of paramagnetic ions in the substrate. Electron paramagnetic resonance spectroscopy has been used primarily by McBride and co-workers and by Motschi to study the local environments of Cu(II) (McBride, 1976, 1978, 1982a, b, 1985, 1986; Bassetti, Burlamacchi and Martini, 1979; Zelewsky and Bemtgen, 1982; Motschi, 1983, 1984, 1987; Clark and McBride, 1984; Harsh, Doner and McBride, 1984; McBride and Bouldin, 1984; McBride, Fraser and McHardy, 1984), Mn(II) (Hronsky *et al.*, 1978, 1979; Burlamacchi and Martini, 1980; Hronsky and Rakos, 1980; Martini, 1981; Bleam and McBride, 1985), and V(IV) (Martini, Ottaviani and Seravalli, 1975; Motschi, 1987) at oxide–water interfaces. The EPR studies of Cu(II) on a series of aluminum oxides and clay minerals by McBride and co-workers have been used to infer oxidation state, coordination, hydrolysis, and orientation of the adsorbate; to distinguish between different sorption mechanisms (mononuclear versus multinuclear complexes and three-dimensional precipitates); and to detect multiple adsorption sites. In an EPR study of Mn(II) adsorption on γ-AlOOH (boehmite) and goethite, Bleam and McBride (1985) found permissive evidence for magnetically ordered Mn arrays on boehmite and of 'Mn clusters' on goethite. Similar conclusions about the formation of Mn(II) 'clusters' on silica gel have been reached in EPR studies by Hronsky *et al.* (1978, 1979) and by Hronsky and Rakos (1980). In contrast, other EPR studies of Mn(II) and Cu(II) on silica gel (Burlamacchi and Martini, 1980; Martini, 1981) were interpreted as indicating that these ions are loosely bound to the surface as monomeric, hexaquo complexes. These examples illustrate the fact that EPR measurements of adsorption complexes at mineral–water interfaces can provide very useful molecular-level information on specific paramagnetic ions, but the conflicting results for Mn(II) on silica gel point out the somewhat indirect nature of the structural information provided by past EPR studies. This type of element-specific microscopic information is provided by only a few spectroscopic methods, including XAS.

Synchrotron-based XAS was first used to study sorption complexes at mineral–water interfaces by Hayes *et al.* (1987) and has proven to be a powerful means of obtaining information on the geometrical and electronic structure, oxidation state and composition of such complexes (Brown *et al.*, 1988; Brown and Parks, 1989; Brown, 1990; Manceau *et al.*, 1992). To date, about two dozen XAS studies of this type have been completed, and several of these will be reviewed below. Because practical applications of XAS require the use of synchrotron X-rays, we begin with a brief overview of synchrotron radiation in the context of surface studies and of synchrotron-based surface methods. This is followed by a review of the fundamentals of XAS and a discussion of experimental methods, with an emphasis on its applications to mineral–water interface geochemistry.

5.4.3 Synchrotron radiation and surface methods

In the 20 years since synchrotron radiation sources have been available as multi-user experimental facilities, they have revolutionized the way in which we study the structure of materials (Winick and Doniach, 1980; Koch, Eastman and Farge, 1983) and have had a particularly significant impact on studies of surfaces and interfaces (Bachrach, 1992). Synchrotron radiation is the light emitted by accelerated, relativistic electrons or positrons and is produced in electron or positron storage rings. The main attributes of synchrotron radiation are:

- its very high photon flux relative to sealed or rotating-anode X-ray tubes (10^3 to 10^5 greater flux) (Figure 5.5);
- its broad spectral range (a few eV to >30 keV) (Figure 5.5) and tunability with appropriate monochromators;

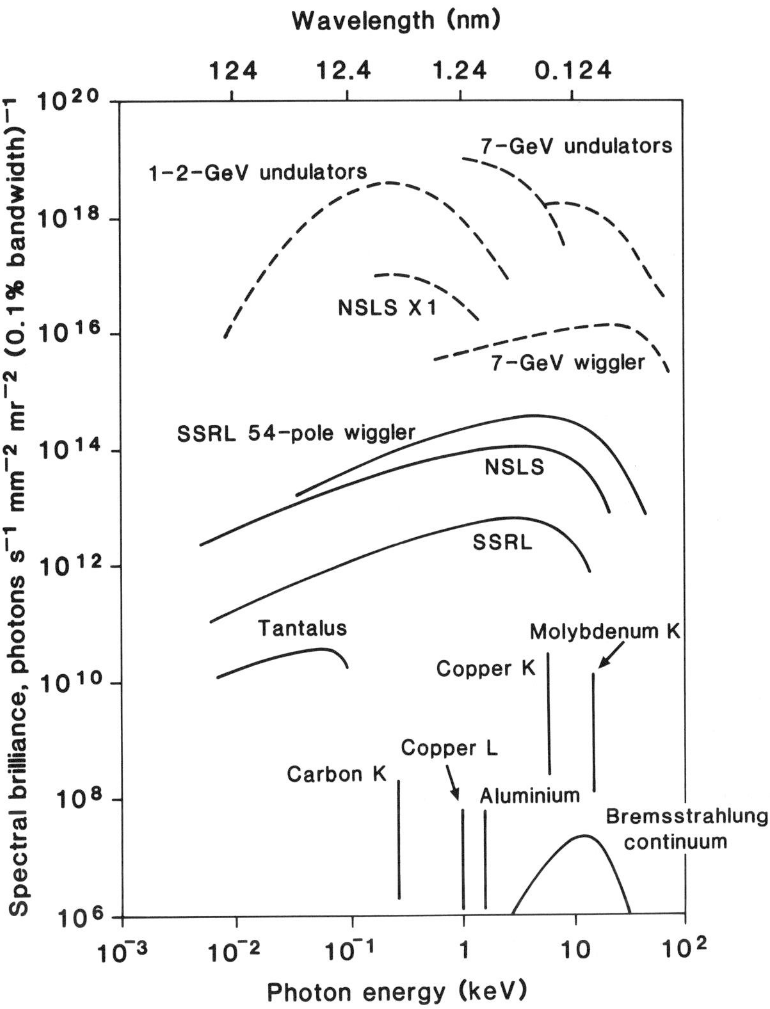

Figure 5.5 Plots of spectral brilliance vs. photon energy (keV) for various synchrotron radiation sources (SSRL – Stanford Synchrotron Radiation Laboratory; NSLS – National Synchrotron Light Source, Brookhaven National Laboratory; Tantalus – University of Wisconsin) compared with brilliances of characteristic X-ray lines from sealed X-ray tubes with various target materials (C, Cu, Al, Mo) and including the Bremsstrahlung from a sealed X-ray tube with tungsten target. The solid lines labelled SSRL, NSLS and Tantalus represent bending magnet beam lines, whereas the solid line labelled SSRL 54-pole wiggler and the dashed lines represent insertion device beam lines at SSRL, NSLS, the Advanced Light Source at Lawrence Berkeley Laboratory (1–2 GeV undulators) and the Advanced Photon Source at Argonne National Laboratory (7 GeV wiggler and 7 GeV undulators). (From Lear and Weber, 1987.)

- its high vertical collimation (a few milliradians), which permits surface-sensitive X-ray spectroscopic and scattering studies to be performed at or below the critical angle of a solid (a few tenths of a degree for metal oxides and silicates);
- its high polarization in the plane of the storage ring, which permits detailed studies of adsorbate geometries, including orientation effects, on single-crystal surfaces.

The two key attributes of synchrotron radiation, *vis-à-vis* microscopic studies of cations and oxyanions adsorbed at mineral–water interfaces, are its extreme intensity and its tunability over a broad energy range, which is required for element-specific spectroscopic probes such as XAS. In order to study adsorbed species at relatively low concentrations (corresponding to less than monolayer coverage or a few μmoles of the adsorbate per m^2 of a metal oxide), an extremely intense X-ray beam is required to generate a detectable X-ray signal from the adsorbed species. High X-ray flux is also required because of the need to characterize surface complexes in the presence of bulk water. Water attenuates the incident X-ray beam as well as the transmitted, diffracted and fluorescence X-ray signal.

A number of surface methods utilizing synchrotron radiation have been developed and applied to a variety of surfaces (Bachrach, 1992). These include photoemission spectroscopy (PES) (Lindau and Spicer, 1980; Allen, 1992; Eberhardt, 1992; Flodström, Nyholm and Johansson, 1992), Surface EXAFS (SEXAFS) spectroscopy (Stöhr, 1988; Rowe, 1992), photon-stimulated desorption (PSD) (Rehn and Rosenberg, 1992), photoelectron and Auger electron diffraction (PD and AED) (Fadley, 1992), grazing-incidence X-ray scattering (GIXS) (Fuoss, Liang and Eisenberger, 1992), X-ray standing wave (XSW) methods (Bedzyk, Bommarito and Schildkraut, 1989; Abruña, Bommarito and Acevedo, 1990; Bedzyk *et al.*, 1990; White, 1991; Woicik *et al.*, 1992) and fluorescence-yield X-ray absorption spectroscopy (FYXAS) (Hayes *et al.*, 1987; Brown, Parks and Chisholm-Brause, 1989; Brown, 1990). All but GIXS, XSW, and FYXAS require UHV environments, so they cannot be used for *in situ* studies of adsorbates at solid–bulk water interfaces. The synchrotron-based method that has become most useful for such studies is FYXAS. We will explore this method in some detail, with an emphasis on its application to adsorbates at mineral–water interfaces.

5.4.4 Fundamentals of X-ray absorption spectroscopy (XAS)

X-ray absorption spectroscopy was first developed as a quantitative structural method in the early 1970s (Sayers, Stern and Lytle, 1971) and has since been applied to a broad range of problems in biology, chemistry, geochemistry, materials science, mineralogy, solid-state physics and surface science. The method and its applications have been discussed in a number of publications (e.g. Teo and Joy, 1981; Hayes and Boyce, 1982; Stern and Heald, 1983; Teo, 1986; Wong, 1986; Brown *et al.*, 1988; Koningsberger and Prins, 1988; Brown and Parks, 1989; Lytle, 1989; Bachrach, 1992; Stöhr, 1992). The discussion that follows focuses mainly on those aspects of the technique relevant to structural studies of adsorbates at mineral–water interfaces. X-ray absorption spectroscopy is a bulk technique, but because of its element-selectivity and its sensitivity to low element concentrations, it can be used to provide unique structural and compositional information on surface complexes when the element of interest is present dominantly at a surface. These attributes also make XAS useful when the adsorbate occurs on a compositionally complex substrate, such as a natural mineral. XAS is commonly subdivided into two methods known as EXAFS (extended X-ray absorption fine structure) or XAFS (X-ray absorption fine structure) spectroscopy, and XANES (X-ray absorption near edge structure) or NEXAFS (near-edge X-ray absorption fine structure) spectroscopy. EXAFS and XANES arise from different physical processes and provide complementary information (Brown *et al.*, 1988).

Contributions to the XAS signal come dominantly from the two or three closest shells of atomic neighbours surrounding the absorbing atom (typically less than ~6 Å radius in oxides), owing to the short electron mean free path in most materials; thus it is a local structural probe. Because of the short-range nature of the signal, XAS is applicable to crystalline as well as non-crystalline materials, including interface regions such as mineral–water interfaces, which may lack long-range order. X-ray absorption spectroscopy is both a spectroscopy, in which the electron is excited from one energy level to another, and a scattering method, in which the ejected photoelectron is scattered from atoms surrounding the absorber. Because the electron remains within the solid, a UHV environment is not required for an XAS experiment, in which changes in X-ray absorption or emission are detected. Its sensitivity to variations in interatomic distances derives from the interference between outgoing and backscattered photoelectron waves, whereas its sensitivity to electronic structure, including oxidation state of the absorber, results from electronic excitations from deep core levels of the absorbing atom to valence and continuum levels. X-ray absorption spectroscopy is also sensitive to the types and number of neighbours surrounding the absorber, and thus it provides information on the composition of an adsorption complex. Careful analysis of EXAFS spectra, using amplitude and phase-shift functions derived from appropriate model compounds or theory, is capable of yielding average interatomic distances accurate to better than ±0.02 Å and average coordination numbers accurate to ±15–20%, assuming that systematic errors have been minimized in the experiment and data analysis and that static and thermal disorder are small. One important caveat is that the structural information provided by XAS is summed over all environments of a given absorber element. Thus XAS is capable of detecting different environments of an element only if these environments are sufficiently different to produce resolvable features in an EXAFS spectrum. In practice, metal ions in different sites of an oxide or silicate with the same coordination number (e.g. octahedral sites) can be distinguished if the difference in average M–O distance is $\geqslant 0.1$ Å and the k-range of the EXAFS data is sufficiently large, where k is the photoelectron wave-vector ($k = [0.262(E - E_0)]^{1/2}$; E is the photoelectron energy, and E_0 is the continuum threshold energy of the photoelectron). However, when different octahedral sites in the same mineral exhibit a range of distances (e.g. the M1 and M2 sites in olivines), such distinction may not be possible. Also, different coordination geometries of a cation in a material (e.g. tetrahedral vs. octahedral) can be distinguished in many cases.

An EXAFS spectrum is generated by exposing a sample to an incident monochromatic beam of synchrotron X-rays of appropriate energy, which is scanned through a range of energies below and above the absorption edge of the element of interest. Various processes occur when X-rays interact with matter, including X-ray scattering (both elastic and inelastic), production of optical phonons, production of photoelectrons and Auger electrons, production of fluorescence X-ray photons, and positron–electron pair production. In the X-ray energy range (0.5 to 100 keV), photoelectron production is the dominant process and is the primary cause of X-ray attenuation by matter. When the energy of the incident X-ray beam ($h\nu$) is less than the binding energy (E_b) of a core electron on the element of interest, little absorption takes place. However, when $h\nu \approx E_b$, electron transitions to unoccupied, bound energy levels occur, contributing to the main absorption edge and producing features below the main edge, referred to as the pre-edge spectrum (Figure 5.6). As $h\nu$ increases beyond E_b, electrons can be ejected to continuum (unbound) levels and remain in the vicinity of the absorber for a short time with excess kinetic energy. In the energy region extending from just above to about 50 eV above E_b, the electrons are multiply scattered among neighbouring atoms (Figure 5.7(a)), which produces the XANES or NEXAFS spectrum. When $h\nu$ is about 50 to 1000 eV above E_b, electrons are ejected from the absorber, are singly or multiply scattered from first- or second-neighbour atoms back to the absorber, and then leave the vicinity of the absorber (Figure 5.7(b)). The EXAFS spectrum is due to interference between outgoing and backscattered photoelectrons, which modulates the atomic absorption coefficient (Figure 5.6). Multiple scattering of the ejected electrons may produce important contributions to the EXAFS spectrum, depending on the geometrical arrangement of nearest and more-distant neighbours around an element and the electron scattering amplitude of these neighbours. This contribution should be assessed in order to avoid possible misinterpretation of the EXAFS signal. Multiple-scattering codes such as FEFF (Rehr and Albers, 1990; Rehr *et al.*, 1991; Rehr, Albers and

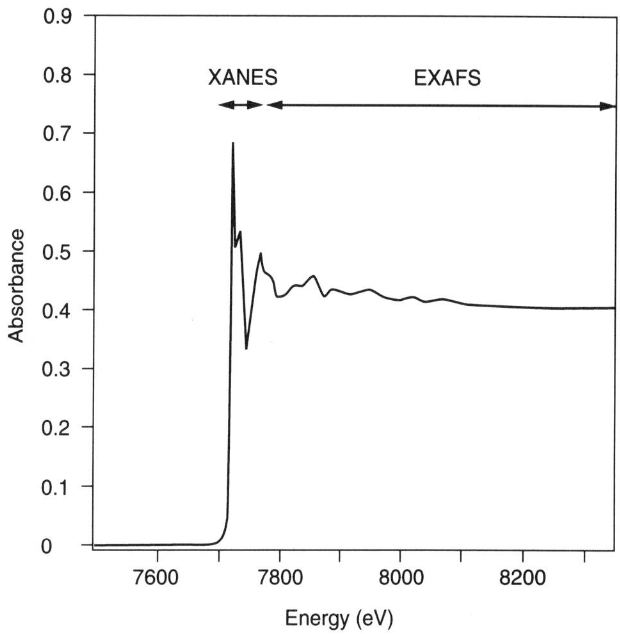

Figure 5.6 Co K-edge X-ray absorption spectrum of $CoCO_3$ recorded in the transmission mode showing the XANES and EXAFS regions. The pre-edge region is from a few eV to $\approx 10\,eV$ below the main absorption edge and shows a small pre-edge feature due to a $1s \rightarrow 3d$ bound-state electron transition. (From Xu, 1993.)

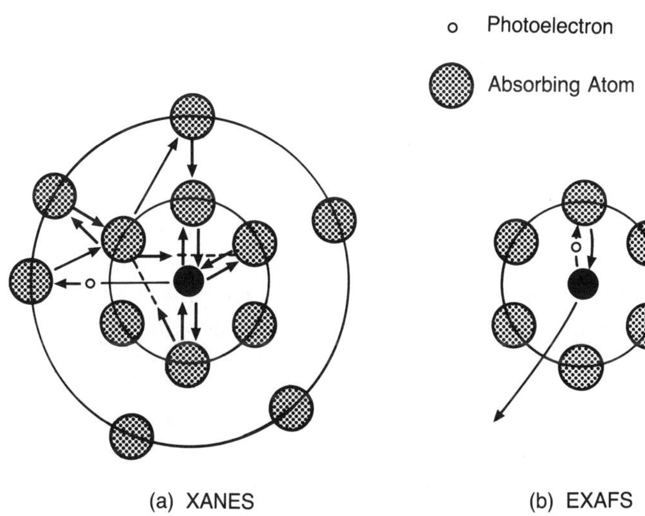

Figure 5.7 Electron scattering processes leading to (a) XANES and (b) EXAFS (see text for details). (After Brown *et al.*, 1988.)

Zabinsky, 1992) are now available for this purpose, and work quite well, even when the structure is quite complex (O'Day *et al.*, 1994a).

Depending on the absorber and the energy of incident X-rays striking the sample, the photoelectrons can be from any atomic shell (K, L_I, L_{II}, L_{III}, etc.). For heavier adsorbate elements such as U, however, the binding energy of K-shell electrons is too high to be accessible with existing synchrotron sources (UK-edge energy = 115.6 keV). In these cases, XAS at the L_{III} edge is commonly measured. With currently available monochromator crystals and synchrotron radiation sources, XAS studies are not feasible at edge energies > 25–30 keV. Thus the K-edges of elements of atomic number (Z) $\geqslant$ 50 (Sn) are not commonly studied using XAS methods. However, this situation is changing with the commissioning of third-generation synchrotron radiation sources like the Advanced Photon Source (at Argonne National Laboratory, in the USA) and the European Synchrotron Radiation Facility (at Grenoble, in France), which have higher critical energies and higher brightnesses than second-generation sources.

5.4.5 Experimental methods

XAS experiments may be carried out in several modes that differ in the type of particle detected [typically, X-rays transmitted through the sample (transmission experiment), fluorescent X-rays emitted as a result of absorption of the incident X-ray beam (fluorescence-yield experiment), or emitted photoelectrons (electron-yield experiment) are measured]. In a transmission XAS experiment, the incident (I_0) and transmitted (I_1) X-ray intensities are recorded as a function of increasing incident X-ray energy (E) to produce an absorption spectrum, which is typically plotted as $\ln(I_0 I_1)$ vs. E (in eV) (Figure 5.6). The relationship between these intensities and the linear absorption coefficient μ (in cm^{-1}) of a sample of thickness x (in cm) is $\ln(I_0/I_1) = \mu x$. In a fluorescence-yield experiment, the X-ray fluorescence from the sample, I_f, can be measured and ratioed with I_0 as I_f/I_0, which is proportional to μ for dilute samples. Fluorescence X-ray emission arises from de-excitation of the core hole generated by absorption of an incident X-ray photon. Thus I_f is directly proportional to the number of absorption events for elements at low concentrations (in practice <1 wt % of the sample). The primary advantage of the fluorescence-yield method relative to the transmission method is its greater sensitivity to elements at low concentrations in highly absorbing matrices.

A relatively simple, efficient and inexpensive detector for fluorescence-yield experiments is the large solid angle, gas-filled ion chamber detector developed by Lytle *et al.* (1984). A schematic drawing of the experimental arrangement for fluorescence-yield XAS measurements employing this type of ion chamber is shown in Figure 5.8. Appropriate filters are necessary in X-ray fluorescence measurements to reduce background caused by elastic

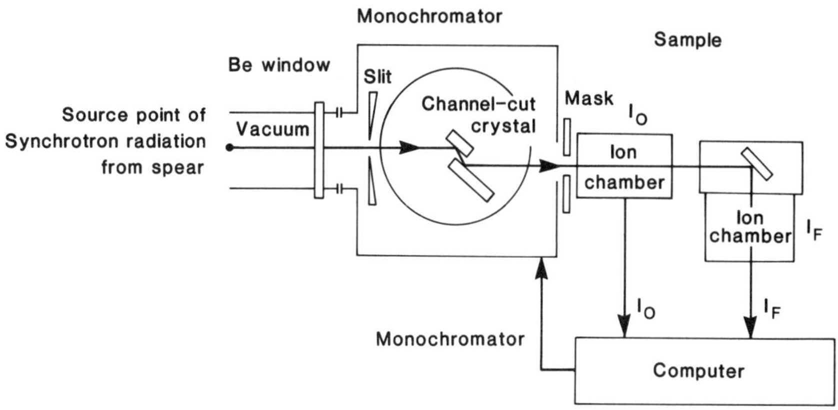

Figure 5.8 Schematic illustration of the experimental set-up used for fluorescence-yield XAS measurements. (After Wong, 1986.)

and Compton scattering. This can be accomplished by using a ($Z - 1$) or ($Z - 2$) filter material with Z one or two less than that of the absorber element. Alternatively, a solid-state detector capable of energy discrimination may be used in place of a gas-filled ion chamber. We have been using a 13-element Ge detector array and associated electronics for the past few years and have found it to provide superior signal-to-noise data for very dilute samples relative to the ion chamber detector. Adsorbate coverages, about an order-of-magnitude lower than those possible with an ion chamber detector, can be studied using the multi-element solid state detector. However, a solid state array detector is significantly more expensive (currently >\$200 000) than the ion chamber detector (<\$10 000) and is limited to about 40 000 cps per detector channel (required to remain in the linear response region) with currently available counting electronics. Our experience has shown that count rates for dilute samples often exceed this limit when the adsorbate is on a highly scattering substrate or one that contains an element that is strongly fluoresced by the X-ray beam (for example, when Co is adsorbed on an Fe-oxide). In this example, the FeKα emission can be electronically discriminated against; however, the FeKα signal may result in a very high total count rate relative to the CoKα signal, making such an experiment quite difficult.

Fluorescence-yield XAS is the method of choice in studies of adsorbates at low concentrations at mineral–water interfaces, but it has some limitations. For example, low Z adsorbates such as Na, Al and Ca cannot be characterized using currently available *in situ* XAS methods because X-rays with energies lower than ~5 eV are strongly absorbed by even a small amount of water. Thus such studies are limited in practice to adsorbate elements with $Z \geqslant 22$ (Ti). A limitation is also imposed by sorbents containing elements that are strongly fluoresced by the X-ray beam, as mentioned above. When an ion-chamber detector is used in such a case, high count rates are not a problem, but an appropriate filter and soller slits must be placed between the sample and the detector to reduce unwanted fluorescence that degrades S/N. Such background fluorescence is not a problem when the sorbent contains only low Z elements such as Si, Al and O. For example, Co adsorbed on Al_2O_3 or SiO_2 will cause fluorescence of Al $K\alpha$, Si $K\alpha$, and O $K\alpha$ X-rays, but these X-rays are strongly absorbed by water and air, and thus are not detected. On the other hand, light atoms are difficult to detect as second neighbours of the absorber, so a compromise in choice of systems for study using XAS is often necessary. The electron-yield method is not particularly useful for *in situ* XAS studies of adsorbed species, even though it is quite surface-sensitive, because bulk water strongly attenuates electrons.

Sorption samples suitable for *in situ* fluorescence-yield XAS experiments may be of two forms: wet, high surface area (<1 to $>100\,m^2\,g^{-1}$) powdered solids or flat single-crystal surfaces on which a particular type of cation or oxyanion is adsorbed. The wet, powdered sample is typically in the form of a paste with a volume of $\sim 1\,cm^3$ to as small as a few μm^3, depending on element concentration, X-ray energy, and X-ray flux. It is loaded into a mylar-windowed sample holder made of low Z materials (aluminium or Teflon). The single-crystal sorption sample can be examined in grazing-incidence geometry, in which the incident X-rays strike the sample below the critical angle, α, of the material, resulting in total external reflection (for oxides and silicates α is of the order of a few tenths of a degree). Such measurements require a more specialized sample holder mounted on a high-precision two-circle goniometer (Abruña, 1991b). In either case, the adsorbed element of interest is usually at concentrations ranging from a few μmolar to mmolar (moles of adion per litre of sample volume), which corresponds to less than monolayer coverage (typically a few μmoles of the adsorbate per m^2 of the adsorbent for samples with surface areas of the order of 10 to $100\,m^2\,g^{-1}$). The major constraint on minimum concentration limit and sample size is adequate spectral signal-to-noise ratio for extracting structural data. The ranges quoted above are typical in our experience using wiggler beam lines. When adsorbed species on single-crystal samples are examined, the surfaces may be common cleavage planes, which can be oriented along different crystallographic directions in the X-ray beam to produce polarized X-ray absorption spectra and direction-dependent absorber–scatterer distance information (Waychunas and Brown, 1990). This application takes advantage of the natural polarization of synchrotron X-rays.

5.4.6 *Analysis and interpretation of XANES and EXAFS spectra*

Discussions of the origin and interpretation of XANES spectra can be found in Bianconi (1988) and Stöhr (1992). Although XANES spectra cannot be analysed in as quantitative a fashion as EXAFS spectra with currently available theories, unique information on the local structural environment and bonding of an element can be obtained (Bianconi, 1988; Stöhr, 1992). Furthermore, the near-edge region of an X-ray absorption spectrum is significantly more intense than the EXAFS region. Thus, when an element is present at concentration levels too low to produce a useful EXAFS spectrum, as is often the case in environmental samples containing contaminants at the low ppm or ppb levels, the XANES spectrum may provide qualitative or semi-quantitative information on the local coordination environment and oxidation state of the contaminant element. Such information is derived by comparison of features in the XANES spectrum of the sample under investigation with features in XANES spectra of appropriate, well-characterized reference compounds (Waychunas, Apted and Brown, 1983; Wong *et al.*, 1984). For first-row transition elements, pre-edge spectral features caused by $1s \rightarrow 3d$ electronic transitions contain useful information about the symmetry of the absorber's local environment. Specifically, the presence of a peak 4–5 eV below the main absorption edge, with an intensity >6% of the edge jump, usually indicates a regular site without a centre of symmetry or a very irregular local coordination geometry (Waychunas, Apted and Brown, 1983; Roe *et al.*, 1984). The absence of such a feature in XANES spectra from first-row transition elements indicates that the absorber occupies a relatively undistorted site characterized by a centre of symmetry, such as an octahedron. This type of structural information is independent of and complementary to that provided by EXAFS analysis. Several examples of XANES spectra are shown in Figure 5.9, which will be discussed below in the context of Co adsorption on γ-Al_2O_3, rutile, and kaolinite.

An EXAFS spectrum consists of an oscillatory signal whose frequency is approximately inversely proportional to the distance between the absorber and neighbouring backscatterer atoms and whose amplitude is directly proportional to the number and backscattering cross-sections of the neighbouring atoms, modified by several amplitude reduction factors discussed below. When several nearest-neighbour atoms are present at different distances around the absorber, multiple frequencies may be present in the EXAFS signal. An ion specifically adsorbed at a mineral–water interface will often generate multiple frequencies in its EXAFS spectrum, and analysis of these frequencies can lead to unique structural and compositional information about the chemisorbed species and its mode of sorption.

Analysis of an EXAFS spectrum involves extracting interatomic distances, coordination number, and identity of first, second and more distant shells of

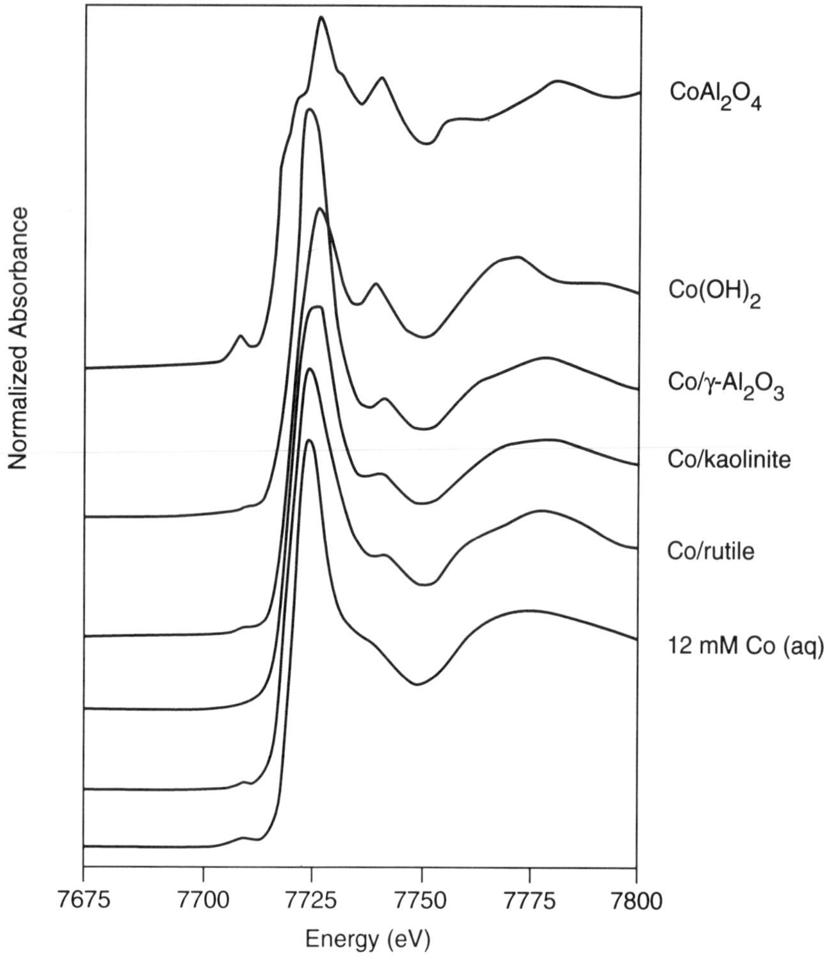

Figure 5.9 Co K-edge XANES spectra of $CoAl_2O_4$ spinel, crytalline $Co(OH)_2$, three samples with aqueous Co(II) adsorbed on γ-Al_2O_3, kaolinite ($Al_2Si_2O_5(OH)_4$) and rutile (TiO_2), and a 12 mM aqueous $Co(NO_3)_2$ solution. (From Chisholm-Brause *et al.*, 1990a.)

→

Figure 5.10 Illustration of the steps involved in analysis of an EXAFS spectrum, using the Co K-XAS of solid $Co(OH)_2$ as an example. Subtraction of background (dashed line in EXAFS spectrum) yields the EXAFS function ($\chi(k) \times k^3$) vs. k (Å^{-1}). Fourier transformation of the background-subtracted EXAFS function yields a radial structure function (Fourier transform magnitude vs. R (Å)), which contains peaks at different distances associated with various shells of neighbouring atoms around Co. Fourier filtering of the different shells, and back Fourier transformation yields the frequency contribution of each neighbouring shell of atoms around Co. In this example, the first peak in the radial structure function is caused by backscattering of the photoelectron from six nearest-neighbour O atoms at 2.08 Å from Co, and the second peak by backscattering from six second-neighbour Co atoms at 3.17 Å from the central Co atom. (After O'Day, 1992.)

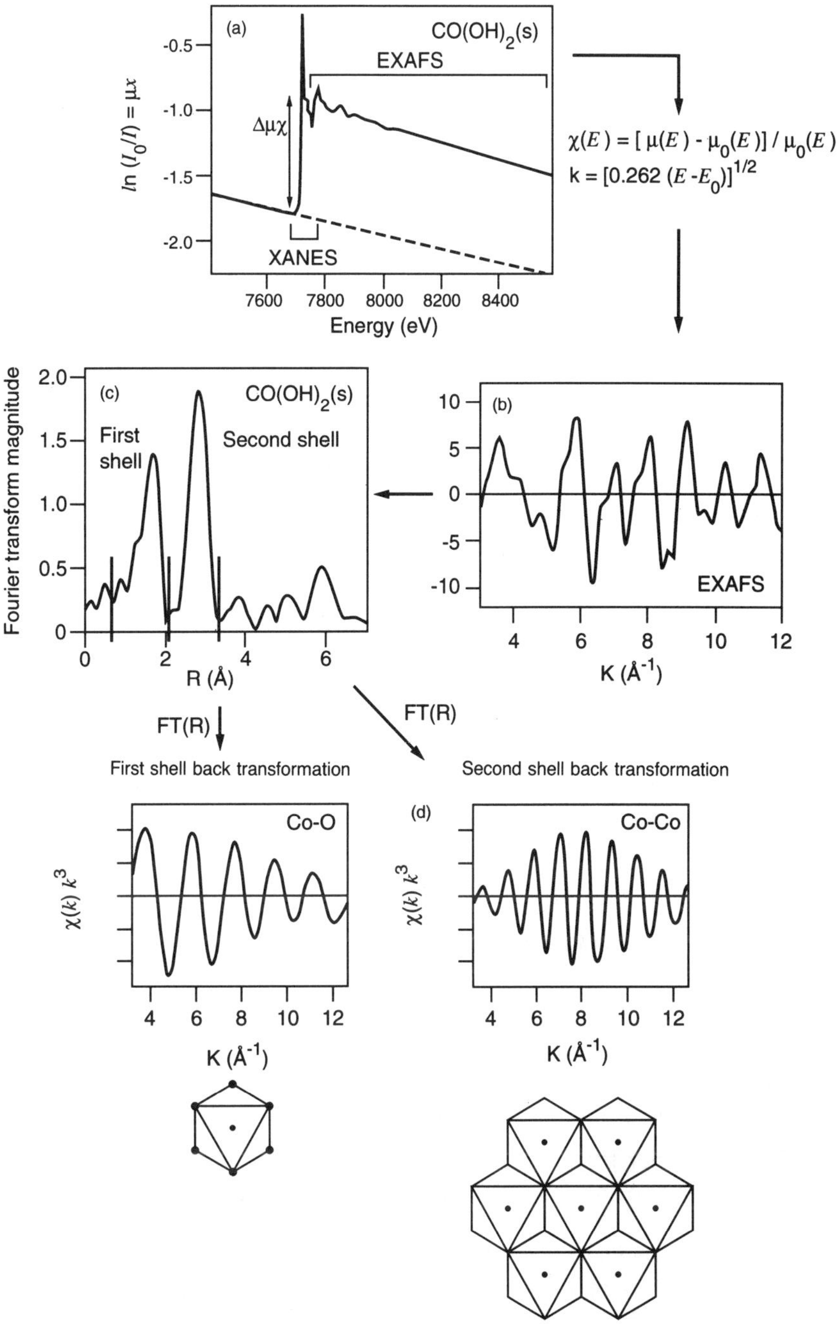

(a)
CO(OH)$_2$(s)
EXAFS
XANES
$ln\ (I_0/I) = \mu x$
$\Delta\mu\chi$
Energy (eV)
$\chi(E) = [\mu(E) - \mu_0(E)] / \mu_0(E)$
$k = [0.262\ (E - E_0)]^{1/2}$
(b)
EXAFS
K (Å^{-1})
(c)
CO(OH)$_2$(s)
First shell
Second shell
Fourier transform magnitude
R (Å)
FT(R)
FT(R)
First shell back transformation
Second shell back transformation
Co-O
(d)
Co-Co
$\chi(k)\ k^3$
K (Å^{-1})

neighbours around an absorber (Teo, 1986; Brown *et al.*, 1988; Sayers and Bunker, 1988) and consists of several steps illustrated in Figure 5.10. In step one (Figure 5.10(a)), a polynomial function is fitted to the pre-edge region of the spectrum and subtracted, leaving zero background in the pre-edge region. The next step involves fitting a spline function through the EXAFS region (not shown in Figure 5.10). The order of the spline function must not be so great as to fit the oscillations. Instead, it should produce a fit approximately through the inflection points of successive oscillations. At this point, a Victoreen function is usually fitted through the data using McMaster coefficients (Brown *et al.*, 1988; Sayers and Bunker, 1988) to model the theoretical fall of absorbance with increasing photon energy and the EXAFS function is given by $\chi(E) = [\mu\text{data} - \mu\text{spline}]/\mu\text{Victoreen}$. This step also normalizes absorbance to a per atom basis. After this procedure, the edge step (or step height), $\Delta\mu x$, should be proportional to the amount of absorbing element present. At this point (Figure 5.10(b)), the energy values are converted to the corresponding k values using the equation shown in Figure 5.10. Fourier transformation of the EXAFS function in k-space form produces a radial structure function (Figure 5.10(c)) which gives pair correlations involving the absorbing element in Å. These distances are not corrected for phase shift, therefore they are typically 0.3–0.5 Å too short. At this point, the Fourier transform (FT) may be filtered as shown in Figure 5.10(c) to isolate individual shells of neighbouring atoms around the absorbing element, and backtransformed to give the individual EXAFS frequencies from the various shells of the neighbouring atoms. These backtransforms can then be fitted to an equation of the form below, and structural parameters extracted, including distances which are corrected for phase shift.

$$\chi(\mathbf{k}) = \sum_i [S_i(k)N_i/kR_i^2]F_i(k)\exp[-2\sigma_i^2k^2]\exp[-2R_i/\lambda(k)] \sin[2kR_i + \phi_i(k)] \quad (5.5)$$

where $F_i(k)$ is the backscattering amplitude of the N neighbouring atoms of type i at distance R_i, modified by three amplitude-reduction factors:

1. a Debye–Waller factor, σ_i^2, which accounts for thermal vibration (in the harmonic approximation) and static disorder (assuming a Gaussian pair distribution);
2. the term $S_i(k)$, which accounts for amplitude reduction due to many-body relaxation effects of the absorber and multielectronic excitations (Stern and Heald, 1983);
3. the term $\exp[-2R_i/\lambda(k)]$, which accounts for inelastic losses in the scattering process, where λ is the electron's mean free path.

The frequency of each oscillation produced by backscatterers in shell i is given by $\sin[2kR_i + \phi_i(k)]$, where $\phi_i(k)$ is a phase-shift function specific to a particular absorber–backscatterer pair. The sum in equation (5.5) is over all shells of neighbouring atoms around the absorber contributing to the EXAFS, which in practice does not extend beyond the third or fourth shell of neighbouring atoms for sorption samples. Many past analyses of EXAFS spectra have utilized a plane-wave, single-scattering approximation, but full curved-wave, multiple-scattering codes are now available and should be used for maximum accuracy in interatomic distances and coordination numbers.

A convenient means of analysing the frequencies of EXAFS oscillations involves Fourier transformation of the experimental $\chi(k)$ function, which produces a radial structure function (rsf). The real part (modulus) of the Fourier transform consists of a sum of radial peaks located at R_i, which correspond to the distances between the absorber and successive shells i of neighbouring atoms, uncorrected for phase shift. The rsf also contains compositional information because the amplitudes of its peaks are dependent on the number and backscattering amplitudes of the atoms in a particular shell i. Figure 5.11 illustrates the significant difference between the

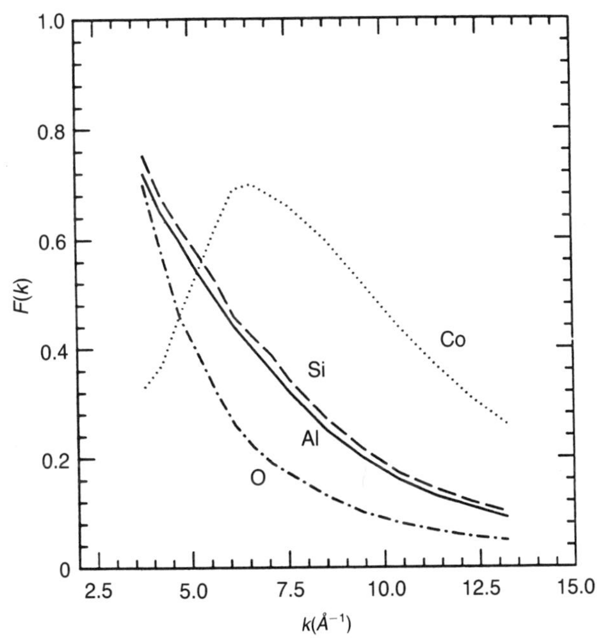

Figure 5.11 Calculated backscattering amplitude function, $F(k)$, from Co, Si, Al, and O vs. photoelectron wave-vector k ($Å^{-1}$).

backscattering amplitude, $F(k)$, for Co vs. Si or Al vs. O as a function of wave vector k. The differences in scattering amplitude between Co and Si, Al or O are sufficient to distinguish Co from the other atoms in an EXAFS analysis of Co adsorbed on SiO_2, Al_2O_3 or $Al_2Si_2O_5(OH)_4$ (kaolinite). Similarly, O can be distinguished easily from Co or from Si or Al. However, the scattering amplitudes of Si and Al are too close to be distinguished. Compositional information also comes from the phase shift, which is dependent on the types of absorber and backscattering atoms.

To derive accurate interatomic distances, coordination numbers, and atom identities from an EXAFS spectrum, the relationships between frequency and distance and between amplitude and the number and identity of coordinating atoms must be calibrated as a function of k (Figure 5.11). This may be done using experimental EXAFS data from appropriate model compounds or by calculating phase-shift and amplitude functions from first principles (Teo and Lee, 1979; Binstead *et al.*, 1988; McKale *et al.*, 1988; Rehr *et al.*, 1991). Model compounds must have known structures and contain the absorber and nearest-neighbour backscatterers of interest. In practice, calibration using model compounds is accomplished by obtaining an EXAFS spectrum of the element of interest in a model compound with nearest neighbours similar to those in the unknown element. While it is usually possible to find suitable model compounds for most metal–oxygen pairs, few good model compounds are available with appropriate metal–second-neighbour metal pairs. In these cases, theoretical amplitude and phase-shift functions must be calculated using FEFF or some other theory. Our recent experience has shown that FEFF-derived functions can be used to generate EXAFS spectra that precisely match experimental spectra of complex crystalline solids with well-known structures (O'Day *et al.*, 1994a). Use of FEFF also allows an assessment of multiple-scattering contributions to the EXAFS, which can be important for certain adsorbate geometries (O'Day *et al.*, 1994b).

Several fitting procedures can be used to extract structural information from an EXAFS spectrum. In one procedure which derives phase-shift and amplitude functions from model or reference compounds, the peak in the rsf of the model compound corresponding to a selected shell of neighbours around the absorber is isolated (also referred to as Fourier filtering or windowing) and back-Fourier transformed to produce the corresponding $\chi(k)$ contribution, $\chi(k)_{\mathrm{Fil}}$, for that shell of neighbours. A parameterized form of equation (5.5), $\chi(k)_{\mathrm{P}}$, is fitted to $\chi(k)_{\mathrm{Fil}}$ after fixing interatomic distances, R_i, coordination numbers, N_i, and Debye–Waller factor, σ_i^2, at values taken from the known crystal structure of the model compound. Phase-shift and backscattering amplitude functions for each shell of neighbours around the absorber are adjusted in a least-squares fitting procedure that minimizes a reduced chi-squared goodness-of-fit parameter, ε^2, summed over all data points i and expressed as

$$\varepsilon^2 = 1/v \sum_i (\chi(k)_{\mathrm{Fil}} - \chi(k)_{\mathrm{P}})_i^2/s_i^2 \tag{5.6}$$

where v is the number of data points minus the number of fit parameters, $(\chi(k)_{\mathrm{Fil}} - \chi(k)_{\mathrm{P}})$ is the difference between the Fourier-filtered experimental data and an EXAFS function derived using a parameterized form of equation (5.5) for a particular absorber environment for each point i, and s_i^2 is the standard deviation of each experimental point based on counting statistics. Once these phase-shift and amplitude functions are obtained, the rsf for the unknown compound is treated in the same way, except that E_0 (the continuum threshold energy of the photoelectron), R_i, N_i and σ_i^2 are varied, while phase-shift and amplitude functions are those refined from the model compound.

In another procedure which curve fits the entire EXAFS function rather than the filtered, back-Fourier transformed function, the EXAFS of an unknown compound is fitted by adjusting R_i, N_i and σ_i^2 for the nearest-neighbour shell of atoms around the absorber element and using phase-shift and amplitude functions derived from a theory such as FEFF (Rehr *et al.*, 1991) or from reference compounds. When more than one frequency is present in the EXAFS spectrum of an unknown, which indicates more than one shell of backscatterers around the absorber, the first-shell structural parameters derived in the preceding fit procedure are fixed, including E_0, and the second shell distance, number of atoms, and σ^2 values are adjusted to minimize ε^2 (equation (5.6)). In this case, an educated guess must be made about the identity of the second shell of backscatterers, and reasonable starting values of R_i, N_i and σ_i^2 must be used in the curve-fitting procedure. If third or higher shells of neighbours are also present, a similar procedure is used to derive structural information for these shells.

The interatomic distances and coordination numbers derived in either of these procedures are accurate to ± 0.02 Å and ± 15–20%, respectively, assuming that systematic errors have been minimized in the experiment and data analysis. Identities of atoms in first-, second- or more distant-neighbour shells can be determined by trial and error if their atomic numbers differ by more than ~3. An example of this is shown in Figure 5.12 for Co sorbed on TiO_2 (rutile) at a surface coverage of 1.2 μmoles m^{-2} (Chisholm-Brause *et al.*, 1989) using the Fourier-filtered, experimental EXAFS signal of the second-shell peak in the rsf of this sample. A least-squares fit assuming only Co backscatterers shown in Figure 5.12(a) gives two Co atoms at about 3.3 Å and is relatively poor in both phase and amplitude, indicating the presence of other atoms in the second shell. The addition of Ti to the second shell of Co improves the fit significantly (Figure 5.12(b)) and indicates about one Co atom at 3.33 Å and about two Ti atoms at 3.29 Å.

In fitting an EXAFS spectrum using the above procedures, the number of adjustable parameters, N_{adjust}, is limited by the relationship (Brillouin, 1962)

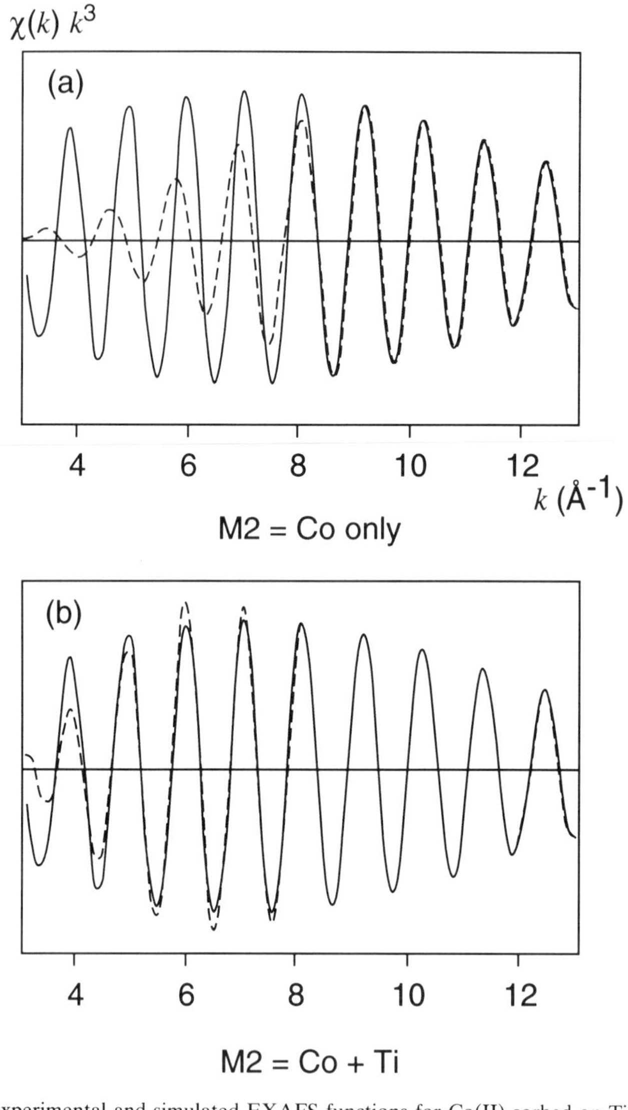

Figure 5.12 Experimental and simulated EXAFS functions for Co(II) sorbed on TiO_2, showing two fits of the experimental data: (a) with only Co second neighbours (M2) around the central Co; (b) with both Co and Ti second neighbours. (After Chisholm-Brause *et al.*, 1990a.)

$$N_{\text{adjust}} \approx 2\Delta r_{\text{filt}} \Delta k / \pi \tag{5.7}$$

where Δr_{filt} is the filtered window width in the radial structure function for the back transform and Δk is the k-range of the forward Fourier transformed data. A typical case for a sorption sample might involve a filtered window width, Δr_{filt}, of 2 Å and a Δk of 9 Å^{-1}, which limits the number of adjustable parameters to about 12. In this case, simultaneous refinement of E_0, N_i, R_i and σ_i^2 for three shells of neighbouring atoms around a cation adsorbed at the mineral–water interface is permitted, in principle. In practice, however, strong correlations between E_0 and R_i and between N_i and σ_i^2 limit the number of adjustable parameters that can be varied simultaneously.

5.5 XAS studies of sorption reactions at mineral–water interfaces

Table 5.1 lists many of the XAS studies of adsorbates at mineral–water interfaces carried out over the past seven years. Most of the samples examined had less than monolayer coverages of an adsorbate, although a few were designed to produce surface precipitates, so coverages were higher than monolayer in these cases. Examination of Table 5.1 shows that many of the metal adsorbates examined so far using XAS bind to oxide, oxyhydroxide or silicate surfaces in an inner-sphere fashion, and that a bidentate attachment mechanism is the dominant one. Although not indicated in Table 5.1, several studies have been done as a function of surface coverage. For example, in several systems (Co(II) sorption on γ-Al_2O_3, α-Al_2O_3, kaolinite, and quartz; Pb(II) sorption on γ-Al_2O_3). Co(II) and Pb(II) form mononuclear complexes at the lowest surface coverages (see discussion below) and multinuclear (MN) complexes at higher surface coverages. At the highest coverages, a three-dimensional metal–hydroxide precipitate forms. In the case of Co(II) sorption on calcite (Xu, 1993), Co forms a co-precipitate on the calcite surface in the form of a $(Ca,Co)CO_3$ solid solution extending to a depth of several hundred Ångströms. Outer-sphere complexes have been observed for selenate sorption on α-FeOOH, and permissive evidence has been found for outer-sphere Co(II) sorption complexes on kaolinite at the lowest surface coverages examined.

The discussion below focuses on several XAS studies carried out in our laboratories, including selenium adsorbed on goethite (α-FeOOH) and Co(II) adsorbed on γ-Al_2O_3, TiO_2 (rutile), α-SiO_2 (quartz) and $Al_2Si_2O_5(OH)_4$ (kaolinite), which serve to illustrate the type of microscopic information obtainable from *in situ* XAS analysis. Recent reviews of sorption complexes on mineral surfaces derived from XAFS studies include Brown (1990) and Manceau *et al.* (1992).

5.5.1 *Experimental protocols*

Key requirements for *in situ* XAS measurements on adsorbates at mineral–water interfaces are

- sufficient adsorbate concentration to produce measurable X-ray fluorescence;

Table 5.1 Summary of selected EXAFS studies of sorption at mineral–water interfaces

Sorbate	Sorbent	d (M–O)	d (M–Ms) (Å)	Mode of attachment*	Reference
SeO_4^{2-}	α-FeOOH	1.65 Å	none	OS	Hayes *et al.* (1987)
SeO_3^{2-}	α-FeOOH	1.70	3.38	IS, bidentate, Mon	Hayes *et al.* (1987)
SeO_3^{2-}	HFO⁺	1.69	3.35	IS, bidentate, Mon	Manceau *et al.* (1992)
SeO_3^{2-}	am-Al_2O_3	1.70	–	IS, ?	Papelis *et al.* (1994)
SeO_3^{2-}	α-Al_2O_3	1.70	–	IS, ?	Papelis *et al.* (1994)
SeO_3^{2-}	γ-$Al(OH)_3$	1.70	–	IS, ?	Papelis *et al.* (1994)
AsO_4^{3-}	α-FeOOH	1.66	3.28	IS, bidentate, Mon	Waychunas *et al.* (1993)
	β-FeOOH	1.66	3.28	IS, bidentate, Mon	
	γ-FeOOH	1.66	3.28	IS, bidentate, Mon	
	Ferrihydrite	1.66	3.25	IS, bidentate, Mon	
	Ferrihydrite	1.66	3.60	IS, monodentate, Mon	
Co(II)	γ-Al_2O_3	2.08	3.23–3.31	IS, bidentate, MN	Chisholm-Brause *et al.*
	TiO_2	2.09	3.29	IS, bidentate, MN	(1990a)
	kaolinite	2.08	2.67–3.43	IS, bidentate, MN	O'Day, Brown and Parks (1991)
	quartz	2.08	3.37–3.44	IS, bidentate, MN	O'Day, Brown and Parks (1991)
	calcite	2.08	3.86–3.89	Co-precipitate	Xu (1993)
	α-Al_2O_3	2.08	3.23–3.90	IS, bidentate, MN	Towle *et al.* (1993)
Pb(II)	γ-Al_2O_3	2.23–2.46	3.72	IS, monodentate, Mon and MN	Chisholm-Brause *et al.* (1990b)
	α-FeOOH	–	–	IS, ?	Roe *et al.* (1991)
	HFO†	2.22–2.42	3.3–3.45	IS, bidentate	Manceau *et al.* (1992)
	HMO†	2.23–2.38	3.14	IS, bidentate	Manceau *et al.* (1992)
Cr(III)	HMO†	–	3.0–4.1	IS, bidentate	Manceau and Charlet (1992)
	HFO†	–	3.0–4.0	IS, bidentate	Charlet and Manceau (1992)
$U(VI)O_2^{2+}$	HFO†	1.8, 2.35	3.1, 3.48	IS, bidentate	Combes (1988)
	smectite	1.8, 2.40	000, 3.8 (U)	IS, ?	Dent, Ramsay and Swanton (1992)
	kaolinite	1.8, 2.3–2.5	000, 3.9 (U)	IS, ?	Thompson *et al.* (1994)
	smectite	1.8, 2.4	–	IS, ?	Chisholm-Brause *et al.* (1994b)
$Np(V)O_2^{+}$	α-FeOOH	1.85, 2.51	~3.0	IS, ?	Combes *et al.* (1992)

d(M–O), metal–oxygen distance. A dash (–) means this distance was unreported or unobserved.
d(M–Ms), distance between the sorbed metal atom (M) and a metal atom in the adsorbent (Ms).
*HMO and HFO, hydrous manganese oxide and hydrous ferris oxide, respectively
†Mode of attachment refers to the type of linkage to the sorbent (outer-sphere (OS), inner-sphere (IS), monodentate, bidentate, monomer (Mon) and multinuclear (MN) surface complex. At low surface coverages in some sorption systems, monomers are sometimes found. When multinuclear complexes are present, M–M distances are observed, but are not reported in this table. The reference cited should be consulted for these distances.

- very low concentration of the element of interest in the bulk mineral and solution phase to minimize the contribution of bulk species to the total fluorescence signal;
- an X-ray excitation energy high enough to minimize the effects of matrix and water absorption of the fluorescent X-rays;
- sufficient differences between the backscattering amplitudes and phases of the adsorbate element (A) and elements (B) in the adsorbent to permit discrimination between A and B among second and more distant neighbours to determine if backscattering is from the adsorbate or the adsorbent.

Such discrimination is essential for determining whether the adsorbate forms an inner-sphere or an outer-sphere complex and/or whether the adsorbate is present as mononuclear or multinuclear species at the mineral–water interface. This requirement constrains the types of sorption systems that can be studied by XAS. For example, it would limit the accuracy of studies of Co sorbed on iron oxides, owing to the similarity of backscattering amplitudes and phase shifts for Co and Fe, which would make these atoms indistinguishable except for differences in Co—O vs. Fe—O distances. Manceau and Charlet (1992) and Charlet and Manceau (1992) used differences in Cr—O, Fe—O, and Mn—O distances as well as metal–metal (second-neighbour) distances to develop geometrical models for distinguishing second-neighbour atoms in their EXAFS study of Cr sorption on Fe and Mn oxyhydroxides. When cation–oxygen and cation–cation interatomic distances for a cation sorbed on an metal oxide or oxyhydroxide differ by more than about 0.05 Å, this procedure can be used.

In general, sorption samples are prepared by equilibrating the sorbent with appropriate solutions containing the sorbate element at concentrations and pH values selected to avoid precipitation (unless desired) and to achieve maximum uptake from solution consistent with a desired surface coverage. After equilibration for a minimum period of time (typically 16–24 h), the samples are centrifuged to remove as much liquid as possible, typically between 80 and 90%. The remaining wet suspensions or pastes are mounted in Teflon cells with thin mylar windows. A portion of the removed supernatant is examined by EXAFS spectroscopy to verify that the remaining sorbate concentration is low. When uptake and solid/liquid ratio in the XAS sample are high, the remaining solute in the wet sample makes no significant contribution to the EXAFS spectrum from adsorbed species. When preparing sorbates on single-crystal surfaces for grazing-incidence XAS study, sorbents are reacted with a solution at a pH chosen to achieve the desired uptake of the cation or oxyanion and the single crystal is mounted in a special polypropylene-covered holder after reaction with solution for EXAFS data collection.

5.5.2 Selenium on α-FeOOH (goethite)

Selenium poses a significant environmental hazard when it occurs in sufficient concentrations (more than a few hundred ppb) in soils and natural waters, such as in the Kesterson National Wildlife Refuge in central California, where it has been found to occur locally at concentrations as high as a few ppm (Tokunaga and Benson, 1992), and in various oxidation states, depending on local redox conditions (Pickering, Brown and Tokunaga, 1994). The oxidation state of Se affects its solubility and sorption behaviour on soil minerals, which in turn affect its transport and dispersal in ground waters. Thus knowledge of the sorption behaviour of Se at the microscopic level is required to understand its speciation and binding in natural soils and ground waters.

Several studies have addressed the macroscopic sorption of the selenate (SeO_4^{2-}) and selenite (SeO_3^{2-}) forms of Se on the common soil mineral goethite (Hayes, 1987; Hayes and Leckie, 1988), the results of which were discussed in section 5.3. Uptake measurements of selenate and selenite as a function of pH and ionic strength were interpreted by Hayes (1987) as indicating outer-sphere and inner-sphere sorption modes, respectively (Figure 5.2). Thus, selenate is thought to bond weakly and selenite is thought to bond strongly to the goethite surface. Figure 5.13 shows EXAFS spectra of

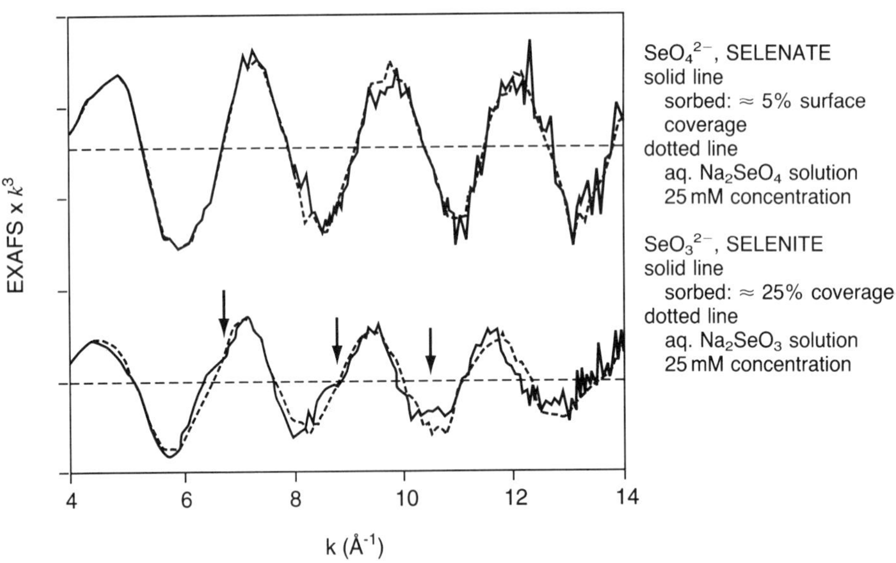

Figure 5.13 Se K-EXAFS spectra of aqueous selenate and selenite ions adsorbed on goethite. Also shown are the spectra of Na_2SeO_4 and Na_2SeO_3 solutions (both 25 mM). Note the second frequency in the spectrum of selenite (indicated by arrows), which is caused by a second shell of neighbours around the central Se. (After Hayes *et al.*, 1987.)

selenate and selenite adsorbed at the goethite–water interface and compares the spectra for the sorption samples with those of selenate and selenite in aqueous solution (Hayes *et al.*, 1987). The sorption samples were prepared using high surface area goethite powder ($\sim 55\,m^2\,g^{-1}$) equilibrated with Na_2SeO_3 or Na_2SeO_4 solutions at pH 2.5, resulting in the equivalent of 5 mM selenite (~25% surface coverage) and 2 mM selenate (~5% surface coverage) adsorbed on goethite. The EXAFS spectrum of selenate adsorbed on goethite shows a single frequency typical of a single shell of nearest-neighbour backscatterers at a single distance. A similar Se K-EXAFS spectrum was obtained from a 25 mM Na_2SeO_4 solution. In contrast, the spectrum of selenite adsorbed on goethite shows two components – an intense, low frequency component and a weaker, higher frequency component – suggesting two distinct shells of backscatterers at different distances. Fourier transforms of these spectra (Figure 5.14(A)) show a clear second-shell contribution in the case of selenite adsorbed on goethite (solid line), but only a first-shell peak in the case of the Na_2SeO_3 solution (dotted line). Fourier transforms for selenate adsorbed on goethite and for Na_2SeO_4 solution are

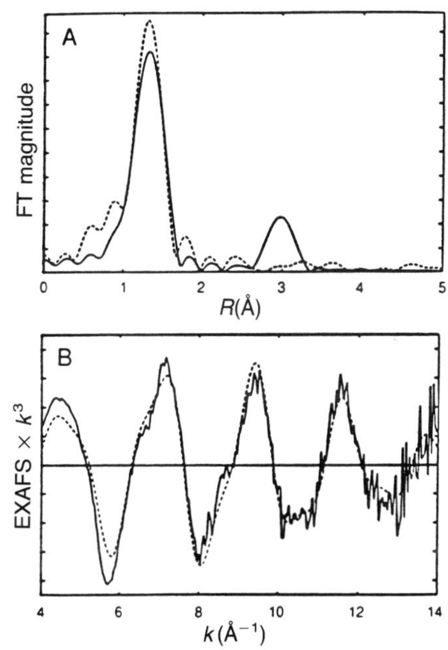

Figure 5.14 (A) Comparison of the Fourier transform of the EXAFS spectrum of aqueous selenite ions adsorbed on goethite (solid line) with that of aqueous Na_2SeO_3 solution (dotted line); (B) comparison of experimental EXAFS spectrum (solid line) for aqueous selenite ions adsorbed on goethite with a spectrum calculated assuming three nearest-neighbour oxygens at 1.7 Å and two Fe atoms at 3.38 Å.

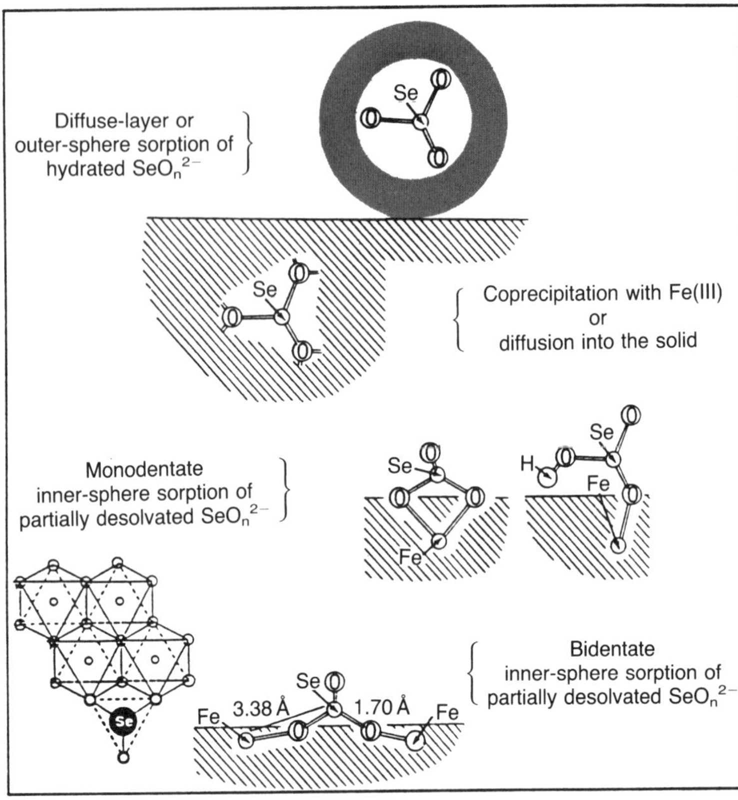

Figure 5.15 Schematic illustration of several possible modes of adsorption of selenium oxyanions at the goethite–water interface. (After Hayes *et al.*, 1987, and Manceau *et al.*, 1993.)

not shown but are the same as that for Na_2SeO_3 solution. Least-squares fitting of the back-transformed spectra using one shell of four oxygens for selenate and two shells for selenite (one shell of three oxygens and one shell of two irons) gave satisfactory fits to the experimental EXAFS. The resulting fit for the EXAFS spectrum of selenite adsorbed on goethite is shown in Figure 5.14(B). The best-fit results in this case are three oxygens at 1.70 Å and two irons at 3.38 Å around selenium, indicating that selenium is bonded in a bidentate fashion to two oxygens at the goethite surface as an inner-sphere, trigonal complex (Figure 5.15). The best fit for selenate adsorbed on goethite yields four oxygens at 1.65 Å and no second-neighbour irons, indicating that selenate is adsorbed as an outer-sphere, tetrahedral complex (Figure 5.15). These structural results provide the first direct molecular-level explanation for the strong binding of selenite and the weak binding of selenate at the goethite–water interface. They also clarify the chemical reactions (see reactions 5.1–5.4) needed to generate macroscopic models of the adsorption of selenite and selenate on goethite, showing that selenite binds as a mononuclear bidentate sorption complex (reaction 5.3) rather than a mononuclear monodentate complex (reaction 5.2) or a multinuclear complex on the goethite surface. Similar results have been obtained for the adsorption of selenite on hydrous ferric oxide (Manceau *et al.*, 1992). Although a less direct method than XAS, an earlier IR spectroscopy study of selenite adsorption on goethite (Parfitt and Russell, 1977) was interpreted as indicating that this anion is adsorbed by a ligand exchange reaction to form a bidentate bridging complex on the goethite surface.

5.5.3 Co(II) on oxides and silicates

Our choice of Co(II) as the adsorbate in a series of XAS studies of metal ion adsorption complexes on various oxides and silicates at ambient temperature

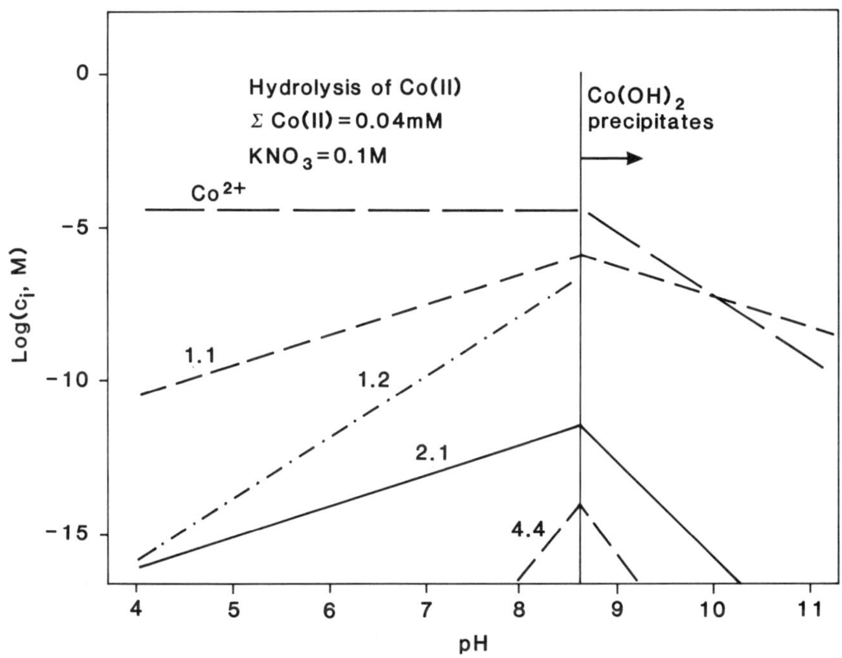

Figure 5.16 Plot of the distribution of Co(II) among various hydrolysis products as a function of pH. The dashed line labelled Co^{2+} represents the hexa-aquo complex of Co^{2+}, whereas the other lines represent various hydrolyzed solution species: 1.1 represents CoOH; 1.2 represents $Co(OH)_2$; 2.1 represents the hydroxy-bridged dimer Co_2OH; and 4.4 represents the tetranuclear species $Co_4(OH)_4$. The vertical line represents the pH at which the solution becomes saturated with respect to the solid hydroxide, resulting in precipitation.

and pressure was dictated by a number of factors. The stability constants for aqueous Co(II) hydroxo complexes and the solubilities of solid Co(II) hydroxide and oxide phases have been critically examined (Baes and Mesmer, 1986). The distribution of Co(II) hydroxo species at total Co(II) concentrations of 10^{-5} M is illustrated in Figure 5.16. Note that there is little or no polymerization of Co(II) in solution at the Co concentrations used in most of the sorption experiments described below, and that sorption occurs well below the pH at which precipitation is predicted to occur in the absence of an adsorbent. Other reasons for choosing Co as an adsorbate include the fact that its co-ordination chemistry (Wells, 1962; Wyckoff, 1963; Cotton and Wilkinson, 1988) and aqueous chemistry (Baes and Mesmer, 1986) are well known. The macroscopic sorption behaviour of Co(II) on aluminium, silicon and titanium oxides is also well known (e.g. James, 1971; James and Healy, 1972; Tewari, Campbell and Lee, 1972; Tewari and Lee, 1975). Additionally, Co is an important trace constituent in some natural waters, and it is a hazardous element from industrial and nuclear wastes. For example, it occurs in metal processing and electroplating wastes, and ^{60}Co is among the elements of concern in radioactive waste products (Krauskopf, 1988). From an experimental standpoint, Co is a good choice for fluorescence-yield XAS measurements because the energies of the Co K-edge and Co Kα fluorescence X-rays are sufficiently high to minimize their absorption by the oxide substrate or water. Furthermore, the Co K near-edge spectrum contains useful information about site geometry. Finally, Co(II) is representative of other first-row transition elements (e.g. Fe and Ni) in its behaviour, so the microscopic results derived for Co(II) sorption complexes provide clues about the sorption behaviour of other cations of similar size and charge. The oxides and silicates chosen as sorbents in our studies are also well characterized and have been the subjects of numerous sorption studies, some of which are cited below.

The macroscopic adsorption behaviour of Co(II) at the γ-Al_2O_3–water and TiO_2 (rutile)–water interfaces has been studied using uptake measurements as a function of pH, total Co concentration and ionic strength (James and Healy, 1972; Tewari, Campbell and Lee, 1972; Chisholm-Brause, 1991). James and Healy (1972) successfully simulated adsorption isotherm data with a macroscopic model that assumes adsorbed Co(II) and its mononuclear hydrolysis products retain their inner hydration shells, a model consistent with outer-sphere binding. They observed an unusually rapid increase in sorption density and surface charge reversal with pH at the highest total Co concentrations studied (cf. Figure 5.4) and suggested surface-induced precipitation of $Co(OH)_2$ as a possible explanation. Tewari and co-workers (Tewari and McIntyre, 1974; Tewari and Lee, 1975) carried out *ex situ* XPS measurements of Co adsorbed at high surface coverage on γ-Al_2O_3 and interpreted similarity in the Co $2p_{3/2}$ binding energies of these spectra with that from solid $Co(OH)_2$ as evidence for surface-induced precipitation of

$Co(OH)_2$, in agreement with James and Healy (1972), though interpretation of the XPS data was later challenged (Briggs and Bosworth, 1977).

In order to provide more detailed microscopic information on the adsorption of Co(II) at oxide–water interfaces, *in situ* XAS studies of Co(II) adsorbed on wet, powdered γ-Al_2O_3 ('Alon' from Cabot Corporation; BET surface area of $117\,m^2\,g^{-1}$) and wet, powdered TiO_2 (synthetic rutile from British Titan Products; BET surface area of $23\,m^2\,g^{-1}$) were carried out by Chisholm-Brause, Brown and Parks (1989, 1991) and Chisholm-Brause *et al.* (1990a). Comparison of Co K-edge XANES spectra for crystalline $CoAl_2O_4$, $Co(OH)_2$, aqueous Co(II) adsorbed on γ-Al_2O_3, kaolinite and TiO_2 (rutile), and 12 mM $Co(NO_3)_2$ solution (Figure 5.9) shows that the local environments of Co(II) adsorbed on γ-Al_2O_3, kaolinite and TiO_2 at similar sorption densities ($\Gamma \approx 1.2\,\mu mol\,m^{-2}$) differ slightly. Furthermore, the spectra for these sorption samples are markedly different from those of crystalline $CoAl_2O_4$ (in which a significant fraction of Co is four-coordinated by oxygens), and crystalline $Co(OH)_2$ and the aquo-complex of Co^{2+} in solution (in both of which Co is six-coordinated by oxygens in relatively regular octahedra). These observations suggest that Co in these sorption samples has a second-neighbour environment distinctly different from that in $Co(OH)_2$. If crystalline $Co(OH)_2$ is a reasonable structural model for a surface precipitate of Co hydroxide, then surface precipitation can be ruled out as the mode of accumulation for Co^{2+} sorbed on γ-Al_2O_3 or TiO_2 (rutile) under the conditions of these experiments, which should be well below thermodynamic saturation with respect to solid $Co(OH)_2$. The lack of similarity of the XANES spectra for the sorption samples relative to that of Co^{2+}(aq) is strong evidence that Co does not form an outer-sphere complex at these oxide–water interfaces. Moreover, the slightly broader main absorption edges for the three sorption samples could be due to a broader range of distances around Co relative to the Co^{2+}(aq) complex. Finally, none of the spectra except that for $CoAl_2O_4$ displays a significant $1s \rightarrow 3d$ pre-edge feature, indicating that Co(II) in each of the sorption samples is in a fairly regular, near-centrosymmetric, local co-ordination environment, such as a slightly distorted octahedron. The Co K-XANES spectrum of $CoAl_2O_4$ spinel shows a relatively strong $1s \rightarrow 3d$ pre-edge feature indicative of a non-centrosymmetric site, which is consistent with a significant fraction of Co in this spinel being tetrahedrally co-ordinated by oxygens.

A more detailed *in situ* XAS study of Co(II) adsorption on γ-Al_2O_3 was carried out by Chisholm-Brause (1991), Chisholm-Brause, Brown and Parks (1991) and Chisholm-Brause *et al.* (1994a). Normalized EXAFS spectra for solid $Co(OH)_2$, aqueous Co(II) at different coverages on γ-Al_2O_3, and 12 mM $Co(NO_3)_2$ solution are shown in Figure 5.17. At the lowest surface loading examined ($\Gamma = 0.08\,\mu mol\,m^{-2}$, which is well below monolayer coverage), the EXAFS spectrum bears some resemblance to the spectrum of the 12 mM Co solution, with one dominant frequency suggestive of a single

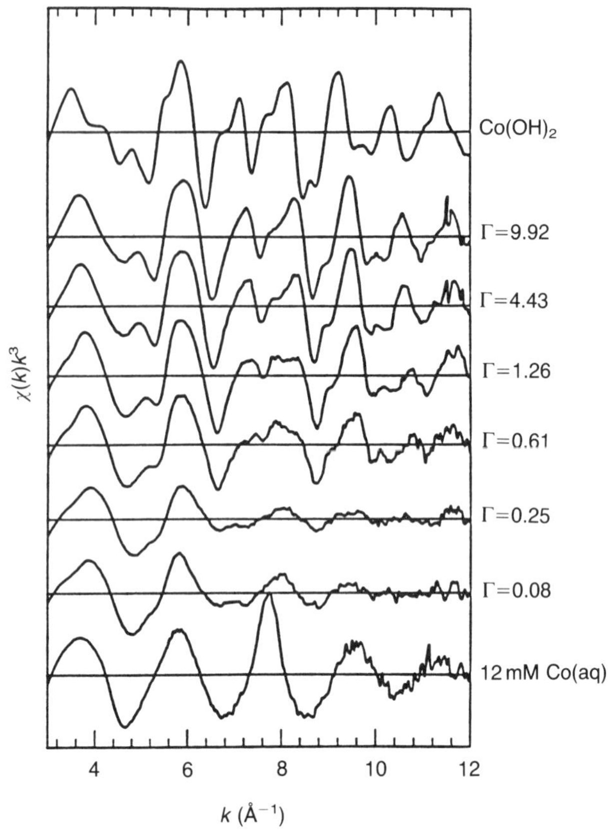

Figure 5.17 Background-subtracted Co K-EXAFS spectra for samples of aqueous Co(II) adsorbed on high surface area γ-Al_2O_3 at various sorption densities (Γ in μmol m^{-2}). Also shown are spectra for solid $Co(OH)_2$ and a 12 mM aqueous $Co(NO_3)_2$ solution. (From Chisholm-Brause, 1991.)

shell of nearest neighbours. The Fourier transform of this spectrum (Figure 5.18) shows a strong first-neighbour feature but no clear evidence of a second shell of atoms around Co (Chisholm-Brause *et al.*, 1994a). As the surface loading of Co is increased, a clear second frequency appears at $k \approx 5\,Å^{-1}$, $7\,Å^{-1}$, and $10\,Å^{-1}$ (Figure 5.17), indicating a second shell of atoms around adsorbed Co (Figure 5.18). At the highest Co loading examined ($\Gamma = 9.92\,\mu mol\,m^{-2}$, which is near monolayer coverage and above bulk solution saturation with respect to a $Co(OH)_2$ solid phase), the EXAFS spectrum is generally similar to that of crystalline $Co(OH)_2$. The Fourier transform of this spectrum shows a strong second-shell feature at a slightly shorter distance than that of crystalline $Co(OH)_2$ and a strong feature at 5–6 Å (Figure 5.18).

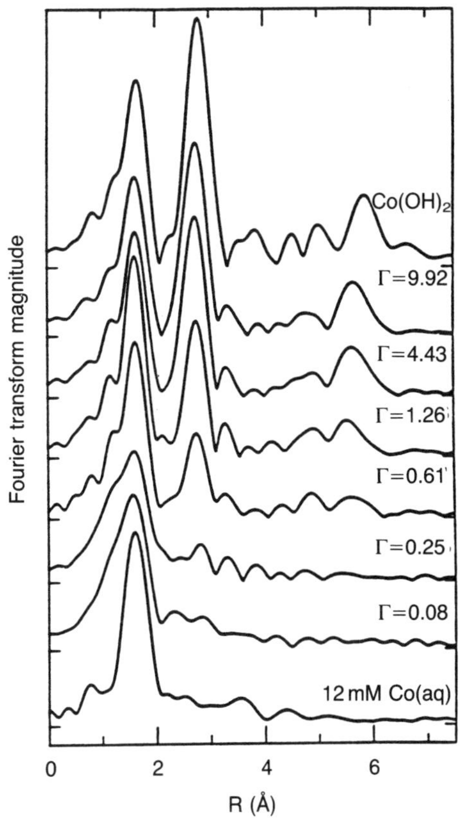

Figure 5.18 Fourier transforms of the EXAFS spectra shown in Figure 5.17.

Least-squares fits of the filtered EXAFS of the first peak in the radial structure function (rsf) from Co on γ-Al_2O_3 samples yielded six nearest-neighbour oxygens at 2.08 ± 0.02 Å for all sorption densities examined (see Figure 5.10 for the steps in this procedure). Fits of the second major peak in the rsf of the $\Gamma = 0.25$ sample indicate ~1 Co atom at 3.21 ± 0.03 Å and ~1 Al atom at 3.31 ± 0.03 Å. These results are consistent with the presence of oxy- or hydroxy-bridged Co dimers on the γ-Al_2O_3 surface at this loading. As sorption density increases, the number of second-neighbour Co atoms increases to ~4 and the number of second-neighbour Al atoms ranges from 1 to 2. In addition, there is a third major peak in the rsf at about 5.7 Å (Figure 5.18) that was fitted to 2–3 fourth-neighbour Co atoms at 6.2 ± 0.05 Å. Crystalline $Co(OH)_2$ by comparison has six second-neighbour Co atoms at 3.17 Å and six fourth-neighbour Co atoms at 6.35 Å. The smaller number of second- and fourth-neighbour Co atoms in the sorption sample relative to $Co(OH)_2$ may be due to multiple-scattering effects as observed for Co(II) sorption on kaolinite at high coverages (O'Day *et al.*, 1994b). These results for the different loadings indicate the Co adsorbs as a monomer at the lowest loadings, forms dimers and more complex oligomers at higher surface loadings, and is present as a three-dimensional $Co(OH)_2$-like precipitate at the highest loadings studied. Additional analysis of the geometries of possible sorption sites on the (111) surface of γ-Al_2O_3, making use of the constraints imposed by the observed Co–Al distance, indicates that the Co octahedral sorption complex most likely bonds in a bidentate fashion, sharing an O–O edge with AlO_6 octahedra in the γ-Al_2O_3 surface (Chisholm-Brause *et al.*, 1990a, 1994a). Similar conclusions have also been reached for Co(II) sorption on α-Al_2O_3 (Towle *et al.*, 1993; Hayes, Katz and Penner-Hahn, 1993).

Co(II) sorption on TiO_2 (rutile) has also been examined using XAS (Chisholm-Brause *et al.*, 1990a), and the results show some similarities as well as significant differences relative to those for Co(II) on γ-Al_2O_3 presented above. Fits of the second-shell feature in the rsf of Co(II) adsorbed on TiO_2 at $\Gamma = 1.2\,\mu mol\,m^{-2}$ yielded ~1 Co atom at 3.33 ± 0.03 Å and ~2 Ti atoms at 3.29 ± 0.03 Å (Chisholm-Brause, 1991). No strong fourth-shell Co peak at 5–6 Å was observed in the rsf, unlike Co adsorbed on γ-Al_2O_3 samples at similar and higher Co loadings. As discussed earlier, inclusion of metal ions of the oxide (Ti) in the second shell around the adsorbed Co, together with second-neighbour Co, produced significantly better fits than only second-neighbour Co atoms (Figure 5.12). This observation leads to the conclusion that Co is bonded directly to the rutile surface as an inner-sphere complex. The presence of second-neighbour Co around adsorbed Co indicates the presence of multinuclear sorption complexes, as was also found for Co on γ-Al_2O_3. However, the radial extent of the oxy- and/or hydroxy-bridged Co adsorbed on TiO_2 appears to be less than that on the γ-Al_2O_3 surface. To the best of our knowledge, this is the first direct structural evidence for multinuclear metal ion sorption complexes on oxide surfaces (Chisholm-Brause *et al.*, 1990a), though 'clustering' of Mn^{2+} on boehmite and goethite has been inferred from EPR measurements (Bleam and McBride, 1985).

Aqueous Co(II) surface complexes of increasing sorption density on kaolinite ($Al_2Si_2O_5(OH)_4$) (Chisholm-Brause *et al.*, 1990a; O'Day, Brown and Parks, 1991; O'Day, 1992; O'Day *et al.*, 1994b, c) and on quartz (α-SiO_2) (O'Day, Brown and Parks, 1991; O'Day, 1992) were studied using *in situ* XAS. These investigations reached the following conclusions:

- At low sorption densities ($<0.5\,\mu mol\,m^{-2}$) on kaolinite, analyses of the EXAFS spectra indicate the presence of inner-sphere, bidentate, mononuclear Co sorption complexes. Solution measurements, however, show an ionic strength dependence to Co uptake at pH < 7, suggesting that some kaolinite sites may adsorb Co as an outer-sphere or more weakly-bound surface complex at low pH in addition to the inner-sphere complexes identified with EXAFS.

- At surface coverages of $\sim 0.5\,\mu mol\,m^{-2}$ on kaolinite, the presence of small numbers of second-neighbour Co atoms ($N < 1$) indicates the formation of small, oxy- or hydroxy-bridged, multinuclear sorption complexes. With increasing Co surface coverage on kaolinite (>0.5–$10\,\mu mol\,m^{-2}$), there is an increase in the average number of second-neighbour Co atoms, indicating an increase in the average size of multinuclear complexes.
- At high surface loadings on both kaolinite and quartz, the EXAFS spectra of adsorbed Co are generally similar, but not identical, to that of either dry, crystalline $Co(OH)_2$, or wet, freshly precipitated $Co(OH)_2$. The Co spectra for these sorption samples are the same whether they were equilibrated with solutions below or above thermodynamic saturation with respect to stable and metastable forms of solid $Co(OH)_2$.
- The EXAFS results for Co(II) sorption on kaolinite, quartz, and γ-Al_2O_3, taken together, support previous macroscopic interpretations (e.g. James and Healy, 1972) that metal ion sorption proceeds from the formation of mononuclear surface complexes at low uptake, to the formation of multinuclear oxo- or hydroxo-complexes of increasing size, to the growth of a three-dimensional phase as surface loading increases and solution compositions approach saturation with respect to solid hydroxide phases.

Several observations suggest that the properties of the solid–water interface influence the type of adsorption complex formed. For example, Co(II) adsorbs more strongly and in a different pH range on TiO_2 (rutile) than on α-SiO_2 (James and Healy, 1972). In addition, Ni(II) precipitates as an amorphous 'active' hydroxide in the presence of α-SiO_2 surfaces, whereas in the presence of α-Al_2O_3, it nucleates as more ordered 'inactive' $Ni(OH)_2$ (Ahmed, 1971). Comparison of XAS results for Co(II) sorption on γ-Al_2O_3 (Chisholm-Brause, 1991; Chisholm-Brause *et al.*, 1994a), kaolinite (O'Day, 1992; O'Day *et al.*, 1994b, c) and quartz (O'Day, 1992), all at similar surface coverages (0.5–$0.8\,\mu mol\,m^{-2}$), shows that smaller or more disordered multinuclear complexes form on γ-Al_2O_3 and kaolinite (characterized by 0.6–2.5 second-neighbour Co atoms at ~ 3.1 Å) and that larger or more ordered multinuclear complexes form on quartz (characterized by ~ 6 second-neighbour Co atoms at ~ 3.1 Å). Small multinuclear Co complexes (characterized by ~ 0.8 second-neighbour Co atoms at ~ 3.3 Å) were found on rutile at a slightly higher sorption density ($1.2\,\mu mol\,m^{-2}$) (Chisholm-Brause *et al.*, 1990a). These results support the general idea that the nature of the oxide surface influences the characteristics of the sorption complex formed.

One of the most significant conclusions from this work is that Co(II) forms multinuclear surface complexes on all of the adsorbents studied even though it does not polymerize in solution at the Co concentrations of the adsorption experiments (Figure 5.16). In contrast, cations such as Pb(II), which do polymerize significantly in solution, form dominantly monomers and small multinuclear sorption complexes (Chisholm-Brause *et al.*, 1990b; Roe *et al.*, 1991). This may indicate that the surface interferes with the tendency for Pb to form hydroxide polymers. These observations suggest that the oxide surface may either help or hinder the formation of surface oxy- or hydroxy-bridged metal polymers and that their formation does not depend strongly on the concentration of such polymers in solution.

The presence of multinuclear rather than mononuclear Co(II) sorption complexes on oxide and silicate surfaces can have a dramatic influence on predictions of sorption and partitioning of Co. Most macroscopic models of sorption of metal ions such as Co(II) assume mononuclear complexes. For example, ligand-exchange reactions of the type

$$\underline{S}OH + Co^{2+}(aq) = \underline{S}OCo^{+} + H^{+}(aq) \tag{5.8}$$

$$\underline{S}OH + Co^{2+}(aq) + H_2O = \underline{S}OCoOH + 2H^{+}(aq) \tag{5.9}$$

describe the formation of mononuclear, monodentate complexes, with the possibility of hydrolysing the complex (reaction 5.9). Suppose, however,

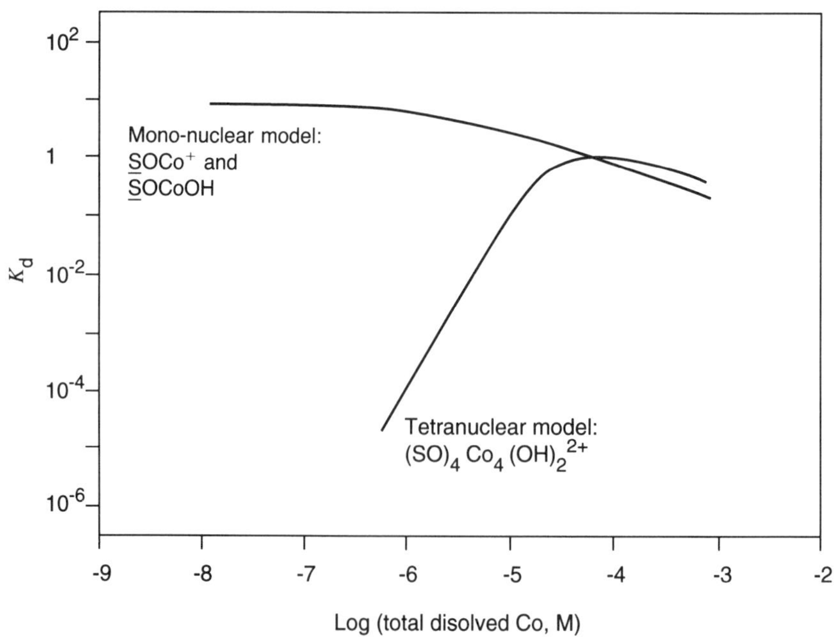

Figure 5.19 Predicted variation in the partition coefficient of Co(II) vs. log(total dissolved Co concentration, *M*) in the aqueous system Co(II) adsorbed on SiO_2. Partition coefficient (K_d) is defined as (concentration of adsorbed Co)/(total dissolved Co).

that an oxy- or hydroxy-bridged tetranuclear surface complex of Co(II) forms instead, as described by the reaction

$$4\underline{S}OH + 4Co^{2+}(aq) + 2H_2O = [(\underline{S}O)_4Co_4(OH)_2]^{2+} + 6H^+ \quad (5.10)$$

Such reactions can be used to predict how the partition coefficient K_d (defined as the concentration of adsorbed Co divided by the total dissolved Co) and uptake of Co on a mineral such as quartz change with decreasing Co concentration. At very low Co concentrations, the mononuclear model predicts very strong uptake and a high K_d, whereas the tetranuclear model predicts very weak uptake and a low K_d (Figure 5.19). If we consider a hypothetical Co contaminant plume moving in a quartz-sand aquifer, Co concentrations would be high near the source and very low far downstream. Use of the mononuclear model for Co uptake would lead to predictions of partitioning that are over six orders of magnitude higher than those assuming a tetranuclear surface complex. This example shows the importance of accurate microscopic information on adsorption complexes in predictions of sorption behaviour using macroscopic models.

5.6 Applications of other synchrotron-based surface methods to adsorbates at mineral–water interfaces

The structure and properties of a mineral surface are almost certainly modified significantly by adsorbates, including water, further complicating models of sorption. For example, Hochella (1990) presents evidence that even a partial monolayer of an adsorbate can dramatically modify the adsorbent surface structure. It is not surprising that the structure of the mineral surface should play some role in determining the structure of specifically adsorbed complexes. However, we do not have much direct knowledge of mineral surface structure or of adsorbate-induced restructuring of mineral surfaces (Brown, 1990, 1994). New types of surface- and interface-sensitive synchrotron radiation methods are now available to help provide such knowledge.

Grazing-incidence EXAFS (GI-EXAFS) spectroscopy (Heald, Keller and Stern, 1984) and X-ray standing wave (XSW) studies (Bedzyk, Bommarito and Schildkraut, 1989; Bedzyk *et al.*, 1990) of adsorbates at mineral–water interfaces hold considerable promise of providing unique information on the nature of these species and the structure and distribution of ions in the interfacial region. Such studies are just beginning, so there is little information to summarize. Towle *et al.* (1993) carried out the first fluorescence-yield, GI-XAS study of Co(II) adsorbed on the $(\bar{1}02)$ surface of α-Al_2O_3 (sapphire) from an aqueous solution at a surface coverage of about 1/30 of a monolayer. This coverage is considerably lower than those feasible on powdered samples of γ-Al_2O_3, α-SiO_2 (quartz), TiO_2 (rutile) and $Al_2Si_2O_5(OH)_4$

(kaolinite) studied by XAS (see earlier discussion), and emphasizes some of the distinct advantages of grazing-incidence (such as lower scattering, higher signal-to-noise, and four times the fluorescence amplitude) relative to powder XAS measurements (Heald, Keller and Stern, 1984). One of the main advantages of GI-XAS studies of sorbates on oriented, single-crystal surfaces is the constraint it places on the orientation of surface complexes and the number of different surface sites available for sorption. In contrast, adsorbates on powder samples can occupy different sites on multiple surfaces in contact with water, thus producing a more complicated XAS signal. Results of the study by Towle *et al.* (1993) indicate significantly fewer Co second neighbours than were found for a powdered sample of α-Al_2O_3 at somewhat higher Co surface coverages. A GI-XAS study of Co-oxide layers deposited on the (0001) surface of α-Al_2O_3 under UHV conditions (Shirai, Asakura and Iwasawa, 1992) has shown that the local environment of Co in the oxide layer is similar to that of Co in the sorption sample.

Surface EXAFS (SEXAFS) (Stöhr, 1992) and photoelectron diffraction (Fadley, 1992) are UHV methods that can provide valuable information on the structure of oxide surfaces, but they have not been used for such measurements to date. Although they require UHV conditions, thus precluding experiments on mineral surfaces in contact with bulk water, these methods can potentially be used to study the effects of a monolayer of water on mineral surfaces that have been exposed to water vapour in the vacuum chamber. These methods are also capable of providing detailed structural information about adsorption complexes at the vacuum–solid interface.

X-ray standing wave (XSW) studies of surfaces were first carried out in the late 1980s (Bedzyk, Bommarito and Schildkraut, 1989), so this is a relatively new synchrotron-based method for studying solid surfaces, adsorbates on surfaces (Woicik *et al.*, 1992), and the distribution of ions in aqueous solutions near the solid surface (Abruña, Bommarito and Acevedo, 1990; Bedzyk *et al.*, 1990). Detectable standing waves are formed by the interference between incident and diffracted beams when an intense X-ray beam is diffracted from a surface. The antinode of the standing wave can be moved in or out relative to a solid surface by changing the angle of X-ray incidence or the energy of the incident beam. When an antinode of sufficient energy is swept through a surface layer or an adlayer, the fluorescence yield from the surface atom or atom in the adlayer will be maximum when the antinode coincides with the centre of the atom, according to the dipole approximation (Bedzyk *et al.*, 1990). Thus the XSW method is particularly sensitive to the z position of the atom. By using standing waves generated from different diffraction planes in a single-crystal substrate, one can triangulate the position of a surface atom very precisely (Woicik *et al.*, 1992). We have recently applied the XSW technique to the surface of CaO (Kendelewicz *et al.*, 1994) with promising results. The clean surface of this model compound was found to undergo very little relaxation, but when hydrated significant surface reconstruction appears to occur. Future applications of XSW methods to mineral surfaces, including adsorbates at mineral–water interfaces, should contribute a great deal of information on the nature of surface complexes as well as the nature of the clean (unreacted) and wet mineral surfaces.

Although they are not useful for studies of mineral surfaces in contact with bulk water, UHV photoemission methods permit studies of oxide surfaces before and after reaction with water and other adsorbates (e.g. Kurtz and Henrich, 1982, 1984; Henrich, 1985; McKay and Henrich, 1987). Such studies provide information on the electronic structure of a clean oxide surface and how it changes after reaction with aqueous sorbates. These methods used in combination with XSW and SEXAFS on the same surfaces (e.g. Woicik *et al.*, 1991a, b) offer significant opportunities for improving our understanding of mineral surfaces.

5.7 Conclusions and future possibilities

In situ synchrotron X-ray studies of adsorbed species on mineral surfaces are beginning to provide unique information about the mode and mechanism of adsorption reactions at mineral–water interfaces. This information is essential for developing accurate and robust macroscopic surface complexation and transport models used to describe many geochemical processes, including the dispersal and fate of aqueous contaminants in the environment. The X-ray flux available from existing hard synchrotron X-ray beam lines is often not sufficient to permit XAS studies of adsorbates on mineral surface and contaminants in soils and natural waters at the low concentration levels often found in natural samples. However, XAS studies of synthetic sorption samples with somewhat higher adsorbate concentrations provide useful clues about the sorption behaviour of chemical species in natural samples. As synchrotron beam lines with higher flux become available, it will be possible to extend XAS studies of the type reviewed in this chapter to natural samples with significantly lower element concentrations. X-ray absorption spectroscopy studies of some natural samples have already begun with promising results (e.g. Pickering, Brown and Tokunaga, 1994).

Acknowledgements

The powder XAS studies of sorption complexes at mineral–water interfaces reviewed in this chapter were generously supported by the US National Science Foundation (Grants EAR-8805440, EAR-8917437, and EAR-9105015), the grazing-incidence XAS studies on single-crystal samples were supported by the US Department of Energy (Grant DE-FG03-93ER14347-A000), and the XSW studies of oxide surfaces recently initiated are supported by NSF Grant EAR-9305028. Catherine Chisholm-Brause

(Los Alamos National Laboratory) was a collaborator on several of these projects and is thanked for her contributions, and François Farges (Institut de Physique du Globe de Paris, Universités Paris 6 et 7) and Lambis Papelis (University of Michigan) helped with data collection. Much of our XAS work has been carried out at the Stanford Synchrotron Radiation Laboratory, which is supported by the US Department of Energy and the National Institute of Health. In particular, we thank Britt Hedman (SSRL) for her advice and help in the experimental work. Finally, the senior author wishes to thank David Vaughan, Richard Pattrick, Michael Henderson and the staff of the Department of Geology at the University of Manchester for making the Mineralogical Society Short Course on Mineralogical Applications of Surface Science enjoyable and fruitful and for their patience in waiting for this contribution.

References

Abruña, H.D. (ed.) (1991a) *Electrochemical Interfaces: Modern Techniques for In-Situ Interface Characterization*, VCH Publishers, New York, 589 pp.

Abruña, H.D. (1991b) X-ray absorption spectroscopy in the study of electrochemical systems, in *Electrochemical Interfaces: Modern Techniques for In-Situ Interface Characterization* (ed. H.D. Abruña), VCH Publishers, New York, pp. 1–54.

Abruña, H.D., Bommarito, G.M. and Acevedo, D. (1990) The study of solid/liquid interfaces with X-ray standing waves. *Science*, **250**, 69–74.

Ahmed, A. (1971) *The Sorption of Aqueous Ni(II) Species on Alumina and Quartz*. PhD Dissertation, Stanford University, Stanford, CA, 113 pp.

Allen, J.W. (1992) Resonant photoemission of solids with strongly correlated electrons, in *Synchrotron Radiation Research, Advances in Surface and Interface Science*, Vol. 1 (ed. R.Z. Bachrach), Plenum Press, New York, pp. 253–323.

Atkinson, R.J. (1969) PhD Thesis, University of Western Australia, Perth.

Atkinson, R.J., Posner, A.M. and Quirk, J.P. (1972) Kinetics of isotopic exchange of phosphate at the α-FeOOH (goethite) – aqueous solution interface. *J. Inorg. Nucl. Chem.*, **34**, 2201–11.

Bachrach, R.Z. (1992) *Synchrotron Radiation Research, Advances in Surface and Interface Science*, Vols 1 and 2. Plenum Press, New York.

Baes, C.F., Jr and Mesmer, R.E. (1986) *The Hydrolysis of Cations*. Robert E. Krieger Publishing Co., Malabar, Florida, 489 pp.

Bassetti, V., Burlamacchi, L. and Martini, G. (1979) Use of paramagnetic probes for the study of liquid adsorbed on porous supports. Copper (II) in water solution. *J. Amer. Chem. Soc.*, **101**, 5471–7.

Bedzyk, M.J., Bommarito, G.M. and Schildkraut, J.S. (1989) X-ray standing waves at a reflecting mirror surface. *Phys. Rev. Lett.*, **62**, 1376–9.

Bedzyk, M.J., Bommarito, G.M., Caffrey, M. and Penner, T.L. (1990) Diffuse double layer at a membrane–aqueous interface measured with X-ray standing waves. *Science*, **248**, 52–6.

Bianconi, A. (1988) XANES spectroscopy, in *X-ray Absorption: Principles, Applications, Techniques of EXAFS, SEXAFS, and XANES* (eds D.C. Koningsberger and R. Prins), John Wiley & Sons, New York, pp. 573–662.

Binning, G., Rohrer, H., Gerber, C. and Weibel, E. (1982a) Tunnelling through a controllable vacuum gap. *Appl. Phys. Lett.*, **40**, 178–80.

Binning, G., Rohrer, H., Gerber, C. and Weibel, E. (1982b) Surface studies by scanning tunnelling microscopy. *Phys. Rev. Lett.*, **49**, 57–61.

Binstead, N., Campbell, J.W., Gurman, S.J. and Stephenson, P.C. (1988) SERC Daresbury Laboratory Program, Daresbury, Warrington, UK.

Bleam, W.F. and McBride, M.B. (1985) Cluster formation versus isolated-site adsorption. A study of Mn(II) and Mg(II) adsorption on boehmite and goethite. *J. Colloid Interface Sci.*, **103**, 124–32.

Briggs, D. and Bosworth, Y.M. (1977) Comment on 'Adsorption of Co(II) at the oxide-water interface. *J. Colloid Interface Sci.*, **59**, 194.

Brillouin, L. (1962) *Science and Information Theory*, 2nd edn, Academic Press, New York.

Brown, G.E. Jr (1990) Spectroscopic studies of chemisorption reaction mechanisms at oxide/water interfaces, in *Mineral–Water Interface Geochemistry* (eds M.F. Hochella, Jr and A.F. White), *Reviews in Mineralogy*, Vol. 23, Mineralogical Society of America, Washington, DC, pp. 309–63.

Brown, G.E. Jr (1994) Mineralogy in two dimensions: mineral surfaces, sorption reactions, and geochemical implications. *Mineral. Mag.* (submitted).

Brown, G.E. Jr and Parks, G.A. (1989) Synchrotron-based X-ray absorption studies of cation environments in earth materials. *Rev. Geophys.*, **27**, 519–33.

Brown, G.E. Jr, Parks, G.A. and Chisholm-Brause, C.J. (1989) In-situ X-ray absorption spectroscopic studies of ions at oxide–water interfaces. *Chimia*, **43**, 248–56.

Brown, G.E. Jr, Calas, G., Waychunas, G.A. and Petiau, J. (1988) X-ray absorption spectroscopy and its applications in mineralogy and geochemistry, in *Spectroscopic Methods in Mineralogy and Geology* (ed. F. Hawthorne), *Reviews in Mineralogy*, Vol. 18, Mineralogical Society of America, Washington, DC, pp. 431–512.

Burlamacchi, L. and Martini, G. (1980) Molecular dynamics in water solutions adsorbed on porous solids studied by ESR of paramagnetic probes, in *Magnetic Resonance in Colloid and Interface Science* (eds J.P. Fraissard and H.A. Resing), D. Reidel Publishing Co., Boston, pp. 621–8.

Cases, J. and François, M. (1982) Etude des propriétés thermodynamiques de l'eau au voisinage des interfaces. *Agronomie*, **2**, 931–8.

Charlet, L. and Manceau, A. (1992) X-ray absorption spectroscopic study of the sorption of Cr(III) at the oxide/water interface. II. Adsorption, coprecipitation and surface precipitation on ferric hydrous oxides. *J. Colloid Interface Sci.*, **148**, 443–58.

Chisholm-Brause, C.J. (1991) *Spectroscopic and Equilibrium Study of Cobalt(II) Sorption Complexes at Oxide/Water Interfaces*. PhD Dissertation, Stanford University, 118 pp.

Chisholm-Brause, C.J., Brown, G.E. Jr and Parks, G.A. (1989) EXAFS investigation of aqueous Co(II) adsorbed on oxide surfaces in-situ. *Physica B*, **158**, 646–8.

Chisholm-Brause, C.J., Brown, G.E. Jr and Parks, G.A. (1991) In-situ EXAFS study of Co(II) sorption on γ-Al_2O_3 at different adsorption densities, in *XAFS VI, X-ray Absorption Fine Structure VI* (ed. S.S. Hasnain), Ellis Horwood Ltd. Publishers, Chichester, UK, pp. 263–5.

Chisholm-Brause, C.J., Roe, A.L., Hayes, K.F. *et al.* (1989) XANES and EXAFS study of aqueous Pb(II) adsorbed on oxide surfaces. *Physica B*, **158**, 674–6.

Chisholm-Brause, C.J., O'Day, P.A., Brown, G.E. Jr and Parks, G.A. (1990a) Evidence for multinuclear metal-ion complexes at solid/water interfaces from X-ray absorption spectroscopy. *Nature*, **348**, 528–31.

Chisholm-Brause, C.J., Hayes, K.F., Roe, A.L. *et al.* (1990b) Spectroscopic investigation of Pb(II) complexes at the γ-Al_2O_3/water interface. *Geochim. Cosmochim. Acta*, **54**, 1897–909.

Chisholm-Brause, C.J., Brown, G.E. Jr and Parks, G.A. (1994a) X-ray spectroscopic study of Co(II) sorption complexes at the γ-Al_2O_3–water interface (in prep.).

Chisholm-Brause, C.J., Conradson, S.D., Buscher, C.T., Eller, P.G. and Morris, D.E. (1994b) Speciation of uranyl at multiple binding sites on montmorillonite. *Geochim. Cosmochim. Acta* (in press).

Clark, C.J. and McBride, M.J. (1984) Chemisorption of Cu(II) and Co(II) on allophane and imogolite. *Clays Clay Mineral.*, **32**, 300–10.
Combes, J.-M. (1988) *Evolution de la Structure Locale des Polymeres et Gels Ferriques Lors de la Cristallisation des Oxydes de Fer. Application au Piegeage de l'Uranium*. Docteur These, Université Pierre et Marie Curie, Paris, France, 194 pp.
Combes, J.-M., Chisholm-Brause, C.J., Brown, G.E. Jr, *et al.* (1992) EXAFS spectroscopic study of neptunium (V) sorption at the α-FeOOH/water interface. *Environ. Sci. Technol.*, **26**, 376–82.
Conway, B.E. (1981) *Ionic Hydration in Chemistry and Biophysics*. Elsevier, New York.
Cotton, F.A. and Wilkinson, G. (1988) *Advanced Inorganic Chemistry*, 5th edn, John Wiley & Sons, New York, 1455 pp.

Davies, C.W. (1962) *Ion Association*. Butterworths: London.
Davis, J.A. and Kent, D. (1990) Surface complexation modeling in aqueous geochemistry, in *Mineral–Water Interface Geochemistry* (eds M.F. Hochella, Jr and A.F. White), *Reviews in Mineralogy*, Vol. 23, Mineralogical Society of America, Washington, DC, pp. 177–260.
Davis, J.A. and Leckie, J.O. (1978) Surface ionization and complexation at the oxide/water interface. II. Surface properties of amorphous iron oxyhydroxide and adsorption of metal ions. *J. Colloid Interface Sci.*, **67**, 90–107.
Davis, J.A. and Leckie, J.O. (1980) Surface ionization and complexation at the oxide-water interface. 3. Adsorption of anions. *J. Colloid Interface Sci.*, **74**, 32–43.
Dent, A.J., Ramsay, D.F. and Swanton, S.W. (1992) An EXAFS study of uranyl ion in solution and sorbed on silica and montmorillonite clay colloids. *J. Colloid Interface Sci.*, **150**, 45–60.
Dines, T.J., Rochester, C.H. and Thomson, J. (eds) (1992) *Catalysis and Surface Characterization*, Royal Society of Chemistry, Cambridge, UK, 292 pp.
Dukhin, S.S. (1971) Dielectric properties of dispersive systems, in *Surface and Colloid Science*, Vol. 3 (ed. E. Matijevic), pp. 83–165.
Dwyer, D.J. and Hoffman, F.M. (eds) (1992) *Surface Science of Catalysis: In Situ Probes and Reaction Kinetics*, ACS Symposium Series 482, American Chemical Society, Washington, DC, 364 pp.
Dzombak, D.A. and Morel, F.M.M. (1990) *Surface Complexation Modeling, Hydrous Ferric Oxide*. John Wiley & Sons, New York, 393 pp.

Eberhardt, W. (1992) Angle-resolved photoelectron spectroscopy, in *Synchrotron Radiation Research, Advances in Surface and Interface Science*, Vol. 1 (ed. R.Z. Bachrach), Plenum Press, New York, pp. 139–97.

Fadley, C.S. (1992) The study of surface structures by photoelectron diffraction and Auger electron diffraction, in *Synchrotron Radiation Research, Advances in Surface and Interface Science*, Vol. 1 (ed. R.Z. Bachrach), Plenum Press, New York, pp. 421–518.
Flodström, A., Nyholm, R. and Johansson, B. (1992) Surface core level spectroscopy, in *Synchrotron Radiation Research, Advances in Surface and Interface Science*, Vol. 1 (ed. R.Z. Bachrach), Plenum Press, New York, pp. 199–251.
Fripiat, J., Cases, J., François, M. and Letellier, M. (1982) Thermodynamic and microdynamic behavior of water in clay suspensions and gels. *J. Colloid Interface Sci.*, **89**, 378–400.
Fuoss, P.H., Liang, K.S. and Eisenberger, P. (1992) Grazing incidence X-ray scattering, in *Synchrotron Radiation Research, Advances in Surface and Interface Science*, Vol. 1 (ed. R.Z. Bachrach), Plenum Press, New York, pp. 385–419.

Gallei, E. and Parks, G.A. (1972) Evidence for surface hydroxyl groups in attenuated total reflectance spectra of crystalline quartz. *J. Colloid Interface Sci.*, **38**, 650–1.

Giese, R.F. Jr and Costanzo, P.M. (1986) Behavior of water on the surface of kaolin minerals, in *Geochemical Processes at Mineral Surfaces* (eds J.A. Davis and K.F. Hayes), ACS Symposium Series 323, American Chemical Society, Washington, DC, pp. 37–53.

Goodman, B.A. (1986) Adsorption of metal ions and complexes on aluminosilicate minerals, in *Geochemical Processes at Mineral Surfaces* (eds J.A. Davis and K.F. Hayes), ACS Symposium Series 323, American Chemical Society, Washington, DC, pp. 342–61.

Grandjean, J. (1992) NMR Studies of interfacial phenomena, in *Annual Reports on NMR Spectroscopy* (ed. G.A. Webb), Academic Press, New York, pp. 181–217.

Hair, M.L. (1967) *Infrared Spectroscopy in Surface Chemistry*, Marcel Dekker, New York.

Harrick, N.J. (1979) *Internal Reflection Spectroscopy*. Harrick Scientific Corp., Ossining, New York.

Harsh, J.B., Doner, H.E. and McBride, M.B. (1984) Chemisorption of copper on hydroxy-aluminum-hectorite: an electron spin resonance study. *Clays Clay Mineral.*, **32**, 407–13.

Haw, J.F. (1992) *In situ* solid state NMR studies of reactions on catalysts, in *Catalysis and Surface Characterization* (eds T.J. Dines, C.H. Rochester and J. Thomson), Royal Society of Chemistry Cambridge, UK, pp. 1–31.

Haworth, A. (1990) A review of the modelling of sorption from aqueous solutions. *Adv. Colloid Interface Sci.*, **32**, 43–78.

Hayes, K.F. (1987) *Equilibrium, Spectroscopic, and Kinetic Studies of Ion Adsorption at the Oxide/Aqueous Interface*. PhD Dissertation, Stanford University, 260 pp.

Hayes, K.F. and Leckie, J.O. (1987) Modeling ionic strength effects on cation adsorption at hydrous oxide/solution interfaces. *J. Colloid Interface Sci.*, **115**, 564–72.

Hayes, K.F. and Leckie, J.O. (1988) Modeling ionic strength effects on anion adsorption at hydrous oxide/solution interfaces. *J. Colloid Interface Sci.*, **125**, 717–26.

Hayes, K.F., Katz, L.E. and Penner-Hahn, J.E. (1993) XAS study of metal/ion partitioning at water/mineral interfaces. *1992 Stanford Synchrotron Radiation Laboratory Activity Report*, pp. 79–82.

Hayes, K.F., Roe, A.L., Brown, G.E. Jr *et al.* (1987) In-situ X-ray absorption study of surface complexes: selenium oxyanions on α-FeOOH. *Science*, **238**, 783–6.

Hayes, T.M. and Boyce, J.B. (1982) Extended X-ray absorption fine structure spectroscopy. *Solid State Phys.*, **37**, 173–365.

Heald, S.M., Keller, E. and Stern, E.A. (1984) Fluorescence detection of surface EXAFS. *Phys. Lett.*, **103A**, 155–8.

Henrich, V.E. (1985) The surfaces of metal oxides. *Rep. Prog. Phys.*, **48**, 1481.

Hingston, F.J., Posner, A.M. and Quirk, J.P. (1974) Anion adsorption by goethite and gibbsite. II. Desorption from hydrous oxide surfaces. *J. Soil Sci.*, **25**, 16–26.

Hochella, M.F. Jr (1988) Auger electron and X-ray photoelectron spectroscopies, in *Spectroscopic Methods in Mineralogy and Geology* (ed. F. Hawthorne), *Reviews in Mineralogy*, Vol. 18, Mineralogical Society of America, Washington, DC, pp. 573–637.

Hochella, M.F. Jr (1990) Atomic structure, microtopography, composition, and reactivity of mineral surfaces, in *Mineral–Water Interface Geochemistry* (eds M.F. Hochella, Jr and A.F. White), *Reviews in Mineralogy*, Vol. 23, Mineralogical Society of America, Washington, DC, pp. 87–132.

Hochella, M.F. Jr and White, A.F. (eds) (1990) *Mineral–Water Interface Geochemistry. Reviews in Mineralogy*, Vol. 23, Mineralogical Society of America, Washington, DC, 603 pp.

Hochella, M.F. Jr, Eggleston, C.M., Elings, V.B. *et al.* (1989) Mineralogy in two dimensions: scanning tunneling microscopy of semiconducting minerals with implications for geochemical reactivity. *Amer. Mineral.*, **74**, 1233–46.

Hronsky, V. and Rakos, M. (1980) ESR study of Mn^{2+} ion adsorption on Al_2O_3. *Czech. J. Phys.*, **B30**, 1061–7.

Hronsky, V., Rakos, M., Belak, J., Juhar, J. and Kazar, D. (1978) EPR study of Mn^{2+} ion

adsorption on silica gel. *Czech. J. Phys.*, **B28**, 1277–86.
Hronsky, V., Rakos, M., Belak, J., Juhar, J. and Kazar, D. (1979) EPR study of Mn^{2+} ions on silica gel, in *Magnetic Resonance and Related Phenomena* (eds E. Kundla, E. Lippmaa and T. Saluvere), Springer-Verlag, New York, pp. 310–35.

James, R.O. (1971) *The Adsorption of Hydrolysable Metal Ions at the Oxide–Water Interface.* PhD Dissertation, University of Melbourne, Melbourne, Australia.
James, R.O. and Healy, T.W. (1972) Adsorption of hydrolyzable metal ions at the oxide/water interface. I. Co(II) adsorption on SiO_2 and TiO_2 model systems. *J. Colloid Interface Sci.*, **40**, 42–52.

Kendelewicz, T., Woicik, J.C., Liu, P., Brown, G.E. Jr and Spicer, W.E. (1994) X-ray standing wave, SEXAFS, and photoemission study of the clean surface of CaO (in prep.).
Kinney, D.R., Chuang, I-S. and Maciel, G.E. (1993) Water and the silica surface as studied by variable-temperature high-resolution ^{1}H NMR. *J. Amer. Chem. Soc.*, **115**, 6786–94.
Kinniburgh, D.G. and Jackson, M.L. (1981) Cation adsorption by hydrous metal oxides and clay, in *Adsorption of Inorganics at Solid–Liquid Interfaces* (eds M.A. Anderson and A.J. Rubin), AnnArbor Science Publishers Inc, Ann Arbor, MI, pp. 91–160.
Kiselev, A.V. (1971) The effect of the geometrical structure and the chemistry of oxide surfaces on their adsorption properties, in *Discuss. Faraday Soc.* No. 53, *Surface Chemistry of Oxides*, pp. 14–32.
Kiselev, V.F. and Krylov, O.V. (1985) *Adsorption Processes on Semiconductor and Dielectric Surfaces I, Springer Series in Chemical Physics* 32, Springer-Verlag, New York, 287 pp.
Koch, E.E., Eastman, D.E. and Farge, Y. (1983) Synchrotron radiation – a powerful tool in science, in *Handbook on Synchrotron Radiation*, Vol. 1 (ed. E.E. Koch), North-Holland Publishing Co., Amsterdam, pp. 1–63.
Koningsberger, D.C. and Prins, R. (eds) (1988) *X-ray Absorption: Principles, Applications, Techniques of EXAFS, SEXAFS, and XANES*, John Wiley & Sons, New York, 673 pp.
Krauskopf, K.B. (1988) *Radioactive Waste Disposal and Geology*. Chapman and Hall, New York, 145 pp.
Kuhn, W.K., He, J.-W. and Goodman, D.W. (1992) Infrared reflection–absorption spectroscopy: new technique for in situ determination of local surface structure, in *Surface Science of Catalysis: In Situ Probes and Reaction Kinetics* (eds D.J. Dwyer and F.M. Hoffman), ACS Symposium Series 482, American Chemical Society, Washington, DC, pp. 71–84.
Kurtz, R.L. and Henrich, V.E. (1982) Chemisorption of H_2O on the surface of Ti_2O_3: role of d-electrons and ligand geometry. *Phys. Rev.*, **B26**, 6682.
Kurtz, R.L. and Henrich, V.E. (1984) The influence of O_2 and H_2O on the surface electronic structure of corundum transition-metal oxides. *J. Vac. Sci. Technol.*, **A2**, 842.

Langmuir, I. (1916) The constitution and fundamental properties of solids and liquids. *J. Amer. Chem. Soc.*, **38**, 2221–95.
Lear, R.D. and Weber, M.J. (eds) (1987) Synchrotron radiation. *Energy and Technology Review*. Lawrence Livermore National Laboratory, Nov.–Dec. 1987 issue, p. 33.
Lindau, I. and Spicer, W.E. (1974) *J. Electron Spectros. Relat. Phenom.*, **3**, 409.
Lindau, I. and Spicer, W.E. (1980) Photoemission as a tool to study solids and surfaces, in *Synchrotron Radiation Research* (eds H. Winick and S. Doniach), Plenum Press, New York, pp. 159–221.
Little, L.H. (1966) *Infrared Spectra of Adsorbed Species*, Academic Press, New York.
Lytle, F.W. (1989) Experimental X-ray absorption spectroscopy, in *Applications of Synchrotron Radiation* (eds H. Winick *et al.*), Gordon and Breach Science Publishers, New York, pp. 135–223.
Lytle, F.W., Greegor, R.B., Sandstrom, D.R., *et al.* (1984) Measurement of soft X-ray absorption spectra with a fluorescent ion chamber detector. *Nucl. Instr. Meth.*, **226**, 542–8.

Manceau, A. and Charlet, L. (1992) X-ray absorption spectroscopic study of the sorption of Cr(III) at the oxide–water interface I. Molecular mechanism of Cr(III) oxidation on Mn oxides. *J. Colloid Interface Sci.*, **148**, 425–42.

Manceau, A., Charlet, L., Boisset, M.C., Didier, B. and Spadin, L. (1992) Sorption and speciation of heavy metals on hydrous Fe and Mn oxides. From microscopic to macroscopic. *Appl. Clay Sci.*, **7**, 201–23.

Manceau, A., Charlet, L., Boisset, M.C., Didier, B. and Spadin, L. (1993) Sorption and speciation of heavy metals on hydrous Fe and Mn oxides. *Appl. Clay Sci.* (in press).

Martini, G. (1981) The state of water adsorbed on silica gel as determined by ESR of transition metal ion probe. *J. Coll. Int. Sci.* **80**, 39–48.

Martini, G., Ottaviani, M.F. and Seravalli, G.L. (1975) Electron spin resonance study of vanadyl complexes adsorbed on synthetic zeolites. *J. Phys. Chem.*, **79**, 1716–20.

McBride, M.B. (1976) Origin and position of exchange sites in kaolinite: an ESR study. *Clays Clay Mineral.*, **24**, 88–92.

McBride, M.B. (1978) Copper(II) interactions with kaolinite: factors controlling adsorption. *Clays Clay Mineral.*, **26**, 101–6.

McBride, M.B. (1982a) Cu^{2+}-adsorption characteristics of aluminum hydroxide and oxyhydroxides. *Clay Clay Mineral.*, **30**, 21–8.

McBride, M.B. (1982b) Hydrolysis and dehydration reactions of exchangeable Cu^{2+} on hectorite. *Clays Clay Mineral.*, **30**, 200–6.

McBride, M.B. (1985) Influence of glycine on Cu^{2+} adsorption by microcrystalline gibbsite and boehmite. *Clays Clay Mineral.*, **33**, 397–402.

McBride, M.B. (1986) Paramagnetic probes of layer silicate surfaces, in *Geochemical Processes at Mineral Surfaces* (eds J.A. Davis and K.F. Hayes), ACS Symposium Series 323, American Chemical Society, Washington, DC, pp. 362–88.

McBride, M.B. and Bouldin, D.R. (1984) Long-term reactions of copper (II) in a contaminated calcareous soil. *Soil Sci. Soc. Amer. J.*, **48**, 56–9.

McBride, M.B., Fraser, A.R. and McHardy, W.J. (1984) Cu^{2+} interaction with microcrystalline gibbsite: evidence for oriented chemisorbed copper ions. *Clays Clay Mineral.*, **32**, 12–18.

McCafferty, E., Pravdic, V. and Zettlemoyer, Z.C. (1970) Dielectric behavior of adsorbed water films on the α-Fe_2O_3 surface. *Trans. Faraday Soc.*, **66**, 1720–31.

McKale, A.G., Veal, B.W., Paulikas, A.P., Chan, S.K. and Knapp, G.S. (1988) Improved ab initio calculations for extended X-ray absorption fine structure spectroscopy. *J. Amer. Chem. Soc.*, **110**, 3763–8.

McKay, J.M. and Henrich, V.E. (1987) Adsorption and interaction of O_2, H_2O, and CO on the NiO(100) surface. *J. Vac. Sci. Technol.*, **A5**, 722.

Motschi, H. (1983) Cu(II) bound to hydrous surfaces. EPR measurements to characterize surface coordination. *Naturwissenschaften*, **70**, 519–20.

Motschi, H. (1984) Correlation of EPR-parameters with thermodynamic stability constants for copper (II) complexes – Cu(II) as a probe for the surface complexation at the water/oxide interface. *Colloids and Surfaces*, **9**, 333–47.

Motschi, H. (1987) Aspects of the molecular structure in surface complexes: spectroscopic investigations, in *Aquatic Surface Chemistry* (ed. W. Stumm), John Wiley & Sons, New York, pp. 111–25.

O'Day, P.A. (1992) *Structure, Bonding, and Site Preference of Cobalt(II) Sorption Complexes on Kaolinite and Quartz from Solution and Spectroscopic Studies*. PhD Dissertation, Stanford University, 208 pp.

O'Day, P.A., Brown, G.E. Jr and Parks, G.A. (1991) EXAFS study of aqueous Co(II) sorption complexes on kaolinite and quartz surfaces, in *XAFS VI, Sixth International Conference on X-ray Absorption Fine Structure* (ed. S.S. Hasnain), Ellis Horwood Ltd Publishers, New York, pp. 260–2.

O'Day, P.A., Brown, G.E. Jr and Parks, G.A. (1994b) X-ray absorption spectroscopy of Co(II) multinuclear surface complexes and surface precipitates on kaolinites. *J. Colloid Interface Sci.* (in press).
O'Day, P.A., Parks, G.A. and Brown, G.E. Jr (1994c) Molecular structure and binding sites of cobalt (II) surface complexes on kaolinite from X-ray absorption spectroscopy. *Clays Clay Mineral.* (in press).
O'Day, P.A., Rehr, J.J., Zabinsky, S.I. and Brown, G.E. Jr (1994a) Extended X-ray absorption fine structure (EXAFS) analysis of disorder and multiple-scattering in complex crystalline solids. *J. Amer. Chem. Soc.* (in press).
O'Melia, C.R. (1987) Particle–particle interactions, in *Aquatic Surface Chemistry* (ed. W. Stumm), John Wiley & Sons, New York, pp. 385–404.

Papelis, C., Hayes, K.F. and Leckie, J.O. (1988) HYDRAQL: a program for the computation of chemical equilibrium composition of aqueous batch systems including surface-complexation modeling of ion adsorption at the oxide/solution interface. *Technical Report* 306, Environmental Engineering and Science, Stanford University, 130 pp.
Papelis, C., Brown, G.E. Jr, Parks, G.A. and Leckie, J.O. (1994) X-ray absorption spectroscopic studies of cadmium and selenite adsorption on aluminum oxides. *Langmuir* (submitted).
Parfitt, R.L. (1978) Anion adsorption by soils and soil materials. *Advances in Agronomy*, **30**, 1–50.
Parfitt, R.L. and Russell, J.D. (1977) Adsorption of hydrous oxides. IV. Mechanisms of adsorption of various ions on goethite. *J. Soil Sci.*, **28**, 297–305.
Parfitt, R.L., Russell, J.D. and Farmer, V.C. (1976) Confirmation of the surface structure of goethite and phosphated goethite by infrared spectroscopy. *J. Chem. Soc. Faraday Trans.*, **172**, 1082–7.
Parks, G.A. (1990) Surface energy and adsorption at mineral–water interfaces: an introduction, in *Mineral–Water Interface Geochemistry* (eds M.F. Hochella, Jr and A.F. White), *Reviews in Mineralogy*, Vol. 23, Mineralogical Society of America, Washington, DC, pp. 133–75.
Pemberton, J.E. (1991) Surface enhanced Raman scattering, in *Electrochemical Interfaces: Modern Techniques for In-Situ Interface Characterization* (ed. H.D. Abruña), VCH Publishers, Inc., New York, pp. 195–266.
Peri, J.B. (1965) Infrared and gravimetric study of the surface hydration of γ-alumina. *J. Phys. Chem.*, **69**, 211–19.
Peri, J.B. and Hannan, R.B. (1960) Surface hydroxyl groups on γ-alumina. *J. Phys. Chem.*, **64**, 1526–30.
Pickering, I.J., Brown, G.E. Jr and Tokunaga, T. (1994) X-ray absorption spectroscopy of selenium-contaminated soils (submitted to *Nature*).
Poisignon, C., Cases, J.M. and Fripiat, J.J. (1978) Electrical polarization of water molecules adsorbed by smectites: an infrared study. *J. Phys. Chem.*, **82**, 1855–60.

Rehn, V. and Rosenberg, R.A. (1992) Photon-stimulated desorption, in *Synchrotron Radiation Research, Advances in Surface and Interface Science*, Vol. 1 (ed. R.Z. Bachrach), Plenum Press, New York, pp. 327–81.
Rehr, J.J. and Albers, R.C. (1990) Scattering-matrix formulation of curved-wave multiple-scattering theory: applications to X-ray absorption fine structure. *Phys. Rev.*, **B41**, 8139–49.
Rehr, J.J., Albers, R.C. and Zabinsky, S.I. (1992) High-order multiple-scattering calculations of X-ray absorption fine structure. *Phys. Rev. Lett.*, **69**, 3937–41.
Rehr, J.J., Mustre de Leon, J., Zabinsky, S.I. and Albers, R.C. (1991) Theoretical X-ray absorption fine structure standards. *J. Amer. Chem. Soc.*, **113**, 5135–40.
Rhodin, T.N. and Gadzuk, J.W. (1979) Electron spectroscopy and surface chemical bonding, in

The Nature of the Surface Chemical Bond (eds T.N. Rhodin and G. Ertl), North Holland Publishing Co., Amsterdam, pp. 113–273.

Roe, A.L., Schneider, D.J., Mayer, R.J. *et al.* (1984) X-ray absorption spectroscopy of iron-tyrosinate proteins. *J. Amer. Chem. Soc.*, **106**, 1676–81.

Roe, A.L., Hayes, K.F., Chisholm-Brause, C.J. *et al.* (1991) X-ray absorption study of lead complexes at α-FeOOH/water interfaces. *Langmuir*, **7**, 367–73.

Rowe, J.E. (1992) Surface EXAFS, in *Synchrotron Radiation Research, Advances in Surface and Interface Science*, Vol. 1 (ed. R.Z. Bachrach), Plenum Press, New York, pp. 117–36.

Russell, J.D., Paterson, E., Fraser, A.R. and Farmer, V.C. (1975) Adsorption of carbon dioxide on goethite surfaces and its implication for anion adsorption. *J. Chem. Soc. Faraday Trans.*, **171**, 1623–30.

Sayers, D.E. and Bunker, B.A. (1988) Data analysis, in *X-ray Absorption: Principles, Applications, Techniques of EXAFS, SEXAFS, and XANES* (eds D.C. Koningsberger and R. Prins), John Wiley & Sons, New York, pp. 211–53.

Sayers, D.E., Stern, E.A. and Lytle, F.W. (1971) New technique for investigating noncrystalline structures: Fourier analysis of the extended X-ray absorption fine structure. *Phys. Rev. Lett.*, **27**, 1204–7.

Scherson, D.A. (1991) Mössbauer spectroscopy, in *Electrochemical Interfaces: Modern Techniques for In-Situ Interface Characterization* (ed. H.D. Abruña), VCH Publishers, New York, pp. 411–77.

Schindler, P.W. and Stumm, W. (1987) The surface chemistry of oxides, hydroxides, and oxide minerals, in *Aquatic Surface Chemistry* (ed. W. Stumm), John Wiley & Sons, New York, pp. 83–110.

Sherwood, P.M.A. (1985) Photoelectron spectroscopic studies of electrode and related structures. *Chem. Soc. Rev.*, **14**, 1–44.

Shirai, M., Asakura, K. and Iwasawa, Y. (1992) Anisotropic structure analysis for cobalt oxides on α-Al_2O_3(0001) by polarized total-reflection fluorescence extended X-ray absorption fine structure. *Catalysis Lett.*, **15**, 247–54.

Sposito, G. (1984) *The Surface Chemistry of Soils*. Oxford University Press, New York, 234 pp.

Sposito, G. (1986) Distinguishing adsorption from surface precipitation, in *Geochemical Processes at Mineral Surfaces* (eds J.A. Davis and K.F. Hayes), ACS Symposium Series 323, American Chemical Society, Washington, DC, pp. 217–28.

Stern, E.A. and Heald, S.M. (1983) Basic principles and applications of EXAFS, in *Handbook on Synchrotron Radiation*, Vol. 1b (ed. E.E. Koch), North Holland Publishing Co., New York, pp. 955–1014.

Stöhr, J. (1988) SEXAFS: everything you always wanted to know about SEXAFS but were afraid to ask, in *X-ray Absorption: Principles, Applications, Techniques of EXAFS, SEXAFS, and XANES* (eds D.C. Koningsberger and R. Prins), John Wiley & Sons, New York, pp. 443–571.

Stöhr, J. (1992) *NEXAFS Spectroscopy*, Springer-Verlag, New York, 433 pp.

Stole, S.M., Popenoe, D.D. and Porter, M.D. (1991) Infrared spectroelectrochemistry: a probe of the molecular architecture of the electrochemical interface, in *Electrochemical Interfaces: Modern Techniques for In-Situ Interface Characterization* (ed. H.D. Abruña), VCH Publishers, New York, pp. 339–410.

Stumm, W. (1992) *Chemistry of the Solid–Water Interface*, John Wiley & Sons, New York, 428 pp.

Teo, B.K. (1986) *EXAFS: Basic Principles and Data Analysis*, Springer-Verlag, Berlin, 349 pp.

Teo, B.K. and Joy, D.C. (eds) (1981) *EXAFS Spectroscopy*, Plenum Press, New York, 275 pp.

Teo, B.K. and Lee, P.A. (1979) *Ab initio* calculation of amplitude and phase function for

Extended X-ray Absorption Fine Structure (EXAFS) spectroscopy. *J. Amer. Chem. Soc.*, **101**, 2815–30.

Tewari, P.H. and Lee, W. (1975) Adsorption of Co(II) at the oxide–water interface. *J. Colloid Interface Sci.*, **52**, 77–88.

Tewari, P.H. and McIntyre, N.S. (1974) Characterization of adsorbed cobalt at the oxide–water interface. *AIChE Symp. Series*, **71**, 134–7.

Tewari, P.H., Campbell, A.B. and Lee, W. (1972) Adsorption of Co^{2+} by oxides from aqueous solution. *Can. J. Chem.*, **50**, 1642–8.

Texter, J., Klier, K. and Zettlemoyer, A.C. (1978) Water at surfaces, in *Progress in Surface and Membrane Science*, Vol. 12 (eds D.A. Cadenhead and J.F. Danielli), pp. 327–403.

Thompson, H., Parks, G.A. and Brown, G.E. Jr (1994) XAS study of U(VI) adsorption at kaolinite–water interfaces (in prep.).

Tjedor-Tjedor, M.I. and Anderson, M.A. (1986) In situ attenuated total reflectance Fourier transform infrared studies of the goethite (α-FeOOH)–aqueous solution interface. *Langmuir*, **2**, 203–10.

Tokunaga, T.K. and Benson, S.M. (1992) Selenium in Kesterson Reservoir ephemeral pools formed by groundwater rise: I. A field study. *J. Environ. Qual.*, **21**, 246–51.

Towle, S.N., Bargar, J.R., Brown, G.E. Jr *et al.* (1993) Grazing-incidence EXAFS spectroscopy of metal ion sorption on single-crystal α-Al_2O_3. (Abstract) 205th American Chemical Society National Meeting, Denver, CO.

Waite, T.D. (1990) Photo-redox processes at the mineral–water interface, in *Mineral–Water Interface Geochemistry* (eds M.F. Hochella, Jr and A.F. White), *Reviews in Mineralogy*, Vol. 23, Mineralogical Society of America, Washington, DC, pp. 559–603.

Waychunas, G.A. and Brown, G.E. Jr (1990) Polarized X-ray absorption spectroscopy of metal ions in minerals: identification of near edge electronic transitions and scattering resonances and application to site geometry determinations. *Phys. Chem. Minerals*, **17**, 420–30.

Waychunas, G.A., Apted, M.J. and Brown, G.E. Jr (1983) X-ray K-edge absorption spectra of Fe minerals and model compounds: near edge structure. *Phys. Chem. Minerals*, **10**, 1–9.

Waychunas, G.A., Rea, B.A., Fuller, C.C. and Davis, J.A. (1993) Surface chemistry of ferrihydrite: Part I. EXAFS studies of the geometry of coprecipitated and adsorbed arsenate. *Geochim. Cosmochim. Acta*, **57**, 2251–69.

Wells, A.F. (1962) *Structural Inorganic Chemistry*, 3rd edn, Clarendon Press, Oxford, 1055 pp.

White, A.F. (1990) Heterogeneous electrochemical reactions associated with oxidation of ferrous oxide and silicate surfaces, in *Mineral–Water Interface Geochemistry* (eds M.F. Hochella, Jr and A.F. White), *Reviews in Mineralogy*, Vol. 23, Mineralogical Society of America, Washington, DC, pp. 467–509.

White, J.H. (1991) Elucidation of structural aspects of electrode/electrolyte interfaces with X-ray standing waves, in *Electrochemical Interfaces: Modern Techniques for In-Situ Interface Characterization* (ed. H.D. Abruña), VCH Publishers, New York, pp. 131–54.

Winick, H. and Doniach, S. (eds) (1980) *Synchrotron Radiation Research*, Plenum Press, New York, 750 pp.

Woicik, J.C., Kendelewicz, T., Miyano, K.E. *et al.* (1991a) Structure of the Si(111) root3 × root3 – Sb interface by surface X-ray absorption fine structure and photoemission. *J. Vac. Sci. Technol.*, **A9**, 1956–61.

Woicik, J.C., Kendelewicz, T., Miyano, K.E. *et al.* (1991b) Determination of the Sb/Si(111) interfacial structure by back reflection standing wave and surface extended X-ray absorption fine structure techniques. *Phys. Rev.*, **B44**, 3475–8.

Woicik, J.C., Kendelewicz, T., Miyano, K.E. *et al.* (1992) X-ray standing wave measurement of the InP(110) surface relaxation. *Phys. Rev. Lett.*, **68**, 341–4.

Wong, J. (1986) Extended X-ray absorption fine structure: a modern structural tool in materials science. *Mat. Sci. Eng.*, **80**, 107–28.

Wong, J., Lytle, F.W., Messmer, R.P. and Maylotte, D.H. (1984) K-edge absorption spectra of selected vanadium compounds. *Phys. Rev.*, **B30**, 5596–6510.

Wyckoff, R.W.G. (1963) *Crystal Structures*, Vol. 1, 2nd edn, Wiley Interscience, New York, 467 pp.

Xu, N. (1993) *Spectroscopic and Solution Chemistry Studies of Cobalt(II) Sorption Mechanisms at the Calcite–Water Interface*. PhD Dissertation, Stanford University, 143 pp.

Yates, D.E. and T.W. Healy (1975) Mechanism of anion adsorption at the ferric and chromic oxide/water interfaces. *J. Colloid Interface Sci.*, **52**, 222–8.

Zelewsky, A. von and Bemtgen, J.-M. (1982) Formation of ternary copper(II) complexes at the surface of silica gel as studied by EPR spectroscopy. *Inorg. Chem.*, **21**, 1771–7.

Zeltner, W.A., Yost, E.C., Machesky, M.L. *et al.* (1986) Characterization of anion binding on goethite using titration calorimetry and cylindrical internal reflection–Fourier transform infrared spectroscopy, in *Geochemical Processes at Mineral Surfaces* (eds J.A. Davis and F.K. Hayes), ACS Symposium Series 323, American Chemical Society, Washington, DC, pp. 142–61.

Zettlemoyer, Z.C., Micale, F.J. and Klier, K. (1975) Adsorption of water in well-characterized solid surfaces, in *Water: A Comprehensive Treatise*, Vol. 5 (ed. F. Franks), pp. 249–91.

CHAPTER SIX

Surface chemistry during the dissolution of oxide and silicate materials

W.H. Casey

6.1 Introduction

All geochemists are surface chemists, some without even knowing it. The growth of minerals, the corrosion of volcanic glasses and the mineral transformations in rocks all relate in some critical stage to forces at mineral–solution interfaces. Even the vesiculation of lava reflects the interfacial tension between the vapour and the liquid. In this chapter the role that aqueous surface chemistry plays in controlling the rates of mineral dissolution is reviewed.

Why dissolution? The experimental conditions are commonly so precisely controlled as to invalidate quantitative comparison with natural weathering, where the important variables are affected by climate, hydrology or mineral microstructure. The experiments are useful, however, because they provide insight into reaction mechanisms and are, therefore, generally applicable. In this chapter, it is argued that the mechanisms by which metals are released from a dissolving mineral surface are profoundly similar to the mechanisms of ligand exchange around dissolved metal complexes. The forces that control rates of plagioclase dissolution also control the rates of borosilicate glass corrosion and, at a very fundamental level, the rates of depolymerization of dissolved multimers (e.g. $[Al_{13}O_4(OH)_{24}(H_2O)_{12}]^{7+}$) in soil solutions. The consistency can be used to predict reactivity trends and to guide the interpretation of experimental data.

6.2 Rates far from equilibrium

6.2.1 Dissolution as a ligand exchange reaction

The oxygens that bridge metal ions at the surface of a mineral are exchangeable ligands in the same sense as ligands in the inner-coordination sphere of a dissolved metal complex. New ligands enter the coordination sphere as chemical conditions change. The rates of these reactions are controlled

Mineral Surfaces. Edited by D.J. Vaughan and R.A.D. Pattrick. Published in 1995 by Chapman & Hall, London. ISBN 0 412 56340 1

Table 6.1 Measured metal–oxygen distances (Å) for hydrated cations, oxide minerals and orthosilicate minerals. Distances for most dissolved species are given in Burgess (1988). Distances in the oxide and orthosilicate minerals are compiled in Casey (1991) and Bish and Burnham (1984), respectively. Distances for $Ga(H_2O)_6^{3+}$ and $Be(H_2O)_4^{2+}$ are inferred from Raman spectral data (Kanno, 1988)

Ion		Oxide		Silicate	
$Ca(H_2O)_6^{2+}$	2.39–2.46	CaO	2.405	Ca_2SiO_4	2.346–2.392
$Mg(H_2O)_6^{2+}$	2.10	MgO	2.11	Mg_2SiO_4	2.101–2.127
$Be(H_2O)_4^{2+}$	1.67	BeO	1.649	Be_2SiO_4	1.645
$Zn(H_2O)_6^{2+}$	2.08–2.17	ZnO	1.95	Zn_2SiO_4	1.92
$Mn(H_2O)_6^{2+}$	2.18–2.20	MnO	2.22	Mn_2SiO_4	2.185–2.227
$Co(H_2O)_6^{2+}$	2.05–2.08	CoO	2.13	Co_2SiO_4	2.123–2.134
$Ni(H_2O)_6^{2+}$	2.04–2.10	NiO	2.095	Ni_2SiO_4	2.076–2.102
$Al(H_2O)_6^{3+}$	1.87–1.97	Al_2O_3	1.86–1.97		
$Ga(H_2O)_6^{3+}$	2.02	Ga_2O_3	^{IV}Ga: 1.80–1.85 ^{VI}Ga: 1.95–2.08		
$In(H_2O)_6^{3+}$	2.15	In_2O_3	1.93–2.18		

exclusively by short range forces between the metal, coordinated oxygens, and the incoming ligand. Because the coordination numbers and metal–oxygen bond distances of metals in minerals are commonly similar to those in the hydrated ion (Table 6.1), it is logical to begin an examination of mineral dissolution kinetics with study of particularly simple ligand-exchange reactions, such as solvent exchange around dissolved metals.

Rates of solvent exchange under identical conditions differ considerably for different metals (Figure 6.1). Rates vary with the ionic potential (z/r) of the metal for alkaline-earth metals and for metals of the sp-block of the periodic table. Rates for first-row transition metals (Ni^{2+}, Co^{2+}, etc.) vary with the number of d-electrons and are largely independent of ionic radius. Metals with ionic bonds to oxygen are more reactive than metals that covalently bond to oxygen, and reactivity changes dramatically with the metal valence state. Rates of solvent exchange around the $Fe(H_2O)_6^{2+}$ complex, for example, are a factor of 1000 times more rapid than around $Fe(H_2O)_6^{3+}$ (Figure 6.1). Most importantly, rates are dramatically affected by acid–base reactions at the waters of hydration. Removal of a proton from $Ru(H_2O)_6^{3+}$, for example, causes the first-order rate coefficient to increase from $k = 3.5 \times 10^{-6} s^{-1}$ to $k_{OH} = 5.9 \times 10^{-4} s^{-1}$ or, conversely, the rate of exchange to increase by a factor of ~170 (Table 6.2).

The simplest analogue for mineral dissolution is dissociation of a dissolved metal dimer. An example is the $Fe_2(OH)_2(H_2O)_8^{4+}$ species, where two iron metals are associated via common hydroxide bridges and are surrounded by waters of hydration. This dimer dissociates to form iron monomers, either as hydroxide species, $Fe(H_2O)_5(OH)^{2+}$, or as fully aquated ions, $Fe(H_2O)_6^{2+}$.

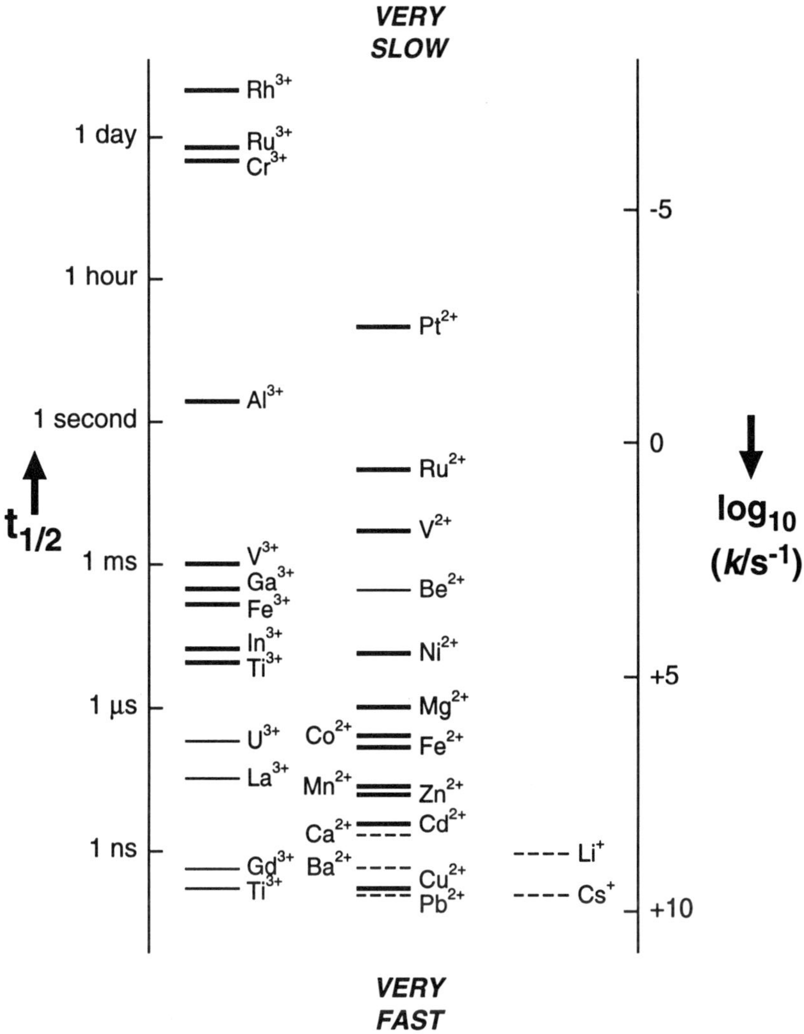

Figure 6.1 The first-order rate coefficient ($\log_{10}(k\,s^{-1})$) for water exchange around the corresponding hydrated metal ion (from Burgess, 1988) and mean residence times ($t_{1/2}$) for water molecules in the inner-coordination spheres of the metal. Data are for 298 K. Octahedral species are indicated by thick lines, non-octahedral species by thin lines, and dashed lines indicate estimates derived from complexation studies. Note that the rates of water exchange around Fe^{3+} and Ru^{3+} are 1000 times slower than for Fe^{2+} and Ru^{2+}; electron exchange resulting in metal oxidation greatly reduces the reactivity of the metal–oxygen bonds.

Table 6.2 First-order rate coefficients (k, in units of s^{-1}) and activation parameters for water exchange on some hexa-aqua and monohydroxypenta-aqua metal ions (Merbach and Akitt, 1990). The ratio k_{OH}/k corresponds to the ratio of rate coefficients for the first hydrolysis product (e.g. $Fe(OH)^{2+}$) relative to the fully hydrated ion (e.g. $Fe(H_2O)_6^{3+}$). The activation parameters $\Delta H^\ddagger$, $\Delta S^\ddagger$ and $\Delta V^\ddagger$ were derived from the temperature and pressure variations of the reaction rates

M^{3+}	k (s^{-1})	$\Delta H^\ddagger$ (kJ mol^{-1})	$\Delta S^\ddagger$ (kJ mol^{-1})	$\Delta V^\ddagger$ (cm^3 mol^{-1})	pK	k_{OH}/k
$Ga(H_2O)_6^{3+}$	4.0×10^2	67.1	+30.1	+5.0	3.9	275
$Ga(OH)^{2+}$	1.1×10^5	58.9	–	+6.2		
$Fe(H_2O)_6^{3+}$	1.6×10^2	64.0	+12.1	−5.4	2.9	750
$Fe(OH)^{2+}$	1.2×10^5	42.4	+5.3	+7.0		
$Cr(H_2O)_6^{3+}$	2.4×10^{-6}	108.6	+11.6	−9.6	4.1	75
$Cr(OH)^{2+}$	1.8×10^{-4}	110.0	+55.6	+2.7		
$Ru(H_2O)_6^{3+}$	3.5×10^{-6}	89.8	−48.2	−8.3	2.7	170
$Ru(OH)^{2+}$	5.9×10^{-4}	95.8	+14.9	+0.9		
$Rh(H_2O)_6^{3+}$	2.2×10^{-9}	89.8	−48.2	−8.3	3.5	19 100
$Rh(OH)^{2+}$	4.2×10^{-5}	95.8	+14.9	−0.9		

The psuedo-first-order rate coefficient can conceivably be resolved into two terms:

$$k = k_1 + k_2(H^+) \tag{6.1}$$

The psuedo-first-order term (k_1, s^{-1}) corresponds to the rate of dissociation into monohydroxide species. The other, psuedo-second-order term (k_2, $M^{-1}s^{-1}$) is proportional to the hydrogen ion (proton) concentration and corresponds to a pathway where a proton first associates with, and weakens, bonds to the bridging hydroxide ion (Wilkens, 1974, p. 81) before dissociation. Although the rate of proton transfer is always fast, the extent of the acid–base reaction varies considerably with solution pH.

One can imagine the case where other ligands affect the dissociation rate. Fluoride ion, for example, could coordinate to the dimer, displace a pre-existing ligand and introduce yet a third pathway for dissociation of the dimer. A dissolved product of this reaction would be an iron-fluoride complex.

Note the fundamental difference in the way that protons and the fluoride ligand affect the reaction. The proton modifies the charge distribution in a pre-existing ligand associated with the metal. The fluoride ion, however, directly associates with the metal and is a wholly new ligand. Thus, there exist **proton-promoted** and **ligand-promoted** pathways for dissociation of the dimer. One could imagine a fourth pathway, whereby a ligand exchanges electrons with the iron before exchange. Any ligand-exchange reaction

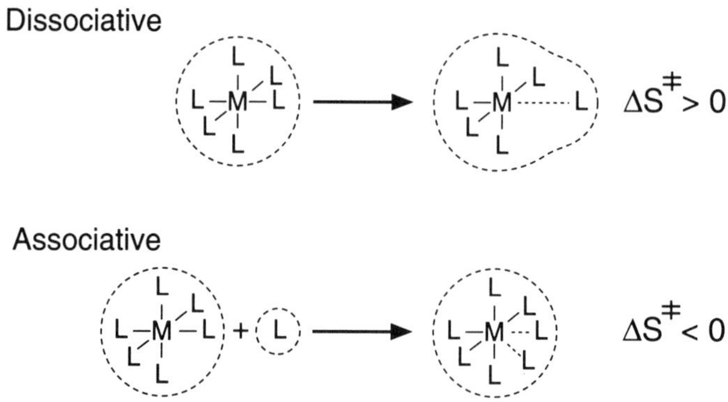

Figure 6.2 Dissociative (top) and associative (bottom) pathways for ligand exchange. A similar, albeit more complicated, figure could be created for destruction of a bridging oxygen at a mineral surface. In this case, a ligand is added to coordinate to one of the two previously bridged metals. (Adapted from Burgess, 1988.)

coupled to electron-exchange would introduce **reductive** and **oxidative** pathways to the overall reaction.

Finally, two extreme pathways can be distinguished. In **dissociative** exchange (Figure 6.2), the metal–ligand bond lengthens to the point of dissociation before entry of the new ligand. The Lewis basicity of the incoming ligand exerts little influence on the reaction rate, the reaction has a positive activation entropy, a small positive activation volume, and a decreased coordination number in the activated state. (Rates are, however, proportional to the concentration of the incoming ligand, which is a surface concentration in the case of dissolution; Hering and Morel, 1990.) At the other extreme, associative exchange requires an increased coordination number in the activated state (Figure 6.2). Rates of associative exchange are considerably affected by the Lewis basicity of the incoming ligand, which must distort the coordination sphere of the metal. These reactions have a small or negative activation volume and negative activation entropies (Table 6.2).

Ion size and bond directionality are particularly important in associative exchange because large ions accommodate the extra ligand in the activated state more easily than small ions. Consider the example of metals in the sp-block of the periodic table. The mechanisms of water exchange into the hydration sphere of $Al(H_2O)_6^{3+}$, $Ga(H_2O)_6^{3+}$ and $In(H_2O)_6^{3+}$ change from dissociative to associative as the ionic radii increase from 0.51 Å to 0.81 Å (Burgess, 1988). Water exchange around $Al(H_2O)_6^{3+}$ is 30 000 times slower than around $In(H_2O)_6^{3+}$. Likewise, the mechanism of ligand exchange changes from associative to dissociative with increasing number of d-electrons in the first row of transition metals. Water exchange around

Mn(II) ($t_{2g}^3e_g^2$) is associative and fast. Water exchange around Ni(II)(($t_{2g}^6e_g^2$) is dissociative and slow (Figure 6.1). With eight valence d-electrons, Ni(II) resists the distortion that accompanies ligand exchange more than metals with no ligand-field stabilization energy.

6.2.2 *Acid–base reactions on mineral surfaces*

The oxide or hydroxide ion bridges at the surface of a mineral also undergo very rapid acid–base reactions with the adjacent solution that impart a net charge on the mineral surface. The reactions are generally written to involve a neutral 'average' hydroxyl functional group:

$$\triangleright SOH_2^+ = \triangleright SOH + H^+ \tag{6.2}$$

$$\triangleright SOH = \triangleright SO^- + H^+ \tag{6.3}$$

where $\triangleright S$ represents a surface atom. The acid–base chemistry of a particular mineral surface is characteristic of its composition and structure, just as the hydrolysis chemistry of a particularly dissolved metal is distinct. Silicic acid (H_4SiO_4) does not hydrolyze at the same pH as the hydrated beryllium ion [$Be(H_2O)_4^{2+}$]; likewise, the hydroxylated surface of silica (SiO_2) is not neutral at the same pH as the surface of bromellite (BeO). This condition of net neutrality is termed the point of zero charge (pzc). Even though we treat the mineral surface as having an average hydroxyl functional group, it is clear that minerals expose several distinct sites to an aqueous solution. Oxygens on the surface of BeO, for example, are associated with one, two or three beryllium ions, depending on how the BeO_4^{4+} structural unit is truncated. These oxygens can be distinguished spectrally (Boehm and Knozinger, 1983) and, by analogy with oxygens in dissolved metal-oxide clusters (e.g. Day, Klemperer and Malthie, 1987; Richens *et al.*, 1989), have distinctly different acid–base properties and reactivities. We ignore the individual sites because models to assign them unique properties are underconstrained.

A concentration-action relation (K) can be defined for the formation of positive or negative surface charge on these generic functional groups. For reaction (6.2), for example,

$$K = \frac{f_{SOH_2^+}\, x_{SOH_2^+}}{(H^+)\, f_{SOH}\, x_{SOH}} \tag{6.4}$$

where f_i and x_i are the rational activity coefficient and mole fraction of the ith surface species, respectively, and (H^+) is the activity of protons in the aqueous phase. The parameter that is most commonly measured, however, is a conditional equilibrium constant

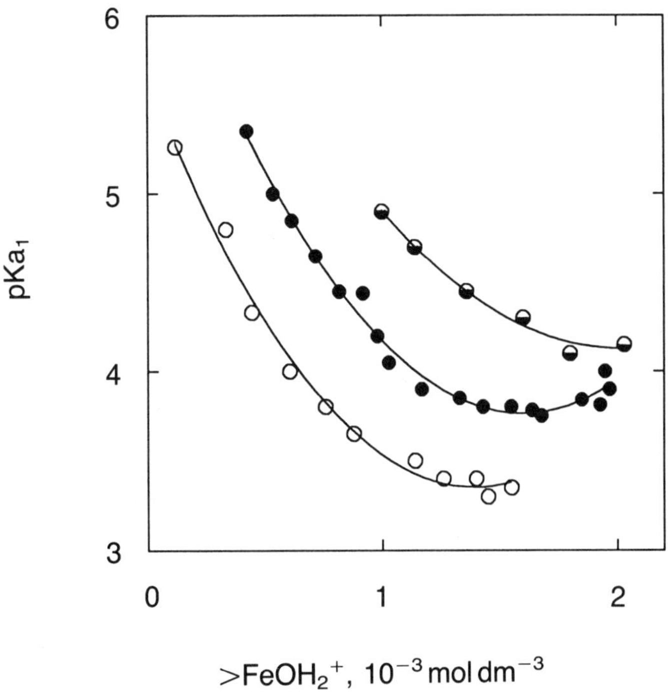

Figure 6.3 The negative logarithm of the conditional equilibrium constant (pKa_1) for protonation of ▷FeOH on hematite (●), magnetite (○) and goethite (◒) (at I = 0.001 M and 25 °C (adapted from Astumian *et al.* 1981). The abscissa corresponds to the concentration of surface sites ($mol\,dm^{-3}$) in the relaxation experiment of Astumian *et al.* (1981). These bulk concentrations can be converted to surface concentrations through knowledge of the cell volume (dm^3), the specific area of the solids ($m^2\,g^{-1}$) and the mass of solid used to make the suspension (g).

$$K' = \frac{K f_{SOH}}{f_{SOH_2^+}} = \frac{x_{SOH_2^+}}{(H^+)\, x_{SOH}} \tag{6.5}$$

which is a function of both temperature and solution composition. The rational activity coefficients, and hence the conditional equilibrium constants, vary by factors of 10–100 with pH, owing to electrostatic interactions among charged sites on the mineral surface (Sposito, 1983, 1984). An example is shown in Figure 6.3, where the negative logarithm of the conditional equilibrium constant for formation of positive charge on goethite, magnetite and hematite changes rapidly and non-linearly with concentrations of positive surface charge ($\triangleright FeOH_2^+$) and, hence, pH.

Empirically, surface charge concentrations for different minerals vary similarly with solution pH (Figure 6.4(A)) once differences in their pzc

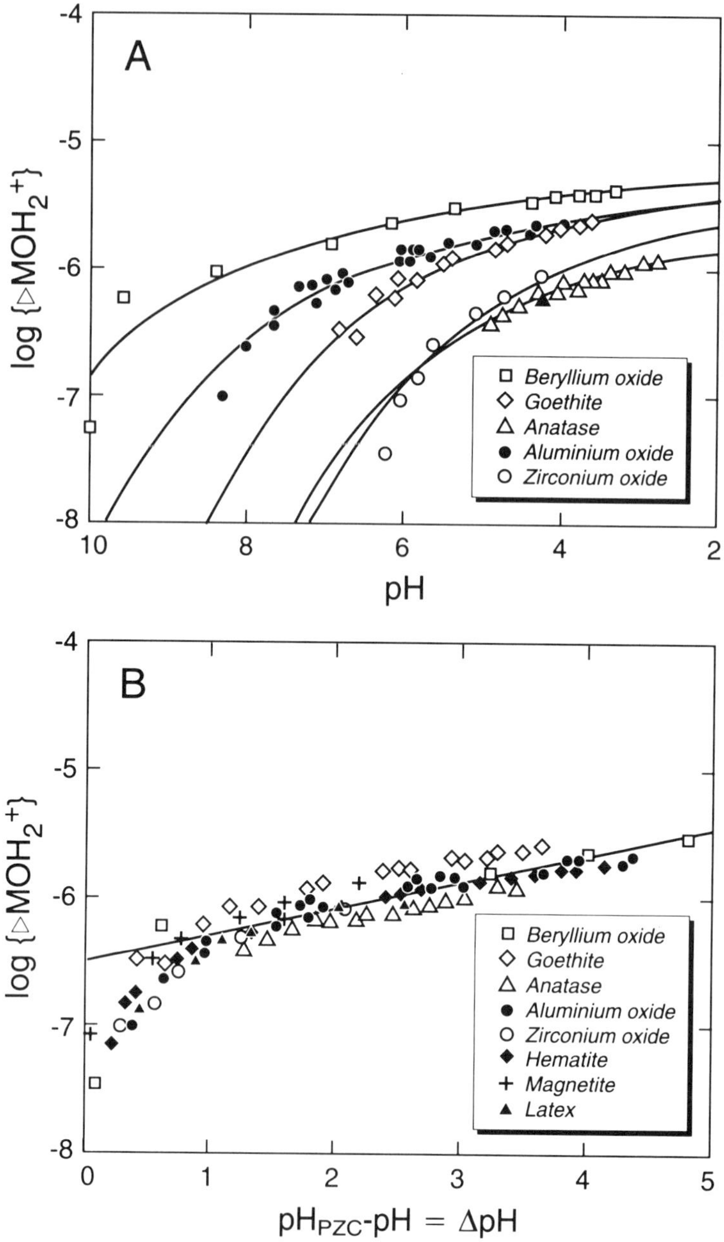

Figure 6.4 (A) Surface protonation isotherms from Wieland, Wehrli and Stumm (1988). Symbols correspond to titration curves at an ionic strength (I) of 0.1 M, except for hematite, where $I = 0.2$ M. The concentrations of protonated sites ($\triangleright SOH_2^+$) is given in mol cm^{-2}. (B) Surface concentrations from Figure 6.4(A) as a function of $\Delta pH = pzc - pH$. (Adapted from Wieland, Wehrli and Stumm, 1988.)

values are taken into account (Figure 6.4(B)). The pH-dependence of charge is apparently controlled by electrostatic interactions among charged surface groups and these interactions are similar on different oxide materials. The situation is a little more complicated for minerals that have a structural charge due to uncompensated cation substitutions (e.g. Al^{3+} for Si^{4+} in clays), but is similar.

6.2.3 *Acid–base reactions and proton-promoted dissolution*

Acid–base reactions influence both the pH- and temperature-dependence of mineral dissolution rates. Rate laws for proton- or hydroxyl-promoted dissolution, therefore, include a term to describe the equilibrium concentration of $\rhd SOH_2^+$ or $\rhd SO^-$. For dissolution under conditions where positive surface charge predominates (pH $\ll$ pzc) and far from equilibrium, one empirical law is (Wieland, Wehrli and Stumm, 1988; Stumm and Wollast, 1990)

$$R_{\text{dis}} = k\,[\rhd SOH_2^+]^n \tag{6.6}$$

where R_{dis} is the experimental dissolution rate (mol cm^{-2} s^{-1}), k is a rate coefficient (mol cm^{-2})$^{(1-n)}$ s^{-1}, $[\rhd SOH_2^+]$ is the concentration of adsorbed protons (mol cm^{-2}), and n is a partial reaction rate order with respect to the concentration of protonated surface groups. It is useful to express equation (6.6) on the mole fraction scale:

$$R_{\text{dis}} = k\,(N_{\text{T}})^n\, x^n_{SOH_2^+} \equiv k'\, x^n_{SOH_2^+} \tag{6.7}$$

where (N_{T}) is the total concentration of protonable sites on the mineral surface.

The partial rate order, n, with respect to charged sites is uncertain. Some workers (Furrer and Stumm, 1986; Wieland, Wehrli and Stumm, 1988) conclude that the rate order, n, in equation (6.7) approximately equals the metal valence state. Thus, $n = 2$ for BeO dissolution, $n = 3$ for δ-Al_2O_3 dissolution and $n = 4$ for SiO_2 dissolution (Figure 6.5). This result can be justified with statistical-mechanical arguments (Wieland, Wehrli and Stumm, 1988), but other workers (e.g. Blum and Lasaga, 1988; Carroll-Webb and Walther, 1988; Brady and Walther, 1992), find first-order dependencies (Figure 6.6) of rate on the concentration of positive or negative charge. The uncertainty about rate order arises because of difficulty in measuring the charge densities on dissolving minerals and particularly on multi-component solids that can be selectively leached. Regardless of arguments about the partial rate order, the relation between the logarithm of rate and pH at pH $<$ pzc can be elucidated by rearranging and differentiating equation (6.5) with respect to pH and combining it with equation (6.7):

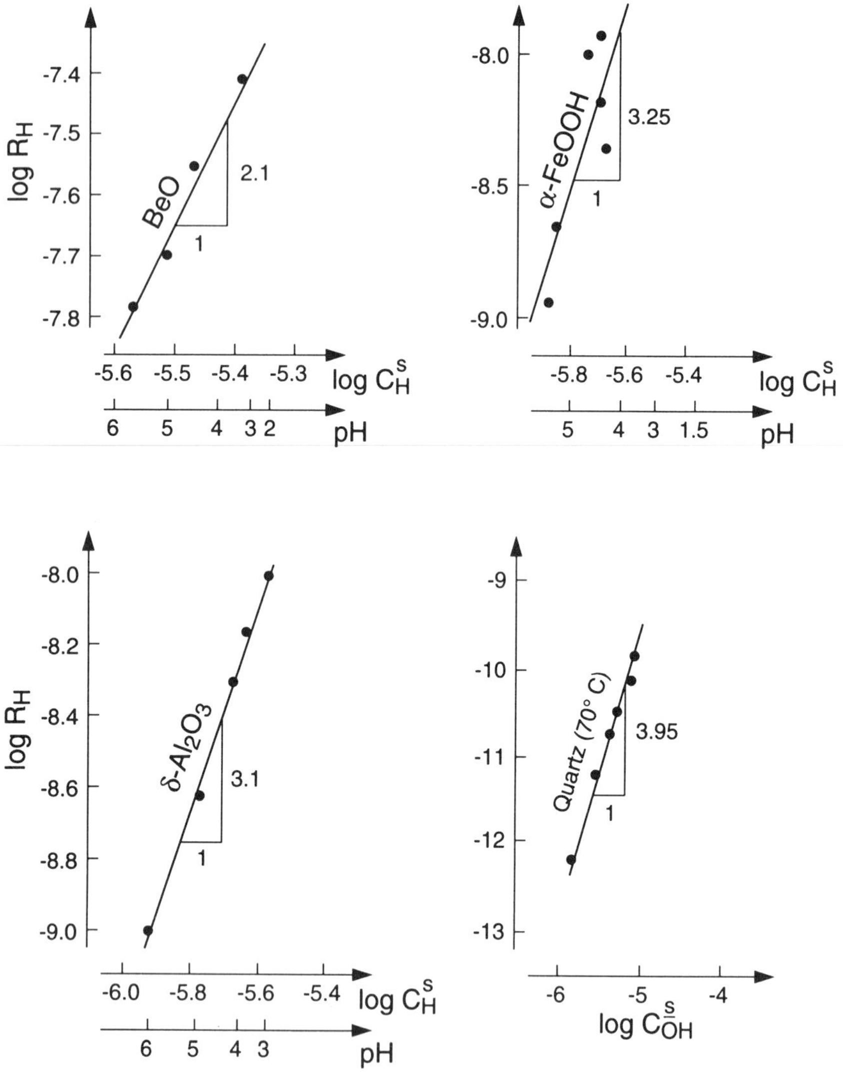

Figure 6.5 Variations in mineral dissolution rates as a function of the concentrations of positively charged sites on the mineral surface (from Stumm and Wollast, 1990). The slope of the lines corresponds to the rate order, *n*, in equation (6.7).

$$\frac{d\log(R_{dis})}{d\,pH} = n(1 - x_{SOH_2^+})\left[\frac{d\log K'}{d\,pH} - 1\right] \tag{6.8}$$

if n and k are independent of pH and $x_{SOH_2^+} + x_{SOH} \approx 1$. Each of the terms in equation (6.8) becomes appreciable at different pH conditions and the net result is a slight nonlinearity with respect to pH.

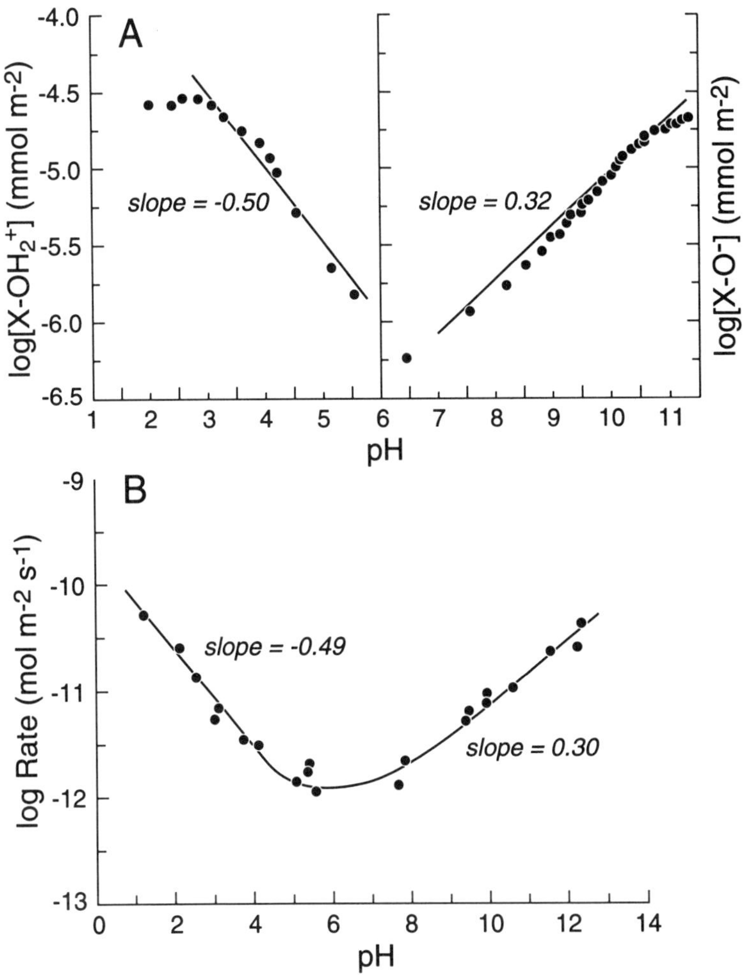

Figure 6.6 (A) Variations in modelled concentrations of positively and negatively charged sites on the surface of albite as a function of solution pH (Blum and Lasaga, 1991). These data are corrected for the amount of sodium released by ion exchange during the titration. (B) Variation in the dissolution rate of albite as a function of solution pH (Chou and Wollast, 1985).

The important points about proton-promoted dissolution are that:

- the rate laws are appropriately expressed in terms of the concentration of surface complexes;
- the surface complexation reactions are generally fast relatively to bond hydrolysis;
- the surface equilibria can be described with conventional thermodynamics;

- equilibrium acid–base reactions at surface functional groups wholly account for the pH-dependence of dissolution rates in simple solutions.

In the next section we show that the mineral surface chemistry also influences the temperature-dependence of corrosion.

6.2.4 Acid–base reactions and the temperature dependence of dissolution

The temperature variation of proton-promoted mineral dissolution rates is generally treated with a version of the Arrhenius relation:

$$R_{\rm dis} = A\ e^{-E_{\rm exp}/RT} \tag{6.9}$$

In equation (6.9), A is a pre-exponential factor (mol cm^{-2} s^{-1}), $E_{\rm exp}$ is the experimental activation energy (kJ mol^{-1}), T is the absolute temperature (K), and R is the molar gas constant (kJ mol^{-1} K^{-1}). Thus calculated, however, $E_{\rm exp}$ values commonly change considerably with pH, as illustrated

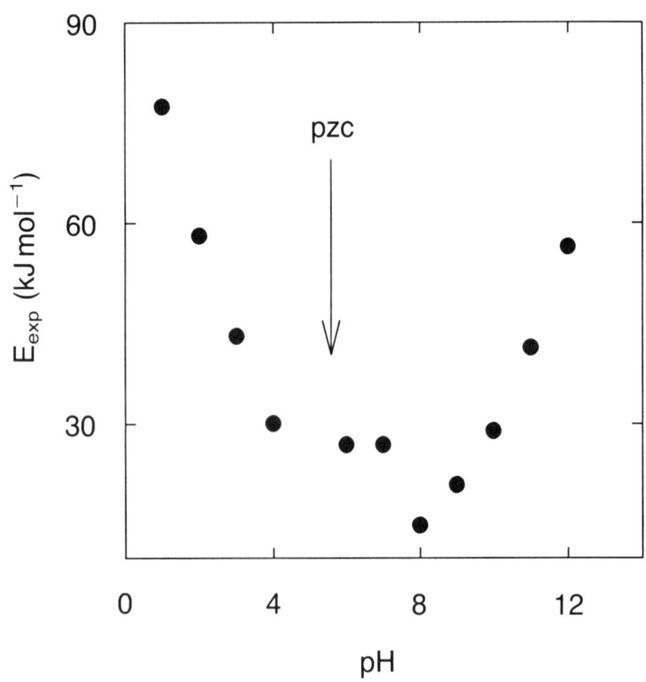

Figure 6.7 The experimental activation energy ($E_{\rm exp}$) for dissolution of kaolinite as a function of pH (Carroll and Walther, 1990). The pzc is approximate.

for kaolinite in chloride solutions in Figure 6.7 (Carroll and Walther, 1990). The E_{exp} value for kaolinite dissolution decreases from $77\,kJ\,mol^{-1}$ at pH 1 to $15.0\,kJ\,mol^{-1}$ at pH 8, then increases to $56.5\,kJ\,mol^{-1}$ at pH 12 (Figure 6.7). The minimum value of E_{exp} occurs at a pH value significantly larger than the estimated point of zero charge (pzc) of kaolinite ($4.5 < pH < 6.5$).

The acid–base equilibrium values vary with temperature and this variation is manifest in the dissolution rates; in other words, the rapid acid–base reactions contribute enthalpies that raise or lower the activation barrier to bond hydrolysis:

$$E_{exp} = \Delta H_{ads} + E_{surf} \tag{6.10}$$

where ΔH_{ads} is the heat of adsorption and E_{surf} is the activation energy of the rate-controlling step (Lasaga, 1981a; Brady and Walther, 1990, 1992). The contribution of enthalpy from acid–base reactions to the experimental activation energy can be separated from other effects:

$$E_{exp} = -R\left[\frac{\partial \ln k'}{\partial(1/T)}\right]_{pH} - nR(1 - x_{SOH_2^+})\left[\frac{\partial \ln K'}{\partial(1/T)}\right]_{pH} \tag{6.11}$$

The first term on the right-hand side of equation (6.11) reflects the variation in the rate coefficient, k', with temperature. Two additional contributions arise from the temperature variation of the conditional equilibrium constant (K' in equation 6.8) at constant pH and ionic strength. The pH variation of K' is usually described by postulating an 'intrinsic' equilibrium coefficient, which refers to a hypothetically chargeless surface with unit mole fraction of charged sites (Sposito, 1983), and a charge-dependent term that accounts for coulombic interactions among charged sites.

The temperature variation of the intrinsic coefficient gives rise to a standard enthalpy of proton adsorption, which relates to K (equation 6.4) via the Van't Hoff equation. Typical enthalpies for metal oxide minerals are estimated through calorimetric measurements (e.g. Mehr *et al.*, 1989) and by measuring the variation in pzc with temperature (Blesa, Maroto and Regazzoni, 1990). For example, the temperature variation in the mineral pzc can be split into standard (intrinsic) enthalpies corresponding to equations (6.2) and (6.3):

$$-\tfrac{1}{2}(\Delta H_{a1}^{int} + \Delta H_{a2}^{int}) = 2.3RT^2 \frac{d(pzc)}{dT} \tag{6.12}$$

Standard enthalpies of proton adsorption are generally negative (an exothermic process) for silicate and oxide minerals and range in absolute magnitude from $15–50\,kJ\,mol^{-1}$ (Blesa, Maroto and Ragazzoni, 1990). This contribution varies in sign depending on whether dissolution rates are affected by positive or negative surface charge. The formation of negatively charged sites, which also enhance the dissolution rates, is endothermic and raises the experimental activation energy.

The charge-dependent factor in equation (6.11) accounts for the dependence of the solid-phase activity coefficient ratio on solution composition. The detailed expression for $f_{SOH}/f_{SOH_2^+}$ depends on the choice of surface speciation model used to describe equation (6.2). Selection of the constant capacitance model (Schindler and Stumm, 1987) is convenient because a consistent set of rate data and surface complexation parameters exist (Wieland, Wehrli and Stumm, 1988). The conditional equilibrium constant in equation (6.8) in this model is

$$K' = [K^{s}_{a1}(\text{int})\, e^{(F\psi_s/RT)}]^{-1} \tag{6.13}$$

where $K^{s}_{a1}(\text{int})$ is an intrinsic equilibrium constant for the dissociation of $\triangleright SOH_2^+$, ψ_s is an electric potential (V) at the mean adsorption plane of the surface species, and F is the Faraday constant (C mol^{-1}). [By convention (Schindler and Stumm, 1987), the equilibrium described by K^{s}_{a1} (int) is the reverse of equation (6.2).] The combination of equations (6.9) and (6.13) yields a model version of equation (6.11)

$$E_{\text{exp}} = -R\left[\frac{\partial \ln k_1'}{\partial(1/T)}\right]_{\text{pH}} - n(1 - x_{SOH_2^-})\Delta H^{s}_{a1}(\text{int}) + n(1 - x_{SOH_2^+})F\psi_s \tag{6.14}$$

where ΔH^{s}_{a1} (int) is a standard proton **desorption** enthalpy (desorption because of a convention in the constant-capacitance model). Note that a contribution arises, owing to electrostatic interactions, that is proportional to ψ_s weighted by the mole fraction of positive charge. This term can potentially contribute several tens of kJ mol^{-1} to E_{exp} (Casey and Sposito, 1992), and the contribution varies non-linearly with pH and the concentration of adsorbed inner-sphere protons. The net result is that E_{exp} values reach a minimum near, but not necessarily at, the pzc (e.g. Figure 6.7).

Because the E_{exp} values vary with pH, so too does the pH-dependence of rate vary with temperature. For many minerals, the variation in E_{exp} with pH causes rates to become increasingly sensitive to pH at elevated temperatures in the range 25–80 °C. Consider the example of kaolinite, where E_{exp} values vary from greater than $60\,\text{kJ mol}^{-1}$ at pH 2.0 to approximately $0\,\text{kJ mol}^{-1}$ at pH 8. A 10 °K increase in temperature would cause rates to increase by a factor of 2.2 if $E_{\text{exp}} = 60\,\text{kJ mol}^{-1}$, by a factor of 1.5 if $E_{\text{exp}} = 30\,\text{kJ mol}^{-1}$ (pH = 3–4), and only by a factor of 1.1 if $E_{\text{exp}} = 10\,\text{kJ mol}^{-1}$ (pH > 5) (Casey, Hochella and Westrich, 1992; Brady and Walther, 1992). The actual variation in E_{exp} with pH and temperature, of course, depends on whether proton adsorption is endothermic or exothermic. One should also remember that there are enthalpy contributions from other sources besides proton adsorption (e.g. leaching) that can reverse these trends predicted from the acid–base equilibria.

6.2.5 *Metal removal from surfaces*

From the similarity in oxygen coordination chemistry, one expects a strong correlation between mineral dissolution rates at well-defined conditions and rates of water exchange about the corresponding dissolved cation. Well-defined conditions mean:

- minerals are compared where divalent metal–oxygen bonds are important to the reaction rate;
- experiments are conducted in similar solutions;
- conditions whereby the valence state of metals at the mineral surface would be altered are excluded.

Such a correlation is shown in Figure 6.8 for the simple orthosilicate minerals at pH 2 and 25 °C. Note that the general reactivity trends for

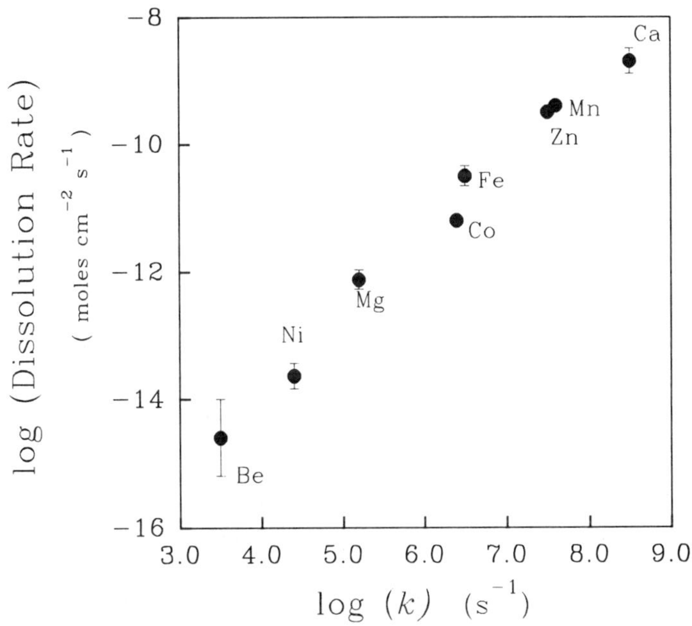

Figure 6.8 Dissolution rates of end-member orthosilicate minerals at pH 2 and 25 °C plotted against the first-order rate coefficient, $k(s^{-1})$, for water exchange from the solvent into the hydration sphere of the corresponding dissolved cation. The mineral and ions are identified by the divalent cation (e.g. Mn^{2+} refers to the dissolution rate of tephroite (Mn_2SiO_4) and rate of solvent exchange around $Mn(H_2O)_6^{2+}$). (From Casey and Westrich, 1992.)

solvent-exchange kinetics (Figure 6.1) are also observed for mineral dissolution. The dissolution rates of silicates containing alkaline earths vary with ionic size of the cation. Calcium can be more quickly removed from a silicate site than magnesium; beryllium is very resistant. For minerals containing first-row transition metals, rates vary with the number of valence d-electrons (Casey and Westrich, 1992). The dissolution rate of liebenbergite (Ni_2SiO_4) is much, much slower than tephroite (Mn_2SiO_4), though the metal–oxygen bond lengths differ only slightly (Table 6.1).

In summary, ligand-exchange kinetics provide a useful, but imperfect, guide to mineral dissolution rates. The guide is useful if the cation in the mineral has a similar valence state and metal–oxygen coordination chemistry as in the dissolved coordination complex, if short range interactions between the divalent metal and oxygens control the reaction, and if the mechanisms of dissolution resemble mechanisms of solvent exchange. Given these caveats, if variables such as ion size, bond hybridization and ligand-field energies influence rates of solvent exchange, they will also influence the rates at which metals are removed from the surface of a mineral. It is critically important to distinguish, however, between the processes by which a single metal is removed from a mineral surface and the overall process of mineral dissolution.

6.2.6 Leaching reactions

The strong correlation between overall dissolution rates and the strengths of bonds between divalent metals and oxygens will ultimately (but not initially) largely disappear for minerals with polymerized silicate structures. Under certain conditions, the dissolution of chain-, sheet- or tecto-silicate minerals can involve selective removal of reactive sites at the mineral surface, leaving a silica-like surface layer. The existence and thickness of this leached layer depends on the solution composition and the mineral structure, but can extend to many thousands of Ångströms. An example of such a leached layer is shown in Figure 6.9. This crystal of labradorite has an ~400 Å leached layer after reaction in a strong acid (0.01 M HCl) for several hundreds of hours.

Murata (1943) identified the controls on these leaching reactions by studying the dissolution of minerals in strong acid. First, there must be Lewis acid sites at the mineral surface with distinctly different reactivities. Second, there must also be sufficient siloxane bridges to maintain structural integrity in the reacted surface once the particularly reactive constituents have been removed. One would not expect a leached layer to form on anorthite ($CaAl_2Si_2O_8$) (without polymerization of silanol groups) because there is no continuously linked silicate fabric once the aluminate (AlO_4^{5-}) groups are removed. From subsequent work, we know also that the reactivity at the different sites changes with solution composition as protons associate with

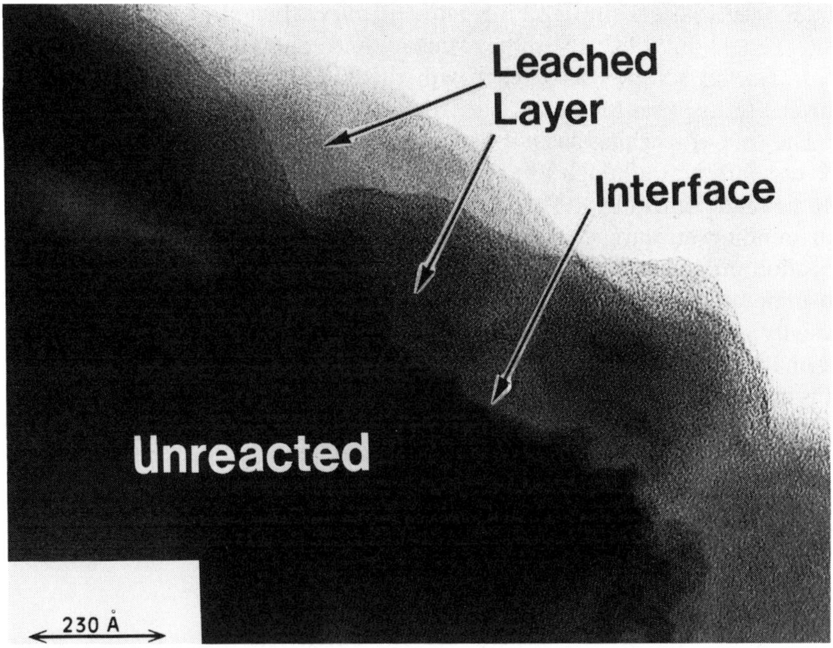

Figure 6.9 High resolution transmission electron micrograph of a plagioclase grain after reaction at pH 2 for 218 h (Casey, Carr and Graham, 1989). The leached layer (~400 Å thick) near the grain consists of porous silica, which is light coloured in the figure. This grain was suspended on to a holey carbon grid for examination and was not ion-etched. (Photo by Dr Jillian Banfield.)

different Lewis acid sites (e.g. Bales and Morgan, 1985; Carroll-Webb and Walther, 1988; Guy and Schott, 1989). Strong acid treatments leach aluminium from aluminosilicate minerals (e.g. Hellmann *et al.*, 1990), while magnesium at the surface of magnesian silicate minerals is leached away at near-neutral and basic pH values (Bales and Morgan, 1985).

Beyond the work of Murata (1943), there has been slow progress in systematizing the controls on leaching. The difficulties have largely been analytical. The properties are usually inferred from surface spectra and rarely directly imaged. Therefore, precipitates, heterogeneities in the surface (Inskeep *et al.*, 1991) or an unfortunate distribution of adsorbed substances can be misinterpreted as being produced by leaching. In addition, there is no satisfying way to rank the reactivities of different sites on the surface of a complicated mixed-oxide mineral. In spite of these difficulties, there have been some clear successes. First, where it is imaged, particularly in glasses, the leached layer is so porous that unreactive solutes probably diffuse through it nearly as rapidly as in bulk solution. It is also clear that leaching is not uniform and that the leached layer interacts with solutes which dramatically affect the leaching rates of minerals and glasses (Buckwalter and Pederson, 1982; Nesbitt, Macrae and Shotyk, 1991). Metals that hydrolyse in solution also react with the silica-leached layer at a mineral surface (e.g. Della Mea *et al.*, 1992).

The role of leaching has not been adequately incorporated into any model of dissolution, although the conceptual framework is well known (e.g. Hume and Rimstidt, 1992). In these models there exists a porous shell surrounding an unreacted core which decreases in volume and area as the reaction proceeds. These 'shrinking-core' models are not yet widely applied to minerals because important parameters in the models, such as the diffusivity of ions in the leached layer and the rates of bond hydrolysis, vary dramatically with time and solution composition, and are virtually unknown. It is also clear that the residual leached surface restructures itself, subsequent to the removal of leachable cations (Casey, Westrich and Arnold, 1988; Casey *et al.*, 1989a).

6.3 Rates near equilibrium

6.3.1 *The law of detailed balance*

In the previous section we examined the kinetics of metal removal from a mineral surface in highly undersaturated solutions. Intuitively, one expects rates near equilibrium to be slower than under these severe conditions. Stated formally, this intuition gives us the law of detailed balance, which arises from the assumption that the activated equilibrium is microscopically reversible. This theory is truly appropriate for reactions that proceed as written at the molecular scale (Lasaga, 1981b); that is, for elementary reactions. Rate laws for such reactions relate directly to the reaction stoichiometry. For a reaction such as

$$A + B = C \tag{6.15}$$

the rate laws are:

$$\overrightarrow{r} = \overrightarrow{k}(A)(B) \qquad \text{forward rate} \tag{6.16a}$$

$$\overleftarrow{r} = \overleftarrow{k}(C) \qquad \text{backward rate} \tag{6.16b}$$

$$r = \overrightarrow{k}(A)(B) - \overleftarrow{k}(C) \qquad \text{net rate} \tag{6.16c}$$

At equilibrium, the net rate is zero and the ratio of rate constants equals the equilibrium constant:

$$\frac{\overrightarrow{k}}{\overleftarrow{k}} = K_{\mathrm{eq}} \tag{6.17}$$

$$r = \vec{k}\,(A)\,(B)\left[1 - \frac{(C)}{(A)\,(B)}\frac{1}{K_{eq}}\right] \tag{6.18}$$

$$r = \vec{k}(A)(B)\left[1 - \frac{Q}{K_{eq}}\right] \tag{6.19}$$

$$r = \vec{k}(A)(B)[1 - e^{(\Delta G_r/RT)}] \tag{6.20}$$

where Q is the activity product for the reaction, ( ) indicate concentrations, and ΔG_r is the reaction free energy. Note in equation (6.20) that the direction of the reaction is determined by the sign of the free energy term and that, conceivably, the rate law for an elementary reaction can be completely constrained from knowledge of the equilibrium constant and the rate law for the reverse reaction. Note also that at $(\Delta G_r/RT) \ll 0$, the back reaction is unimportant to the overall rate. The rate law can be further expanded to include detail, but such expansion is neglected here (see Glasstone, Laidler and Eyring, 1941; Aagaard and Helgeson, 1982).

These rate laws can be extended to overall reactions if the rate-controlling steps are treatable with transition-state theory and if reaction rates along the various paths are constant relative to one another. Essentially, this means a constant disequilibrium state. The reaction rates and stoichiometries of rate-controlling elementary step reaction(s), however, may not equal the overall reaction. To account for possible differences, an average stoichiometric factor m, is added to the steady-state rate law:

$$r = \vec{k}\prod_{i}^{N} a_i^n \left[1 - e^{(m\Delta G_r/RT)}\right] \tag{6.21}$$

where $\Pi_i^N\, a_i^n$ is the activity product of reactants in the overall equation weighted by a rate order, n, and N is the total number of reactants (Boudart and Djega-Mariadassou, 1984; Chapter 3). If the rate is expressed in units of bulk concentration, the rate law must include geometrical factors such as the exposed mineral area and the solution volume. In reality, the assumption of steady state needed to derive equation (6.21) is violated in most experiments, so the stoichiometric factor is reduced to an empirical fitting parameter.

Dissolution rates have long been thought to be relatively insensitive to the concentration of the released solutes at conditions far from equilibrium, though this generalization is justifiably criticized (e.g. Amrhein and Suarez, 1992). Growth rates are very sensitive to solution composition; therefore, rate laws such as equation (6.20) must predict different dependencies of growth and dissolution rates on free energy at conditions where $|\Delta G_r| \gg m/RT$. The variation in rates with driving force for the reaction is shown in Figure 6.10. Rates of dissolution reach a constant value (the dissolution

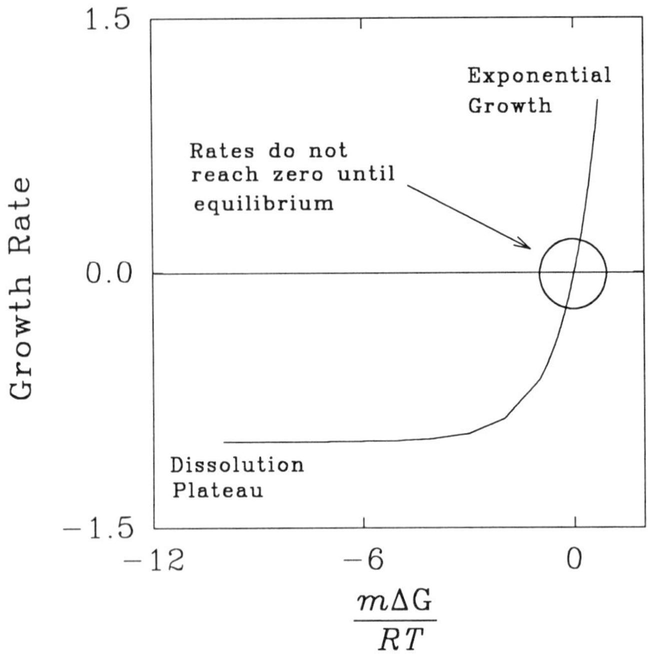

Figure 6.10 The variation in rate with driving force for equation (6.21). The rates are appreciable at all non-zero ΔG_r values. Note that the dissolution rates become constant as the solution becomes increasingly undersaturated, while the growth rates increase exponentially with increased supersaturations.

plateau; Nagy and Lasaga, 1992) at large undersaturations, while rates of growth increase exponentially until solute transport limits the overall rate. The rate laws are appealing because of their versatility. They can be simply evaluated at conditions far from equilibrium and applied to a wide range of conditions (e.g. Lichtner, 1985; Knapp, 1989; Steefel and Lasaga, 1990). Note, for example, that the concentration of surface complexes need not be incorporated into rate laws such as equation (6.21) as long as the speciation steps are much more rapid than dissolution.

6.3.2 Surface microtopography and rate

Measured rates near equilibrium are much more complicated than is implied from equation (6.21). In Figure 6.11, for example, the dissolution rates of gibbsite (Nagy and Lasaga, 1992) and albite (Burch, Nagy and Lasaga, 1993) are shown as a function of ΔG_r. Consider first the results for albite (Figure 6.11(A)). At great undersaturation ($\Delta G_r \approx -7.0\,\text{kcal mol}^{-1}$), the dissolution rate is constant; rates change rapidly, however, in the range $-7 < \Delta G_r < -5\,\text{kcal mol}^{-1}$ and reach a plateau near equilibrium. The reaction rates are

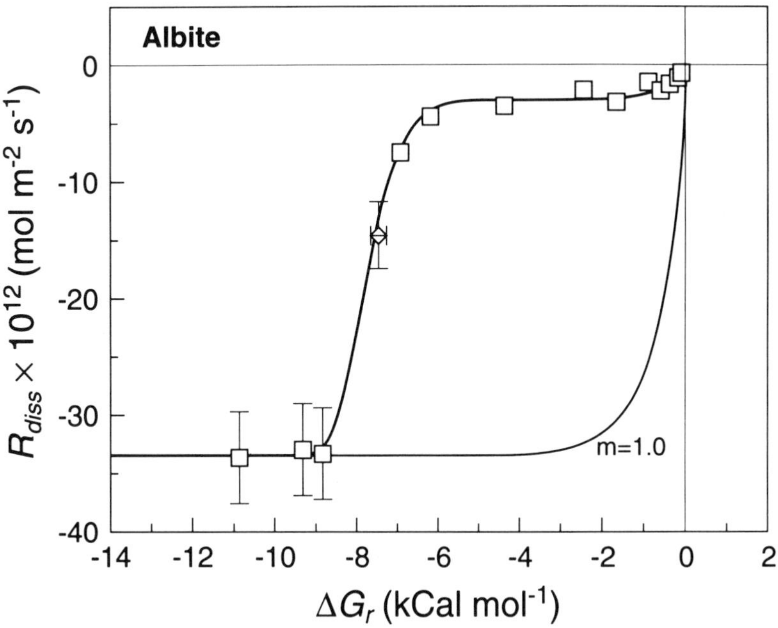

(A)

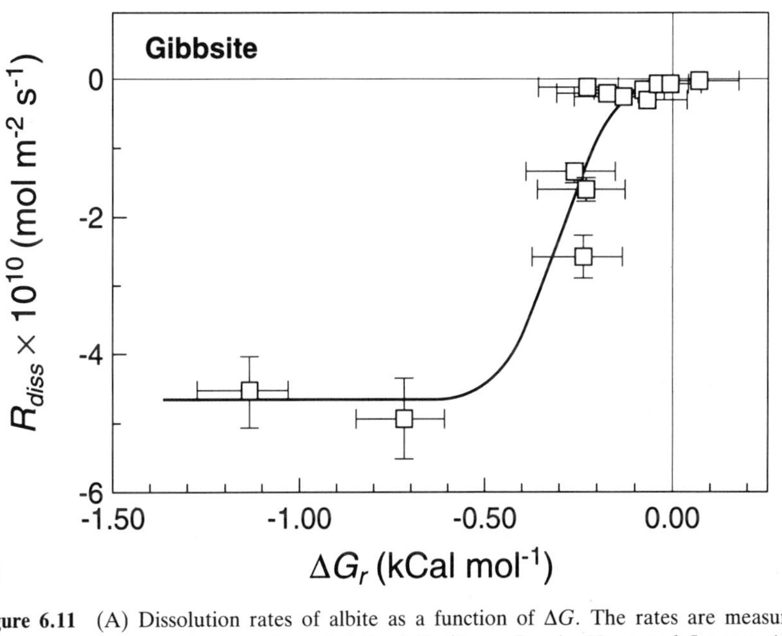

(B)

Figure 6.11 (A) Dissolution rates of albite as a function of ΔG. The rates are measurable but near zero over a large region of $\Delta G_r < 0$. (From Burch, Nagy and Lasaga, 1993.) (B) Dissolution rates of gibbsite as a function of ΔG. Note that the rates reach zero at an appreciable $\Delta G_r < 0$. The equilibrium position ($\Delta G_r = 0$) was measured independently. (From Nagy and Lasaga, 1992.)

virtually constant with undersaturation in the range $-5 < \Delta G_r < 0\,\mathrm{kcal\,mol^{-1}}$ (Figure 6.11(A)). Similar results are observable for gibbsite, though the plateau near equilibrium is less pronounced (Figure 6.11(B)). At undersaturations greater than $-0.50\,\mathrm{kcal\,mol^{-1}}$ from equilibrium ($\Delta G_r = 0$), gibbsite dissolution rates are constant (Figure 6.11(A)). At $-0.50 < \Delta G_r <$ $0.25\,\mathrm{kcal\,mol^{-1}}$, the rates change very rapidly as a function of ΔG_r and again reach a plateau at $-0.15 < \Delta G_r < 0\,\mathrm{kcal\,mol^{-1}}$.

The important point is that the transition in rate for both minerals reflects the energy required to open a dislocation core and form a macroscopic etch pit (cf. Nagy and Lasaga, 1992); this critical energy is different for the two minerals. This second plateau in rates at $-0.15 < \Delta G_r < 0\,\mathrm{kcal\,mol^{-1}}$ is not predicted from a rate law such as equation (6.21), where rates are predicted to vary linearly with ΔG_r near equilibrium with a slope proportional to: $[m/RT]$.

Growth and dissolution rates are non-linear near equilibrium because energy is required to disrupt an atomically smooth surface by precipitating an adatom cluster or a hole. Growth rates are larger on a crystal face with an abundance of edge or kink sites than on an atomically smooth surface. In the absence of a constantly replenished source of adatom clusters or holes, the rates of growth and dissolution are microscopically episodic; rates are controlled by nucleation. The simplest model of growth and dissolution is the **birth and spread** of monolayers (Zhang and Nancollas, 1990). Holes or adatom clusters nucleate on the flat mineral surface and provide a source of edge sites for further reaction. Reaction proceeds rapidly at the edge sites until several expanding features merge. The driving force for the reaction must be larger than some critical value. At this critical value, the energy required to increase the surface area is exactly offset by bulk energy to drive the reaction. (The bulk energy may include strain energy if dislocation outcrops are present.) Rates at energies smaller than this critical value are negligible, unlike rates predicted from equation (6.21), which are appreciable all the way up to $\Delta G_r = 0$ (Figure 6.10).

The results of a Monte Carlo simulation of **birth and spread** growth are shown in Figure 6.12 (from Gilmer, 1977), where the rates at a dislocation-free surface are plotted as a function of a characteristic energy variable (L/kT). At $L/kT < 4.5$, there is no barrier to surface rearrangement and nucleation of two-dimensional clusters on the surface; the rates are appreciable at very small driving forces for the reaction. As the energy is increased, however, the amount of energy required to nucleate sites on the surface of the crystal increases correspondingly. Large supersaturations (for growth) or undersaturations (for dissolution) are necessary before appreciable reaction is measurable. If dissolution rates are limited by the concentration of nucleated holes in the surface, the rate will vary exponentially with ΔG_r once a critical free energy barrier is surmounted.

Growth or dissolution in the birth and spread model is possible only if

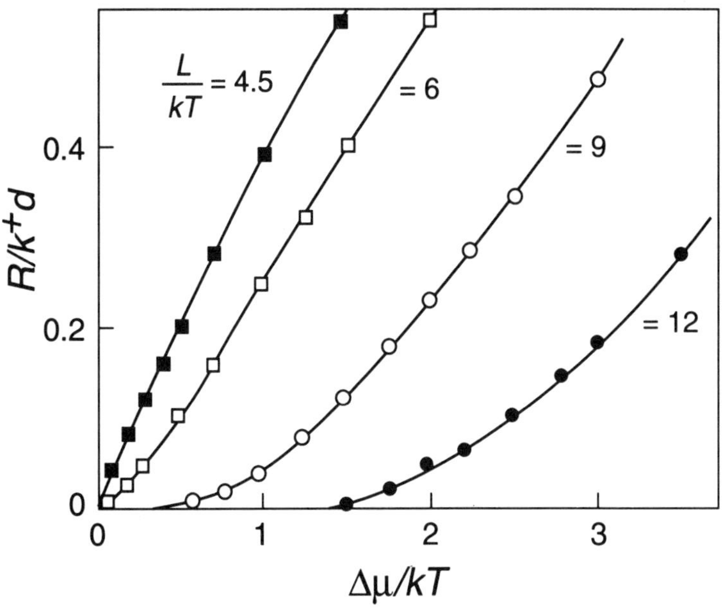

Figure 6.12 Monte Carlo simulations of growth rates for a face on a dislocation-free crystal as a function of $\Delta G_r/RT$ (Gilmer, 1976). The parameter L corresponds to a measure of the crystal binding energy, T is temperature (K), and k is Boltzmann's constant. At $L/kT = 4.9$, there is no activation barrier to formation of two-dimensional crystal nuclei. Note that the rates reach zero at appreciable values of μ/RT for cases where there exists an activation barrier to nucleation of a two-dimensional adatom cluster. (Adapted from Gilmer, 1976.)

nucleation is constant. In experiments, minerals grow at faster rates near equilibrium than are predicted from the 'birth and spread' model (but at slower rates than predicted by equation (6.21)). The high rates were explained by Frank (1951), who demonstrated that the outcrop of spiral dislocations on the surface of a mineral provides an infinite source of nucleation sites. Edge sites are not extinguished in spiral-dislocation growth (e.g. Gilmer, 1976, 1977). As adatoms adsorb, the arms of the spiral rotate and the crystal grows normal to its surface. Growth at spiral dislocations is contrasted with the birth and spread model of dissolution in Figure 6.13(B). The normal abundance of dislocations at crystal surfaces is sufficiently large (10^5–$10^9\,cm^{-2}$) that rates of growth near equilibrium almost always involve dislocations.

Rate laws that emphasize the central role of dislocations in crystal growth are rooted in the paper by Burton, Cabrera and Frank (1951) and are referred to as BCF theory. This theory has been extended to dissolution (Gilmer, 1977; Blum and Lasaga, 1987; Lasaga and Blum, 1986) and has been used to interpret natural and experimental textures (Brantley *et al.*, 1986; Casey, Carr and Graham, 1988). Also, this theory is central to the

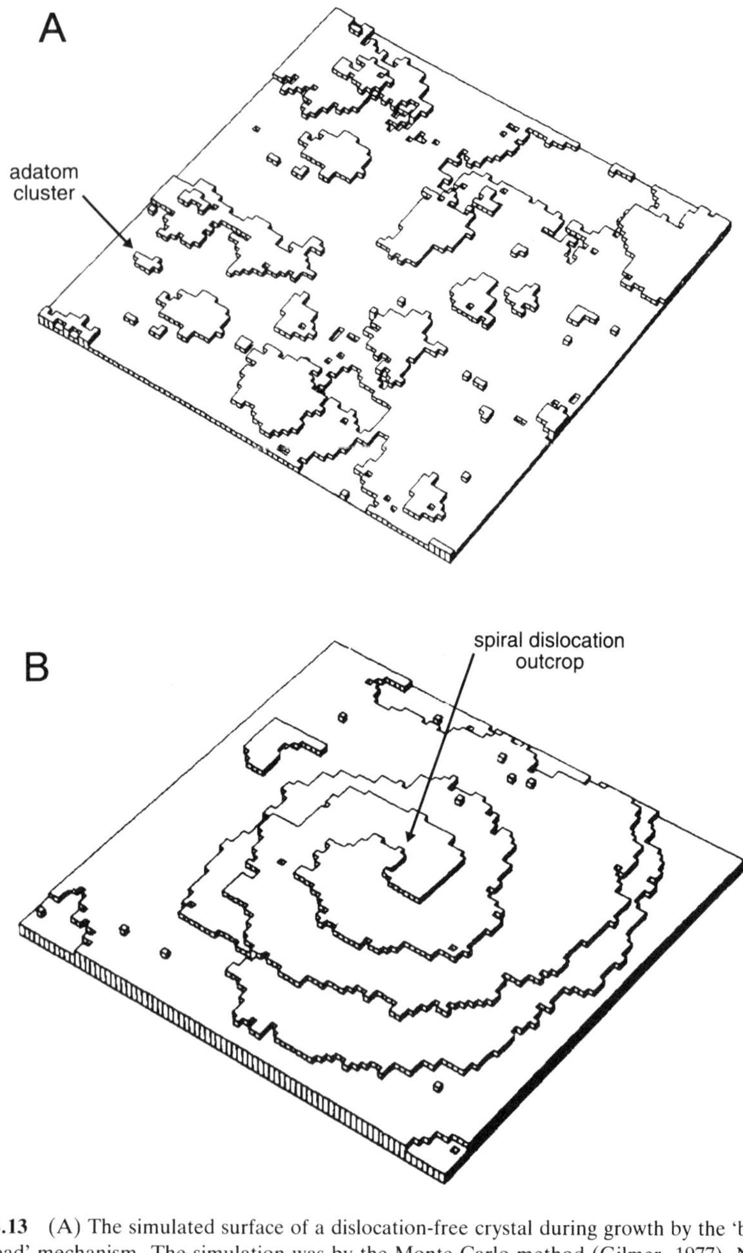

Figure 6.13 (A) The simulated surface of a dislocation-free crystal during growth by the 'birth and spread' mechanism. The simulation was by the Monte Carlo method (Gilmer, 1977). Note that edge sites are abundant on this surface and that several adatom clusters are growing together. (B) Monte Carlo simulation of screw-dislocation growth (Gilmer, 1977; Blum and Lasaga, 1987, and references therein). The particular example has a pair of screw dislocations outcropping on the surface, forming two spiral arms. The dislocations provide a constant source of edge sites for growth. (Adapted from Gilmer, 1977.)

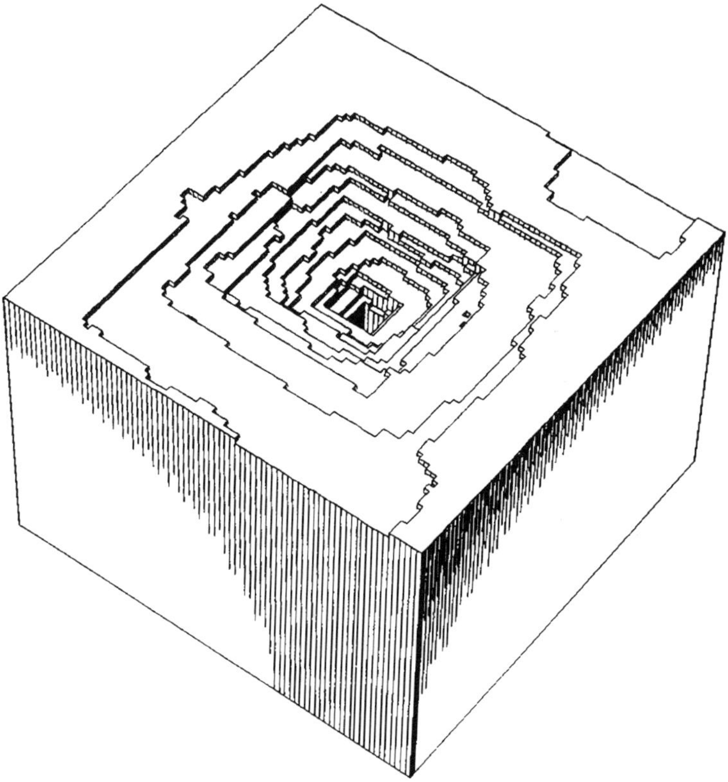

Figure 6.14 The results of Monte Carlo simulation of the excavation of a dislocation outcropping at the mineral surface. Note that this excavated dislocation provides a source of energetically favoured edge sites for dissolution (Lasaga and Blum, 1986; Blum and Lasaga, 1987). (Adapted from Gilmer, 1977.)

interpretation of the rapid transition in rates as a function of ΔG_r (Figure 6.11) (Nagy and Lasaga, 1992; Burch, Nagy and Lasaga, 1993).

The theories identify three terms in an energy balance describing the effect of lattice defects on dissolution rates. The first term addresses the energy released as strained bonds dissolve in the core of the dislocation. The second term accounts for dissolution of edge sites exposed at the excavated dislocation core. A third energy contribution is the positive energy required to increase the area of the crystal during excavation of the dislocation. One can imagine the case where the surface energy required to excavate a dislocation is greater than the energy released by lateral growth of the hole. In this case the strained bonds may be removed, but no macroscopic etch pit (e.g. Figure 6.14) forms. Dislocation densities are unimportant for dissolution in highly undersaturated solutions but are potentially very important near equilibrium.

To accommodate the varied role of surface microtopography on the rates, semi-empirical rate laws of the form

$$r = \vec{k} \prod_{i}^{N} a_i^n [1 - e^{\Delta G_r/RT}]^m \tag{6.22}$$

are being used with increasing frequency to describe rates of growth and dissolution over the entire range of ΔG conditions. These empirical rate laws have been applied to calcite (Lasaga, 1981a, pp. 54–6; Morse, 1983), gibbsite (Nagy and Lasaga, 1992), kaolinite (Nagy, Blum and Lasaga, 1991) and albite (Burch, Nagy and Lasaga, 1993). Unlike the rate law derived from detailed balance, this rate law becomes proportional to: $[|\Delta G_r|/RT]^m$ near equilibrium and the coefficient, m, in equation (6.22) now has no significance in terms of reaction stoichiometry of a single elementary reaction; it is wholly an empirical parameter applied to an overall reaction. The rate must reach zero at equilibrium, but need not be symmetric around $\Delta G_r = 0$. The asymmetry allows for different reaction pathways for growth and dissolution and for a varying influence of surface topography and lattice defects on the reaction kinetics. Rate laws such as equation (6.22) are an improvement over those based upon detailed balance but are too inflexible to incorporate changes in surface speciation. Conceivably, new experiments would have to be conducted to determine the parameter, m, in equation (6.22), as each new set of conditions modifies the surface of the mineral. Such modification might include poisoning of dissolution by adsorption of slowly reacting complexes (such as $Be(H_2O)_4^{2+}$, Figure 6.1).

6.4 Natural weathering

Our success in understanding dissolution and growth in an experimental setting has only been partly transferred to natural conditions. There are some clear successes. Naturally weathered minerals, for example, are commonly selectively etched in a similar fashion to experimentally weathered materials (Berner and Holdren, 1979; Brantley *et al.*, 1986). These etch pits are virtually identical to those produced by simulation (Figure 6.14) and have been shown to control the dependence of the reaction rate on ΔG_r (Blum and Lasaga, 1987; Nagy and Lasaga, 1992). In addition, some simple sheet silicate and hydroxide minerals can be grown and dissolved reversibly (e.g. Nagy and Lasaga, 1992) in ways that are convincingly similar to nature.

There are also some failures. High resolution transmission electron microscopy (HRTEM) has shown that some natural minerals weather in intimate contact with reaction product minerals (e.g. Eggleton, 1975; Banfield, Veblen and Jones, 1990; Banfield, Jones and Veblen, 1991a, b), unlike in well-controlled experiments where a small amount of mineral is dispersed into a large volume of solution. These secondary minerals strongly influence the rate and character of natural weathering (e.g. Eggleton, 1975). The HRTEM images of naturally weathered olivine, for example, show that the reaction proceeds in channels that are so small (~30 Å diameter), and so filled with reaction products (see Banfield, Veblen and Jones, 1990; Banfield, Jones and Veblen, 1991a), that the bulk solution chemistry is a poor guide to the reaction. The hydration volumes (~20–30 $Å^3$) of an ion can take up most of the available pore volume. The development of topotactic secondary minerals, such as smectite and hematite (Banfield, Jones and Veblen, 1991a, Figure 4) or smectite and goethite (Eggleton, 1984), further indicates that growth of the alteration products is influenced by the structure and orientation of olivine and the properties of highly structured water in the pores and grain boundaries.

There is also tantalizing evidence that bacteria influence rates of weathering either directly (Arnold, DiChristina and Hoffmann, 1988; Arnold *et al.*, 1990; Francis and Dodge, 1991) or indirectly by releasing reactive ligands (e.g. Hiebert and Bennett, 1992). Dicarboxylic acids are particularly effective in enhancing rates of dissolution (e.g. Amrhein and Suarez, 1988) because they form a ring complex with metals on the mineral surface. Not only bacteria, but also mycorrizea and plants exude rate-enhancing dicarboxylic acids (Table 6.3).

One of the first steps in evaluating weathering in nature is to constrain the importance of the solute transport process on the overall reaction kinetics. At the simplest level, there are three sequential steps in removing a metal from a dissolving mineral surface. First, reactive solutes (e.g. protons, fluoride, oxalate) migrate through the aqueous solution to the mineral surface. Secondly, the solutes combine with metals and oxygens on the mineral surface in a reactive complex which decomposes into products during the reaction. Thirdly, the dissolved products move away from the mineral surface. The slowest of these three sequential processes controls the rate of the overall reaction because rates of surface reaction are linked to rates of solute transport via the concentration dependence of the rate law. High

Table 6.3 Concentrations ($\mu mol\,l^{-1}$) of dissolved organic acids in eluviated soil horizons of forested and cultivated soils (Hue, Craddock and Adams, 1986). The short dash (–) means the acid was not detectable

Soil Series	pH	Oxalic	Citric	Malic	Malonic	Succinic	Lactic	Formic	Phthalic
Dothan	5.04	22	–	137	–	–	–	90	15
Dothan	5.06	3	12	101	74	125	12	–	–
Dothan	5.05	4	5	117	20	282	52	–	–
Dothan	5.00	0.75	–	–	–	–	–	–	0.05
Lucedale	4.85	3.3	–	3.0	18.0	–	–	8.4	0.15
Malbis	5.40	2.0	–	6.2	4.2	–	–	–	–
Benndale	4.90	3.4	–	3.0	17.5	10.0	–	8.0	0.12
Benndale	4.36	16.5	–	–	–	–	74.5	–	–

concentrations near the mineral surface cause the back reaction to become important via the exponential term in equation (6.21).

Under experimental conditions, most rock-forming silicate minerals are sufficiently unreactive that transport of solutes towards, or away from, the dissolving surface does not control the rate. The rate of the overall reaction is controlled by reactions at the interfaces, i.e. the second sequential step. Under natural conditions, distances for solute transport can be quite large, particularly in subsurface rocks, and the overall rate of dissolution may be controlled by rates of solute transport (Murphy, Oelkers and Lichtner, 1989). Nevertheless, solute transport in most soils is sufficiently rapid that weathering of most silicates is controlled by the kinetics and not solute transport (Berner, 1978).

For natural systems, it is useful to employ dimensionless parameters to describe the relative importance of transport and surface kinetics. The Damkohler number is defined as:

$$\frac{\bar{R}X}{D} = Da \tag{6.23}$$

where X is a characteristic length (cm) for transport (e.g. the thickness of a boundary layer), D is the solute diffusion coefficient ($cm^2\,s^{-1}$) and $\bar{R}$ is a coefficient related to the optimal rate of interface retreat or advance ($cm\,s^{-1}$). When $Da > 1$, the rate of interface retreat is potentially larger than the fluxes of reactive solutes to, and from, the reacting surface. For $Da < 1$, the rates of solute transport are larger than the rates of reaction at the surface; for this case, the overall reaction kinetics are controlled by reaction at the mineral surface. Calculation of actual parameters in the dimensionless group depend on the geometry of transport. For examples, see Knapp (1989), Murphy, Oelkers and Lichtner (1989) and Steefel and Lasaga (1990).

6.5 Conclusions

The dissolution of oxide and silicate minerals is best understood by drawing an analogy between processes by which a metal is removed from a dissolving mineral surface and the processes by which a dissolved metal complex exchanges ligands. This similarity is the key to predicting reactivities and to unravelling the reaction mechanisms. The actual reactive species at the mineral surface have a similar coordination chemistry to common dissolved complexes. Using this framework, we can predict the coordination chemistry to common dissolved complexes. Using this framework, we can predict the temperature dependence of the reaction and the dependence of the reaction on solution composition.

Comparision of laboratory and field rates is most appropriate at the scale of reactivity trends. Not only does the mineral phenacite (Be_2SiO_4) dissolve slowly and irreversibly, for example, but the Be^{2+} cation is difficult to remove from the surface of dissolving beryl ($Be_3Al_2Si_6O_{18}$), even if the overall reaction is controlled by destruction of the six-member silicate rings in the structure. In not all cases, however, is natural mineral dissolution controlled by surface reactions under far-from-equilibrium conditions. Length scales for solute transport can be sufficiently long that growth and dissolution proceeds under near-equilibrium conditions for minerals that can be reversibly grown and dissolved. Even for minerals that dissolve irreversibly, reaction products can accumulate in intimate contact with the primary mineral. There is considerable evidence from HRTEM that this accumulation causes natural weathering reactions to differ considerably from the experimental results.

Acknowledgements

Support from the National Science Foundation via grant EAR-9105318 and a research contract from the Nuclear Regulatory Commission (NRC-04-92-056) are acknowledged. This chapter was greatly improved through the effort of the editors, Prof. David Vaughan and Dr Richard Pattrick, and by helpful suggestions from Dr Kathy Nagy. The author also thanks Drs K. Nagy and T. Burch for permission to use their data on growth and dissolution near equilibrium.

References

Aagaard, P. and Helgeson, H.C. (1982) Thermodynamic and kinetic constraints on reaction rates among minerals and aqueous solutions: I. Theoretical considerations. *Amer. J. Sci.*, **282**, 237–85.

Amrhein, C. and Suarez, D.L. (1988) The use of a surface complexation model to describe the kinetics of ligand-promoted dissolution of anorthite. *Geochim. Cosmochim. Acta*, **52**, 2785–93.

Amrhein, C. and Suarez, D.L. (1992) Some factors affecting the dissolution kinetics of anorthite at 25 °C. *Geochim. Cosmochim. Acta*, **56**, 1815–26.

Arnold, R.G., DiChristina, T.J. and Hoffmann, M.R. (1988) Reductive dissolution of Fe(III) oxides by Psuedomonas sp. 200. *Biotech. & Bioeng.*, **32**, 1081–96.

Arnold, R.G., Hoffmann, M.R., DiChristina, T.J. and Picardal, F.W. (1990) Regulation of dissimilatory Fe(III) reduction activity in Shewanella putrefaciens. *Appl. Environ. Microbiol.*, **56**, 2811–17.

Astumian, R.D., Sasaki, M., Yasunaga, T. and Shelly, Z.A. (1981) Proton adsorption–desorption kinetics on iron oxides in aqueous suspensions, using the pressure-jump method. *J. Phys. Chem.*, **85**, 3832–5.

Bales, R.C. and Morgan J.J. (1985) Dissolution kinetics of chrysotile at pH 7 to 10. *Geochim. Cosmochim. Acta*, **49**, 2281–8.

Banfield, J.F., Jones, B.F. and Veblen, D.R. (1991a) An AEM-TEM study of weathering and diagenesis, Abert Lake, Oregon I: weathering reactions in the volcanics. *Geochim. Cosmochim. Acta*, **55**, 2781–93.

Banfield, J.F., Jones, B.F. and Veblen, D.R. (1991b) An AEM-TEM study of weathering and diagenesis, Abert Lake, Oregon II: diagenetic modification of the sedimentary assemblage. *Geochim. Cosmochim. Acta*, **55**, 2795–810.

Banfield, J.F., Veblen, D.R. and Jones, B.F. (1990). Transmission electron microscopy of subsolidus oxidation and weathering of olivine. *Contrib. Mineral. Petrol.*, **106**, 110–23.

Berner, R.A. (1978) Rate control of mineral dissolution under Earth surface conditions. *Amer. J. Sci.*, **278**, 1235–52.

Berner, R.A. and Holdren, G.R. (1979) Mechanisms of feldspar weathering – II. Observations of feldspars from soils. *Geochim. Cosmochim. Acta*, **43**, 1173–86.

Bish, D. and Burnham, C.W. (1984) Structure energy calculations on optimum distance model structures: application to the silicate olivines. *Amer. Mineral.*, **69**, 1102–9.

Blesa, M.A., Maroto, A.J.G. and Regazzoni, A.E. (1990) Surface acidity of metal-oxides immersed in water: a critical analysis of thermodynamic data. *J. Colloid Interface Sci.*, **40**, 287–90.

Blum, A.E. and Lasaga, A.C. (1987) Monte Carlo simulations of surface reaction rate laws, in *Aquatic Surface Chemistry* (ed. W. Stumm), John Wiley, New York, pp. 255–92.

Blum, A. and Lasaga, A.C. (1988) Role of surface speciation in the low-temperature dissolution of minerals. *Nature*, **331**, 431–3.

Blum, A. and Lasaga, A.C. (1990) The effect of dislocation density on the dissolution rate of quartz. *Geochim. Cosmochim. Acta*, **54**, 283–97.

Blum, A. and Lasaga, A.C. (1991) Role of surface speciation in the dissolution of albite. *Geochim. Cosmochim. Acta*, **55**, 2193–2201.

Boehm, H.P. and Knozinger, H. (1983) Nature and estimation of functional groups on solid surfaces, in *Catalysis Science and Technology* (eds J.R. Anderson and M. Boudart), Springer-Verlag, New York, pp. 39–207.

Boudart, M. and Djega-Mariadassou, G. (1984) *Kinetics of Heterogeneous Catalytic Reactions*. Princeton University Press, Princeton, NJ.

Brady P.V. and Walther, J.V. (1990) Kinetics of quartz dissolution at low temperatures. *Chem. Geol.*, **82**, 253–64.

Brady, P.V. and Walther, J.V. (1992) Surface chemistry and silicate dissolution at elevated temperatures. *Amer. J. Sci.*, **292**, 639–58.

Brantley, S.L., Crane, S.R., Crerar, D.A., Hellmann R. and Stallard R. (1986) Dissolution at dislocation etch pits in quartz. *Geochim. Cosmochim. Acta*, **50**, 2349–61.

Buckwalter, C.Q. and Pederson, L.R. (1982) Inhibition of nuclear waste glass leaching by chemisorption. *J. Amer. Ceram. Soc.*, **65**, 431–97.

Burch, T.E., Nagy, K.L. and Lasaga, A.C. (1993) Free energy dependence of albite dissolution kinetics at 80 °C, pH 8.8. *Chem. Geol.* (in press).

Burgess, J. (1988) *Ions in Solution: Basic Principles of Chemical Interactions*. Ellis Horwood, Chichester.

Burton, W.K., Cabrera, N. and Frank, F.C. (1951) The growth of crystals and the equilibrium structure of their surfaces. *Phil. Trans. Royal Soc. Lond.*, **A243**, 299–358.

Carroll-Webb, S.A. and Walther, J.V. (1988) A surface complex model for the pH-dependence of corundum and kaolinite dissolution rates. *Geochim. Cosmochim. Acta*, **52**, 2609–23.

Carroll, S.A. and Walther, J.V. (1990) Kaolinite dissolution at 25 °C, 60 °C and 80°C. *Amer. J. Sci.*, **290**, 797–810.

Casey, W.H. (1991) On the relative dissolution rates of some oxide and orthosilicate minerals. *J. Coll. Interf. Sci.*, **146**, 586–9.

Casey, W.H. and Sposito, G. (1992) On the temperature dependence of mineral dissolution rates. *Geochim. Cosmochim. Acta*, **56**, 3825–30.

Casey, W.H. and Westrich, H.R. (1992) Control of dissolution rates of orthosilicate minerals by metal-oxygen bonds. *Nature*, **355**, 157–9.

Casey, W.H., Carr, M.J. and Graham, R.A. (1988) Crystal defects and the dissolution kinetics of rutile. *Geochim. Cosmochim. Acta*, **52**, 1545–56.
Casey, W.H., Hochella, M.F. Jr and Westrich, H.R. (1992) The surface chemistry of manganiferous silicate minerals as inferred from experiments on tephroite (Mn_2SiO_4). *Geochim. Cosmochim. Acta*, **57**, 785–93.
Casey, W.H., Westrich, H.R. and Arnold, G.W. (1988) The surface chemistry of labradorite feldspar reacted with aqueous solutions at pH = 2, 3 and 12. *Geochim. Cosmochim. Acta*, **52**, 2795–807.
Casey, W.H., Westrich, H.R., Arnold, G.W. and Banfield, J.F. (1989a) The surface chemistry of dissolving labradorite feldspar. *Geochim. Cosmochim. Acta*, **53**, 821–32.
Casey, W.H., Westrich, H.R., Massis T., Banfield J.F. and Arnold G.W. (1989b) The surface of labradorite feldspar after acid hydrolysis. *Chem. Geol.*, **78**, 205–18.
Chou, L. and Wollast, R. (1985) Steady state kinetics and dissolution mechanisms of albite. *Amer. J. Sci.*, **285**, 963–93.

Day, V.W., Klemperer, W.G. and Malthie, D.J. (1987) Where are the protons in $H_3V_{10}O_{28}^{3-}$? *J. Am. Chem. Soc.*, **109**, 2991–3002.
Della Mea, G., Dran, J.-C., Moulin, V. *et al.* (1992) Migration of colloids in the geosphere and mechanisms of retention on mineral surfaces. *Nucl. Instrum. Methods*, **B64**, 523–7.

Eggleton, R.A. (1975) Nontronite topotaxial after hedenbergite. *Amer. Mineral.*, **60**, 1063–8.
Eggleton, R.A. (1984) Formation of iddingsite rims on olivine: a transmission electron microscope study. *Clays Clay Min.*, **32**, 1–11.

Frank, F.C. (1951) Capillary equilibria of dislocated crystals. *Acta Cryst.*, **4**, 497–500.
Francis, A.J. and Dodge, C.J. (1991) Dissolution of ferrites by Clostridium sp. *Geomicrobiol. Jour.*, **9**, 27–40.
Furrer, G. and Stumm, W. (1986) The coordination chemistry of weathering: I. Dissolution kinetics of δ-Al_2O_3 and BeO. *Geochim. Cosmochim. Acta*, **50**, 1847–60.

Gilmer, G.H. (1976) Growth on imperfect crystal faces. *J. Cryst. Growth*, **35**, 15–28.
Gilmer, G.H. (1977) Computer simulation of crystal growth. *J. Cryst. Growth*, **42**, 3–10.
Glasstone, S., Laidler, K.J. and Eyring, H. (1941) *The Theory of Rate Processes*. McGraw-Hill, New York, 610 pp.
Guy, C. and Schott, J. (1989) Multisite surface reaction versus transport control during the hydrolysis of a complex oxide. *Chem. Geol.*, **78**, 181–204.

Hellmann, R., Eggleston, C.M., Hochella, M.F. and Crerar, D.A. (1990) The formation of leached layers on albite surfaces during dissolution under hydrothermal conditions. *Geochim. Cosmochim. Acta*, **54**, 1267–81.
Hering, J.G. and Morel, F.M.M. (1990) The kinetics of trace metal complexation: implications for metal reactivity in natural waters, in *Aquatic Chemical Kinetics* (ed. W. Stumm), Wiley-Interscience, New York, pp. 145–71.
Hiebert, F.K. and Bennett, P.C. (1992) Microbial control of silicate weathering in organic-rich groundwater. *Science*, **258**, 278–81.
Hue N.V., Craddock, G.R. and Adams, F. (1986) Effect of organic acids on aluminum toxicity in subsoils. *Soil. Sci. Soc. Am. J.*, **50**, 28–34.
Hume, L.A. and Rimstidt, J.D. (1992) The biodurability of chrysotile asbestos. *Amer. Mineral.*, **77**, 1125–8.

Inskeep, W.P., Nater, E.A., Bloom, P.R., Vandervoort, D.S. and Erich M.S. (1991) Characterization of laboratory weathered labradorite surfaces using X-ray photoelectron spectroscopy and transmission electron microscopy. *Geochim. Cosmochim. Acta*, **55**, 787–800.

Kanno, H. (1988) Hydrations of metal ions in aqueous electrolyte solutions: a Raman study. *J. Phys. Chem.*, **92**, 4232–6.
Knapp. R.B. (1989) Spatial and temporal scales of local equilibrium in dynamic fluid–rock systems. *Geochim. Cosmochim. Acta*, **53**, 1955–64.

Lasaga, A.C. (1981a) Rate laws of chemical reactions, in *Kinetics of Geochemical Processes* (eds A.C. Lasaga and R.J. Kirkpatrick), *Reviews in Mineralogy*, **8**, Mineralogical Society of America, Washington, DC, pp. 1–68.
Lasaga, A.C. (1981b) Transition state theory, in *Kinetics of Geochemical Processes* (eds A.C. Lasaga and R.J. Kirkpatrick), *Reviews in Mineralogy*, **8**, Mineralogical Society of America, Washington, DC, pp. 135–69.
Lasaga, A.C. and Blum, A.E. (1986) Surface chemistry, etch pits and mineral–water reactions. *Geochim. Cosmochim. Acta*, **50**, 2363–79.
Lichtner, P.C. (1985) Continuum model for simultaneous chemical reactions and mass transport in hydrothermal systems. *Geochim. Cosmochim. Acta*, **49**, 4009–25.

Mehr, S.R., Eatough, D.J., Hansen, L.D. and Lewis, E.A. (1989) Calorimetry of heterogeneous systems: H^+ binding to TiO_2 in NaCl. *Thermochim. Acta*, **154**, 129–43.
Merbach, A.E. and Akitt, J.W. (1990) High-resolution variable pressure NMR for chemical kinetics. *NMR Basic Principles and Progress*, **24**, 190–232.
Morse, J.W. (1983) The kinetics of calcium carbonate dissolution and precipitation, in *Carbonates: Mineralogy and Chemistry* (ed. R.J. Reeder), *Reviews in Mineralogy*, **11**, Mineralogical Society of America, Washington, DC, pp. 227–64.
Murata, K.J. (1943) Internal structure of silicate minerals that gelatinize with acid. *Amer. Mineral.*, **28**, 545–62.
Murphy, W.M., Oelkers, E.H. and Lichtner, P.C. (1989) Surface reaction versus diffusion control of mineral dissolution and growth rates in geochemical processes. *Chem. Geol.*, **78**, 357–80.

Nagy, K.L. and Lasaga, A.C. (1992) Dissolution and precipitation kinetics of gibbsite at 80 °C and pH 3: the dependence on solution saturation state. *Geochim. Cosmochim. Acta*, **56**, 3093–112.
Nagy, K.L., Blum, A.E. and Lasaga, A.C. (1991) Dissolution and precipitation kinetics of kaolinite at 80 °C and pH 3: the dependence on solution saturation state. *Amer. J. Sci.*, **291**, 649–86.
Nesbitt, H.W., Macrae, N.D. and Shotyk, W. (1991) Congruent and incongruent dissolution of labradorite in dilute, acidic, salt solutions. *J. Geol.*, **99**, 429–42.

Richens, D.T., Helm, L., Pittet, P.-A. *et al.* (1989) Crystal structure of and mechanism of water exchange on $[Mo_3O_4(OH_2)_9]^{4+}$ from X-ray and oxygen-17 NMR studies. *Inorg. Chem.*, **28**, 1394–402.

Schindler, P.W. and Stumm, W. (1987) The surface chemistry of oxides, hydroxides and oxide minerals, in *Aquatic Surface Chemistry* (ed. W. Stumm), John Wiley, New York, pp. 83–107.
Sposito, G. (1983) On the surface complexation model of the oxide–aqueous solution interface. *J. Coll. Interf. Sci.*, **91**, 329–40.
Sposito, G. (1984) *The Surface Chemistry of Soils*, Oxford University Press, Oxford.
Steefel, C.I. and Lasaga, A.C. (1990) Evolution of dissolution patterns, in *Chemical Modeling in Aqueous Systems II* (eds D.C. Melchior and R.L. Bassett), ACS Symposium Series, 416, American Chemical Society, Washington, DC, pp. 212–25.
Stumm, W. and Wollast, R. (1990) Coordination chemistry of weathering: kinetics of surface-controlled dissolution of oxide minerals. *Rev. Geophys.*, **28**, 53–69.

Wieland, E., Wehrli, B. and Stumm, W. (1988) The coordination chemistry of weathering-III: a generalization on the dissolution rates of minerals. *Geochim. Cosmochim. Acta*, **52**, 1969–81.

Wilkens, R.G. (1974) *The Study of Kinetics and Mechanisms of Reactions of Transition Metal Complexes*, Allyn and Bacon, Boston.

Zhang, J.-W. and Nancollas, G.H. (1990) Mechanisms of growth and dissolution of sparingly soluble salts, in *Mineral–Water Interface Geochemistry* (eds M.F. Hochella and A.R. White), *Reviews in Mineralogy*, **23**, Mineralogical Society of America, Washington, DC, pp. 365–93.

CHAPTER SEVEN

Electrochemistry and surface chemistry of sulfide minerals

G.H. Kelsall

7.1 Introduction

The chemistry of mineral sulfide–aqueous solution interfaces is determined both by the interfacial solution pH and the electrode potential, since a range of oxidation states are possible for sulfur and metals. Hence the electrochemical behaviour of sulfide minerals determines their response, *inter alia*, in:

- froth flotation (Woods, 1981), the principal industrial process for concentrating economic minerals from their ores. This involves the exploitation of differences in surface chemistry between the component minerals to effect their separation, usually by selectively imparting surfactant-induced hydrophobicity, followed by flotation of hydrophobic particles with air bubbles, and is often a precursor to some subsequent extractive metallurgical operation;
- hydrometallurgical processes, involving chemical reactions at mineral–aqueous solution (and immiscible organic liquid–aqueous solution) interfaces at ambient temperatures and above, to separate and recover metals;
- electro-hydrometallurgical processes, in which electrical energy is used to dissolve sulfide minerals and/or recovering/reducing metal ions from aqueous solution by electrolysis;
- geological weathering processes, e.g. oxidation of sulfide minerals in oxygenated ground water.

Sulfide minerals are either metallic, or n- or p-type semiconductors, so that after fabrication into electrodes with suitably low resistance ohmic contacts, their chemical behaviour is amenable to investigation by electrochemical techniques. It is their electronic conductivity, which enables them to act as a source/sink for electrons in coupled electrochemical processes, that is responsible for the selectivity against dielectric mineral particles in sulfide froth flotation processes, without which the rapid worldwide expansion in mineral exploitation would not have occurred.

The electrode potential determines both which oxidation and reduction reactions are thermodynamically possible and their kinetics. Faraday's law

Mineral Surfaces. Edited by D.J. Vaughan and R.A.D. Pattrick. Published in 1995 by Chapman & Hall, London. ISBN 0 412 56340 1

relates the amount of (electro-)chemical change (N moles of substance formed) to the electrical charge (Q) passed in an electrochemical reaction involving some oxidized (O) and reduced states (R):

$$O + ze^- \leftrightarrow R \tag{7.1}$$

since the electronic charge is quantized ($e = 1.60219 \times 10^{-19}$ coulombs); hence:

$$N = \frac{Q}{zeL_{\mathrm{Av}}} = \frac{Q}{zF} \tag{7.2}$$

where L_{Av} = Avogadro's number and F is the Faraday constant. As the current (I) is related to the charge (Q) passed by:

$$Q = \int_0^t I\,\mathrm{d}t \qquad \text{or} \qquad I = \frac{\mathrm{d}Q}{\mathrm{d}t} \tag{7.3}$$

the current is also a direct measure of the rate of reaction (R):

$$R = \frac{\mathrm{d}N}{\mathrm{d}t} = \frac{1}{z\mathrm{F}}\frac{\mathrm{d}Q}{\mathrm{d}t} = \frac{I}{zeL_{\mathrm{Av}}} = \frac{I}{zF} \tag{7.4}$$

As with other types of heterogeneous reaction systems, it is usual to normalize these parameters with respect to the (geometrical or real) area (A_e) of the interfacial reaction, so that the current density (i) and charge density (q) are the derived quantities of more fundamental value in investigations of the kinetics, mechanisms and rate-determining steps of electrochemical reactions. From a knowledge of the charge density passed in a particular reaction, the number of monolayers of surface reacted can be calculated, with a measurement precision of better than a tenth of a monolayer.

Electrochemical techniques can also be used to define the kinetics and mechanisms of spontaneous so-called **mixed potential** processes, in which two or more coupled redox reactions occur at equal but opposite rates at the same interface, one component being oxidized and the other reduced. Examples include surfactant and/or mineral oxidation coupled to oxygen reduction in the froth flotation of sulfide mineral particles and the oxidative leaching of sulfide minerals with dissolved oxygen or ferric ions.

The methodology often adopted in investigations of the electrochemical behaviour of sulfide minerals is to use thermodynamic calculations to predict the electrode potential at which the onset of a current response would be expected, enabling assignment of particular reactions to that current response in steady-state current-potential or non-steady-state voltammetric experiments. Hence, brief descriptions of electrochemical thermodynamics and kinetics are given below. The oxidation behaviour of chalcocite (Cu_2S) at pH 9.2 by linear potential sweep voltammetry and the reduction of pyrite (FeS_2) in acidic electrolytes at a ring-disc electrode are used to exemplify some aspects of the electrochemical behaviour of sulfide minerals and two frequently used techniques for their study. The former process involves surface transformations, resulting in adsorbed copper oxide and sulfur products, the properties of which inhibit and control the rate of further oxidation, while the latter involves reductive dissolution of Fe^{2+} ions and H_2S, the rate of which is limited ultimately by the transport of protons from the bulk electrolyte to the electrode surface.

7.2 Electrochemical thermodynamics

A pair of species that are related by one (R) being the reduced (i.e. lower oxidation state) form of the other (O) are known as a redox couple, designated (O/R):

$$O + ze^- \leftrightarrow R \tag{7.1}$$

As the potential difference across a single electrode/electrolyte interface cannot be measured absolutely, the standard hydrogen electrode (SHE) was adopted as an arbitrary zero (at all temperatures) against which the potential of all other redox couples can be referred (Pourbaix 1966; Bard, Parsons and Jordan, 1985). The SHE is a platinized platinum electrode bathed in H_2 gas at 0.1 MPa pressure and immersed into a solution of protons of unity activity. Thus, redox systems have positive potentials if, in principle, they can oxidize H_2 to H^+ ions and negative potentials if they reduce H^+ ions to H_2; adoption of a zero potential for the SHE has the effect of defining the Gibbs energy of formation of protons as zero, as for all elements.

$$H_2 \leftrightarrow 2H^+ + 2e^- \tag{7.5}$$

for which, as explained below, the electrode potential (E) is:

$$E_{H_2/H^+}\,(\mathrm{V}) = 0 - 0.0591\,pH - 0.0296\log p_{H_2} \tag{7.6}$$

The electrode potential of the redox system in reaction (7.1) is defined as the potential difference of the electrochemical cell incorporating a SHE:

$$\mathrm{Pt\ (platinized),\ H_2/H^+ \| O,\ R/Pt} \tag{7.7}$$

where $\|$ refers to a salt bridge of KCl electrolyte, to minimize the liquid junction (diffusion) potential separating the two electrolytes, and / refers to a phase boundary. As Pt may be poisoned by some redox couples and dissolved hydrogen may reduce others, alternative electrodes are normally used as experimental reference electrodes. The one most commonly used in aqueous systems is the saturated calomel electrode (SCE), which is Hg / Hg_2Cl_2/ KCl (sat.), the potential of which is 0.242 V vs. SHE at 298 K when the internal electrolyte is saturated with KCl.

Increasingly positive potentials correspond to increasingly oxidizing conditions, under which most metals or metal sulfides will oxidize either to solution species or (hydr-)oxide phases, the latter giving rise to passivation (inhibition of dissolution, owing to blocking of the surfaces by an adsorbed product(s)). Conversely, decreasing potentials correspond to increasingly reducing conditions, under which most metals will be thermodynamically stable ('immune' to corrosion), but which may cause the reductive decomposition of sulfide minerals.

From the general reaction isotherm

$$\Delta G\ (\mathrm{J\,mol^{-1}}) = \Delta G^0 + RT \ln K \tag{7.8}$$

and the equation

$$\Delta G\ (\mathrm{J\,mol^{-1}}) = -zFE \tag{7.9}$$

for the general reaction

$$pA_q + nH^+ + ze^- \leftrightarrow qB_p + cH_2O \tag{7.10}$$

coupled to an SHE, the Nernst equation (Pourbaix, 1966; Bard, Parsons and Jordan, 1985) can be derived as

$$E_{\mathrm{B/A}}\ (\mathrm{V}) = E^0_{\mathrm{B/A}} + \frac{RT}{zF} \ln\left\{\frac{(A_q)^p\,(H^+)^n}{(B_p)^q\,(H_2O)^c}\right\} \tag{7.11}$$

where the subscripts p and q in reaction (7.10) define the number of atoms associated with each species. Parentheses ( ) denote activities, whereas brackets [ ] indicate concentrations. At 298.15 K, the quotient

$$\frac{2.303RT}{F} = 0.0591\ \mathrm{V} \tag{7.12}$$

Hence,

$$E_{\mathrm{B/A}}\ (\mathrm{V}) = E^0_{\mathrm{B/A}} - \frac{0.0591n}{z} pH + \frac{0.0591}{z} \log\left\{\frac{(A_q)^p}{(B_p)^q}\right\} \tag{7.13}$$

The molar concentration c_A and c_B of the species A_q and B_p are related to their molar elemental concentrations ($c_{T,A}$ and $c_{T,B}$) by:

$$c_{T,A} = qc_A \qquad \text{and} \qquad c_{T,B} = pc_B \tag{7.14}$$

Assuming concentrations can be used in place of activities (i.e. activity coefficients (γ_i) are unity) and taking the activity of water as unity for dilute solutions, the Nernst equation is given by:

$$E_{B/A}\ (V) = E^0_{B/A} - \frac{0.0591n}{z} pH + \frac{0.0591}{z} \log\{c_T^{(p-q)} q^{-p} p^q\} \tag{7.15}$$

Similarly, when reaction (7.10) is chemical ($z = 0$) rather than electrochemical:

$$\log K = \log\{c_T^{(p-q)} q^{-p} p^q\} + npH \tag{7.16}$$

For reactions such as

$$M^{z+} + ze^- \leftrightarrow M \tag{7.17}$$

according to equation (7.15), the Nernst equation is

$$E_{M^{z+}/M}\ (V) = E^0_{M^{z+}/M} + \frac{0.0591}{z} \log(M^{z+}) \tag{7.18}$$

since the activity of a pure solid phase, the metal in this case, is unity by definition. Equation (7.18) predicts that the electrode potential of the metal/metal ion couple decreases by $59/z$ mV for each ten-fold decrease in activity of metal ions.

The Gibbs energy change ΔG^0 for a particular reaction can be calculated from the standard Gibbs energies of formation ΔG_f^0 (when available from data books (Bard, Parsons and Jordan, 1985; Wagman *et al.*, 1982)) of the various species involved, with appropriate stoichiometric factors. Hence, an equilibrium constant for a particular reaction can be calculated from equation (7.8), at equilibrium $\Delta G = 0$, so that for a chemical reaction at 298 K,

$$\log K = -\frac{\Delta G^0\ (\text{kJ mol}^{-1})}{5.70797} \tag{7.19}$$

and for an electrochemical reaction, from equation (7.9),

$$E^0\ (V) = -\frac{\Delta G^0\ (\text{kJ mol}^{-1})}{96.48456} \tag{7.20}$$

7.2.1 Potential–pH diagrams

The electrochemical behaviour of (multi-) element–water systems is most conveniently summarized graphically in the form of potential–pH diagrams (Pourbaix, 1966; Biernat and Robins, 1969; El-Raghy and El-Demerdash, 1989; Yokokawa, *et al.*, 1990; Wadsley, 1992; Kelsall and Thompson, 1993a), examples of which are given below. Such diagrams show predominant species and phases only, for a fixed temperature and total dissolved activity; hence, solubility/speciation diagrams, also exemplified below, provide complementary information, predicting the pH/position of the vertical lines in the potential–pH diagrams. Generally, the area of predominance of solid phases increases with increased total dissolved activity.

All the thermodynamic equations reported here have been calculated from the most recently published, critically assessed thermodynamic data for 298 K (Bard, Parsons and Jordan, 1985).

In addition to the hydrogen reaction (7.5), the oxygen–water reactions are common to all potential-pH diagrams for aqueous systems, and indeed the Nernst equations (7.6) and (7.22) define the boundaries of stability of water itself; above the potential given by equation (7.22) oxygen is predicted to be evolved at suitable electrodes, while below that potential, dissolved oxygen, if present, is predicted to be reduced by the reaction

$$O_2\ (\mathrm{g}) + 4H^+ + 4e^- \leftrightarrow 2H_2O \tag{7.21}$$

for which the Nernst equation is

$$E_{\mathrm{O_2/H_2O}}\ (\mathrm{V}) = 1.2291 - 0.0591\,pH + 0.0148\log p_{O_2} \tag{7.22}$$

The aqueous solubility of oxygen

$$O_2\ (\mathrm{g}) \leftrightarrow O_2\ (aq) \tag{7.23}$$

can be calculated from thermodynamic data (Bard, Parsons and Jordan, 1985) as

$$\log(O_2\ (aq)) = -2.85 + \log p_{O_2} \tag{7.24}$$

As the O_2/H_2O potential–pH line lies above the stability field of all metal sulfides, their spontaneous surface oxidation occurs in the presence of atmospheric or dissolved oxygen, producing, e.g. a sulfur or oxide surface phase, depending on the pH. Such oxidation may occur during fabrication and polishing of sulfide minerals prior to experimental investigations, for instance, involving electrochemical techniques in aqueous solutions. Some oxidation reactions, such as the formation of sulfate, are irreversible, so that even electrochemical reduction of air-formed oxidation products may not be able to regenerate the metal sulfide. If possible, fracture of the sulfide mineral under potential control in a deoxygenated electrolyte is the best means of ensuring a virgin surface prior to electrochemical experiments.

S–H_2O SYSTEMS

The potential-pH diagram of the S–H_2O system (Kelsall and Thompson, 1993a) is shown in Figure 7.1, which indicates that only the oxidation states −2, 0 and +6 are stable; the potentials of $H_2S_2O_8/SO_4^{2-}$, $S_2O_8^{2-}/HSO_4^-$ and $S_2O_8^{2-}/SO_4^{2-}$ couples lie well above those given by equation (7.22), so they are predicted to oxidize water to oxygen. Figure 7.1 also predicts, for example, that SO_4^{2-} ions can be reduced to H_2S/HS^-, whereas many metal electrowinning processes involving sulfuric acid electrolytes are operated at potentials well below those predicted to be required to cause such a reaction. Similarly, the oxidation of HS^- ions is predicted to produce S(VI) species, whereas many lower oxidation species have been reported to form (Kuhn, Charna and Kelsall, 1983).

The metastable behaviour of sulfur–water systems arises from large activation energy barriers in such reactions. Allowance for this can be made by increasing the Gibbs energies of formation of those species that are slow to form; this amounts to making them less and less thermodynamically stable and so ultimately effectively excludes them from consideration in the calculations. If S(VI) species are so treated, the resulting potential-pH diagram

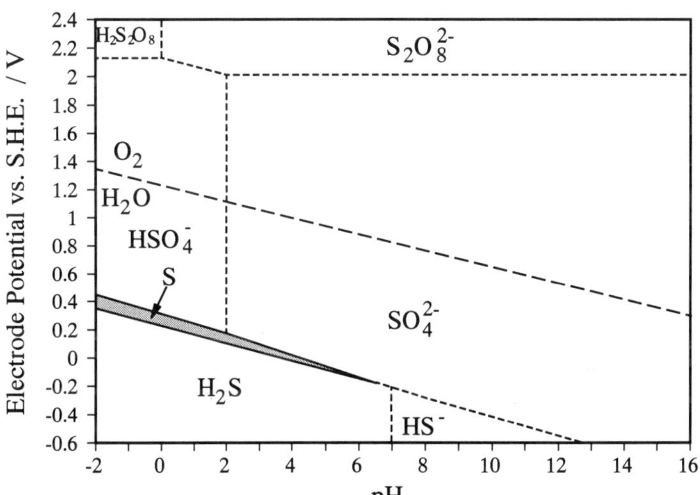

Figure 7.1 Potential-pH diagram for the S–H_2O system at 298 K; total dissolved sulfur activity $= 10^{-2}$, $p_{H_2S} = 0.1$ MPa.

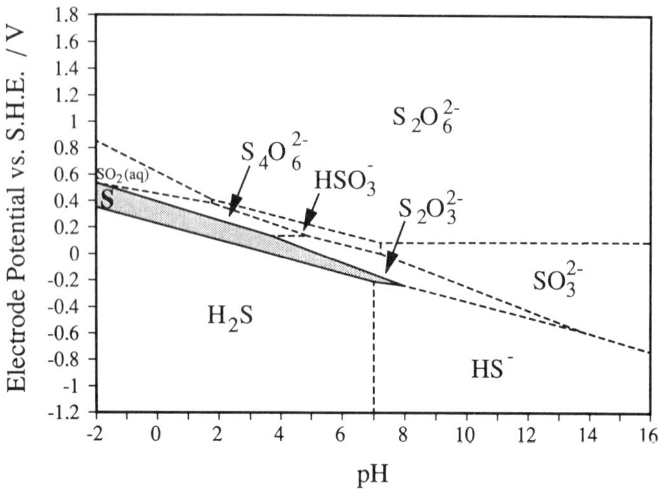

Figure 7.2 Potential-pH diagram for a metastable S–H_2O system at 298 K; total dissolved sulfur activity = 10^{-2}, p_{H_2S} = 0.1 MPa. (Species with oxidation states > 5 excluded from the calculations).

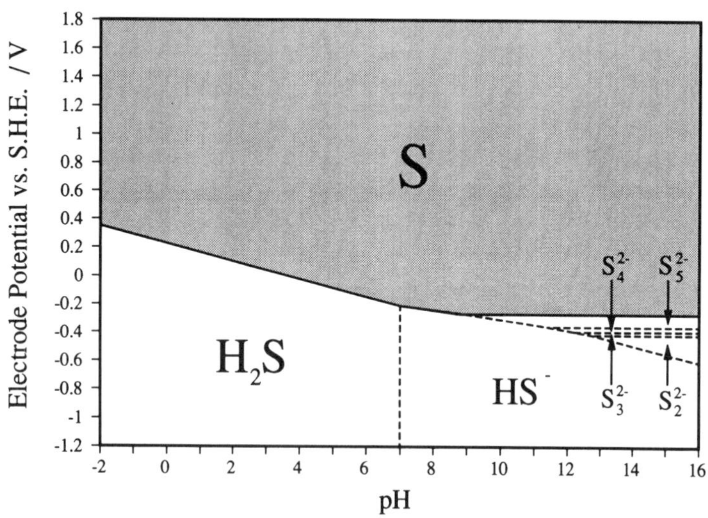

Figure 7.3 Potential-pH diagram for a metastable S–H_2O system at 298 K; total dissolved sulfur activity = 10^{-2}, p_{H_2S} = 0.1 MPa. (Species with oxidation states > 0 excluded from the calculations).

(Figure 7.2) reveals areas of metastability of a range of sulfoxy species, with oxidation states ⩽ 5.

If all sulfoxy species are excluded from the calculations, then polysulfide ions (S_n^{2-}), which have been shown to be intermediates in the oxidation of

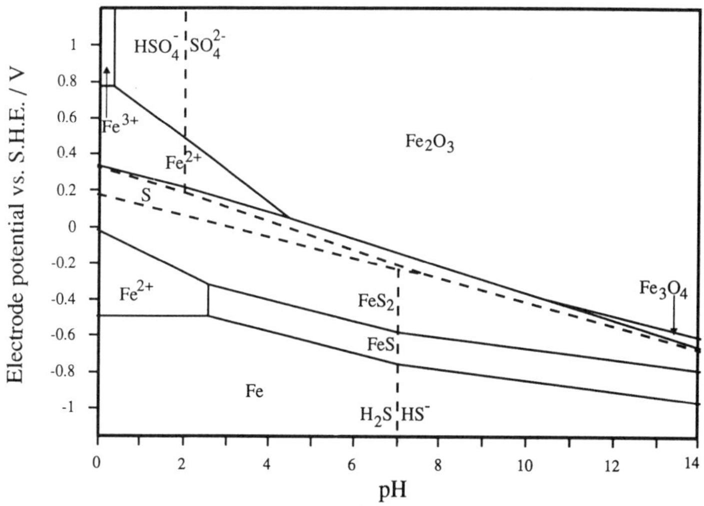

Figure 7.4 Potential-pH diagram for the stable Fe–S–H_2O system at 298 K, p_{H_2S} = 1 kPa. Total dissolved iron and sulfur activities = 10^{-3} and 10^{-1}, respectively (after Marsland, 1992).

HS^- ions (Kelsall and Thompson, 1993b), have areas of predominance under alkaline conditions, as shown in Figure 7.3.

Fe–S–H_2O SYSTEMS

Figure 7.4 shows the potential–pH diagram for the Fe–S–H_2O system, predicting that pyrite (FeS_2), like other metal sulfides, may be decomposed by several types of reactions, e.g.:

1. **Oxidative dissolution/decomposition**, in acid solutions, producing sulfate and soluble metal ions (Biegler and Swift, 1979; Mishra and Osseo-Asare, 1988a, b; Page, 1989; Marsland, Dawe and Kelsall, 1990; Welham, 1992):

$$FeS_2 + 8H_2O \rightarrow Fe^{2+} + 2SO_4^{2-} + 16H^+ + 14e^- \qquad (7.25)$$

   or oxide(s) at higher pH:

$$2FeS_2 + 19H_2O \rightarrow Fe_2O_3 + 4SO_4^{2-} + 38H^+ + 30e^- \qquad (7.26)$$

2. **Non-oxidative dissolution/decomposition**, in acid solutions, leading to H_2S evolution and dissolved metal ions (Marsland, Dawe and Kelsall, 1990; Marsland, 1992). Although formally this does not involve an electrode transfer reaction, the rate of the dissolution reaction is sensitive to potential, through its control of the M_xS_y stoichiometry:

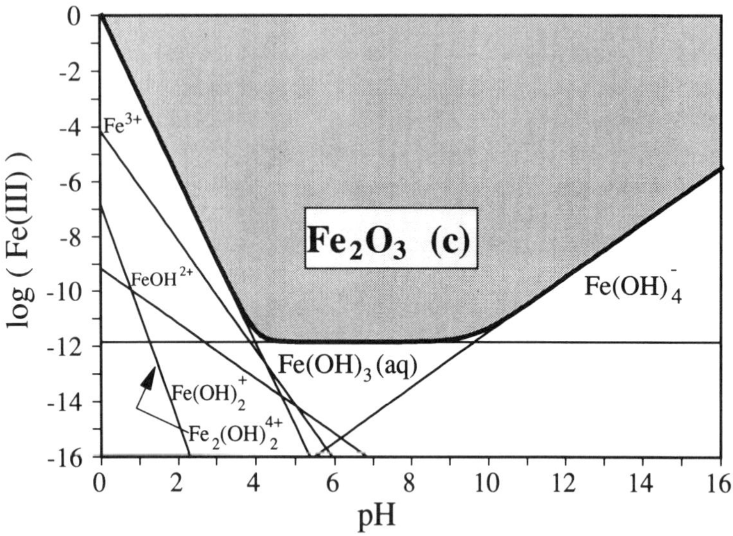

Figure 7.5 Activity-pH diagram for Fe(III) species in equilibrium with Fe_2O_3 (c) at 298 K, the shaded area defining the conditions corresponding to insolubility of that phase.

$$FeS_{1+x} + 2xH^+ + 2xe^- \rightarrow FeS + xH_2S \tag{7.27}$$

$$FeS + 2H^+ \leftrightarrow Fe^{2+} + H_2S \tag{7.28}$$

3. **Reductive decomposition** (Marsland, 1992), in which H_2S is evolved in acid solutions or HS^- ions are formed at higher pH, at lower potentials.

$$FeS_2 + 4H^+ + 2e^- \leftrightarrow Fe^{2+} + 2H_2S \tag{7.29}$$

$$FeS_2 + 2H^+ + 2e^- \leftrightarrow FeS + H_2S \tag{7.30}$$

The low solubility of Fe_2O_3 is indicated in Figure 7.5, from which the pH corresponding to the vertical line representing the Fe^{3+}/Fe_2O_3 equilibrium in Figure 7.4 can be obtained for dissolved iron activities other than 10^{-3}; higher activities expand the Fe_2O_3 predominance area.

Cu–S–H_2O SYSTEMS

Figure 7.6 shows the potential–pH diagram for the stable Cu–S–H_2O system (Woods, Yoon and Young, 1987; Goodenough, Kelsall and Liang, 1988) and predicts, *inter alia*, the oxidative decomposition of Cu_2S by the improbable reaction:

$$Cu_2S + 4H_2O \rightarrow 2Cu + SO_4^{2-} + 8H^+ + 6e^- \tag{7.31}$$

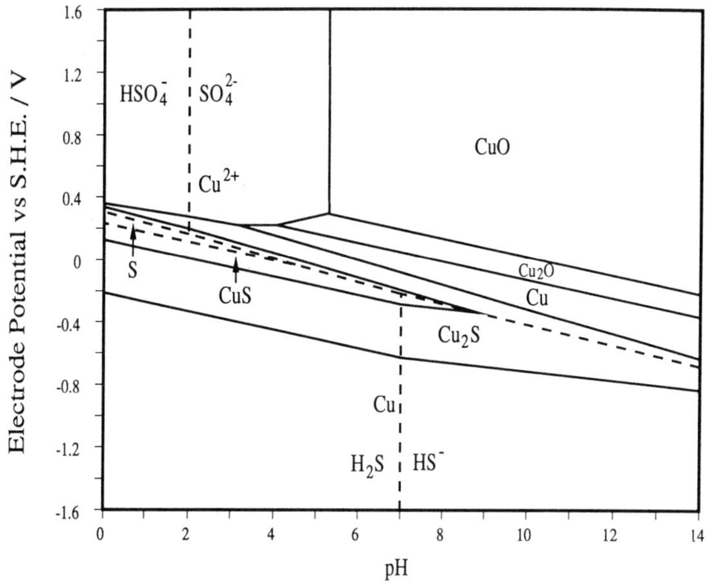

Figure 7.6 Potential-pH diagram for the Cu–S–H_2O system at 298 K with dissolved activities of copper and sulfur of 10^{-4} and 10^{-3}, respectively. (Dashed lines represent equilibria involving the excess sulfur species in the system) (after Liang, 1988).

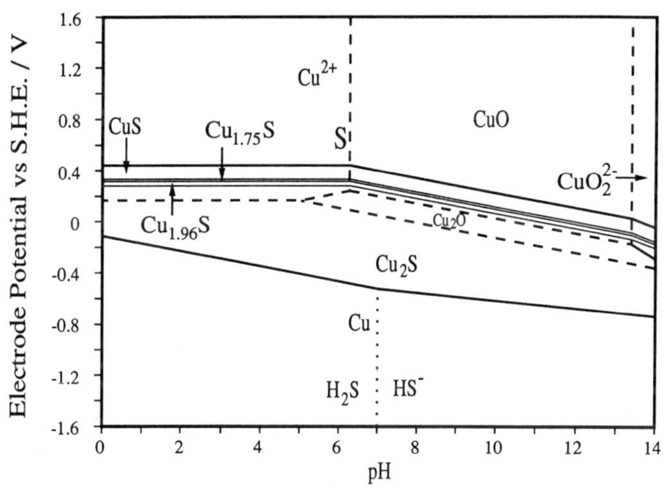

Figure 7.7 Potential-pH diagram for the Cu–S–H_2O system at 298 K with dissolved activities of copper and sulfur of 10^{-6} and 5×10^{-7}, respectively. All sulfur species with oxidation states > 0 excluded from the calculation. (Dashed lines represent equilibria involving the excess copper species in the system) (after Liang, 1988).

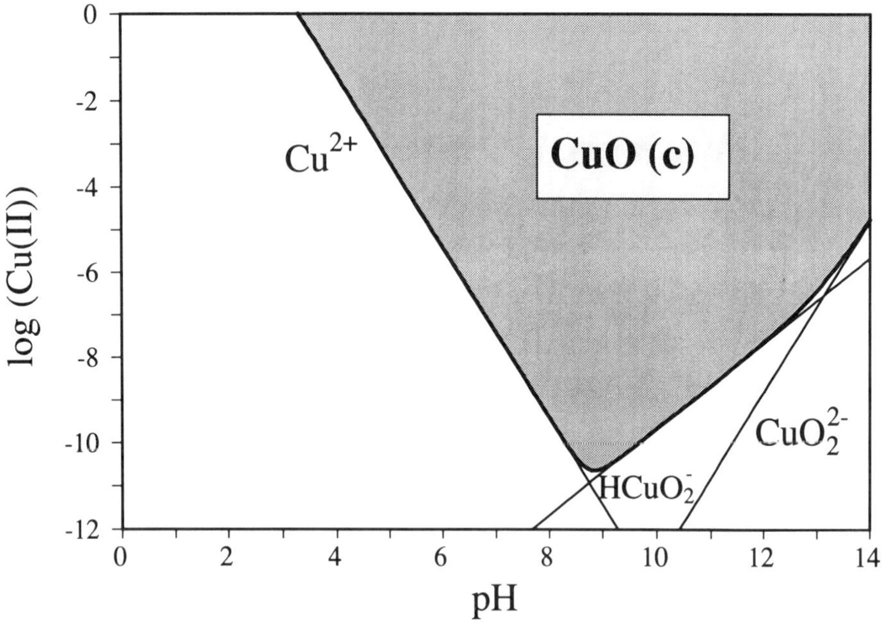

Figure 7.8 Activity-pH diagram for Cu(II) species in equilibrium with CuO (c) at 298 K, the shaded area defining the conditions corresponding to insolubility of that phase.

which would lead to effective loss of sulfur from the system, as sulfate ions cannot be reduced electrochemically under ambient conditions. However, there is no evidence of such a reaction from kinetic experiments with Cu_2S electrodes under ambient conditions (Goodenough, Kelsall and Liang, 1988; Liang, 1988), as discussed below.

Figure 7.7 shows the potential–pH diagram for the metastable Cu–S–H_2O system in which sulfur is constrained to oxidation states $\leqslant 0$ (Goodenough, Kelsall and Liang, 1988). This provides better predictions of the chemical behaviour of the system (Liang, 1988).

Figure 7.8 shows the pH dependence of solubility and speciation for Cu(II)/CuO equilibria (Goodenough, Kelsall and Liang, 1988; Liang, 1988), from which the pH corresponding to the vertical lines representing those equilibria in Figures 7.6 and 7.7 can be obtained for the range of dissolved Cu(II) activities.

7.3 Control of electrode potentials

Many electrochemical systems exhibit maxima in their current–potential behaviour, resulting in a single value of current corresponding to multiple values of electrode potential; hence the importance of controlling the electrode potential and measuring the current response using a potentiostat, rather the converse, using a galvanostat. Though industrial electrochemical processes normally use constant current/voltage control, experimental investigations of the kinetics of electrode reactions normally involve control (and/or programming of the time dependence) of the electrode potential of the so-called working electrode of a three-electrode electrochemical cell and measurement of the current response. This imposes certain instrumental requirements, such as the need for the potentiostat to be capable of delivering the required maximum cell voltage and current, to avoid instrumental limitations on these parameters imposing artefacts on the observed current–potential behaviour.

The cell voltage (U) of an electrochemical cell/reactor, in which a current I is flowing, has at least five components:

1. a thermodynamic potential (E_A) requirement for the anode (A) reaction;
2. an anode overpotential (η_A) which provides the driving force for the anode reaction to depart from its dynamic equilibrium condition;
3. a thermodynamic potential (E_C) requirement for the cathode (C) reaction;
4. an cathode overpotential (η_C) which provides the driving force for the cathode reaction to depart from its dynamic equilibrium condition;
5. an ohmic potential drop (IR) in the electrolyte between anode and cathode:

$$U = E_A + \eta_A - E_C - \eta_C + IR = E_A + \eta_A - E_C - \eta_C + \frac{Id}{\kappa A_e} \quad (7.32)$$

where d is the inter-electrode gap, κ is the electrolyte specific conductivity and A_e is the electrode area. The term electrode potential ($E = \phi_M - \phi_{ref}$) refers to the potential difference between the bulk solid electrode phase (ϕ_M) and the potential in solution adjacent to its surface, as measured by an appropriate reference electrode (ϕ_{ref}).

Usually in electrochemical experiments, the electrolyte ionic strength/conductivity is high, owing to added supporting electrolyte (e.g. sulfuric acid), which does not participate in the electrochemical reactions. Under these conditions, the potential drop across the interface is restricted to the so-called compact double layer or Helmholtz layer between the electrode surface and the outer Helmholtz plane (OHP – the locus of closest approach of hydrated ions). In low ionic strength solutions, some of the potential may also be dropped across the so-called diffuse double layer, giving rise to the zeta potential (ζ), as shown in Figure 7.9 for an anode–electrolyte interface; a similar situation could arise at the cathode. In semiconducting bulk or surface phases, space charge/solid state diffuse double layers may also be

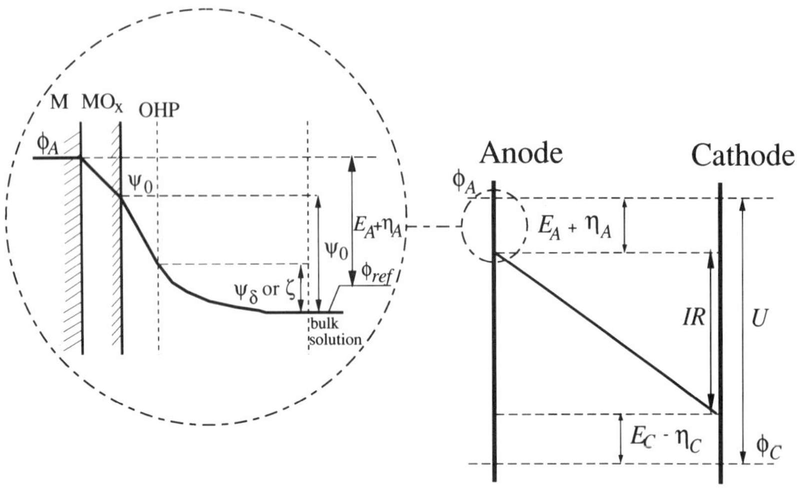

Figure 7.9 Distribution of potential in an electrochemical cell (right) and at an electrode–electrolyte interface (left), when a current I flows between a metal (M) anode at potential ϕ_A and cathode at potential ϕ_C, resulting in a total cell voltage ($U = \phi_A - \phi_C$). The anode and cathode reactions have reversible electrode potentials E_A and E_C, respectively, and the current results in overpotentials η_A and η_C, respectively, at the interfaces and an IR potential drop in the bulk electrolyte. The magnification of the metal anode–electrolyte interface on the left shows the potential decaying from a bulk phase value of ϕ_A through an oxide film (MO_x) with a surface potential ψ_0, across the compact double layer to the outer Helmholtz plane (OHP), with potential ψ_δ or ζ relative to the bulk solution, and through the diffuse double layer. The reversible anode potential E_A is measured with respect to a suitable reference electrode (ϕ_{ref}).

present, e.g. in the surface oxide MO_x shown in Figure 7.9; the surface potential (ψ_0) is also shown for completeness.

A low resistance ohmic electronic contact can be made, *inter alia*, by gold sputtering the back face of the sulfide mineral electrode and then using silver-loaded epoxy resin to connect it to a copper wire, which in turn is connected to an external electronic circuit, such as a potentiostat.

Most experiments to define the kinetics of electrochemical reactions involve control or programming of the electrode potential and the measurement of the current response, using a potentiostat, which incorporates an operational amplifier, as shown in Figure 7.10; transistor-based output stages can be used to achieve output currents of more than, say, 10 mA. These devices are designed to minimize the potential difference between their two inputs, so that application of a potential, e.g. from a wave-form generator, to one of their inputs, effectively applies that potential to the other input (in a potentiostat, connected to a reference electrode), and causes a current from/to its output. That current flows between the working and counter

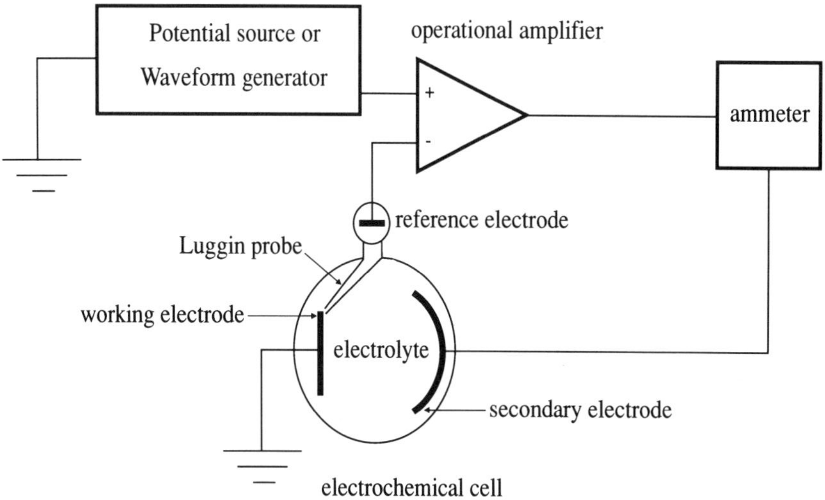

Figure 7.10 Schematic potentiostatic control of the working electrode with respect to a reference electrode in an electrochemical cell, the current flowing between the working and secondary electrodes.

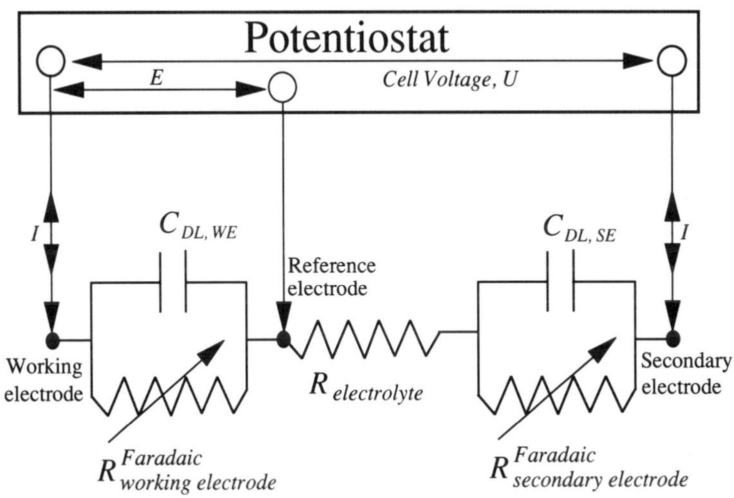

Figure 7.11 Simple equivalent electronic circuit of an electrochemical cell under potentiostatic control, the working and secondary electrode–electrolyte interfaces being represented as (double-layer) capacitances (C) which leak current (I) through (non-linear) Faradaic resistances (R^{Faradaic}), the current flowing between them through the bulk electrolyte resistance ($R_{\text{electrolyte}}$).

electrodes, through the electrolyte; essentially no current flows through the reference electrode, so that, irrespective of any impedance changes at the working electrode, its electrode potential is maintained, provided that the potentiostat is capable of providing the required current and cell voltage between working and secondary electrodes.

The simplest electronic analogue circuit for an electrochemical cell is shown in Figure 7.11. The electrode–electrolyte interfaces are capable of storing charge in their respective double-layer capacitances $C_{DL,WE}$ and $C_{DL,SE}$, but these leak current through their respective (non-linear) Faradaic resistances, $R^{Faradaic}$. The two resistance–capacitance (*RC*) networks representing the electrode–electrolyte interfaces are linked by the resistance of the bulk electrolyte, $R_{electrolyte}$, through which the cell current *I* passes. The potentiostat controls the potential between the working and reference electrodes, while current flows between the working and secondary electrodes, generating a cell voltage *U*.

7.4 Transport processes in electrochemical systems

As electrode reactions involve electron transfer usually at the interface between an electronically conducting solid and ionically conducting electrolyte, reactants and/or products have to be transported to and/or from such interfaces (Bockris and Reddy, 1973; Newman, 1973; Pletcher and Walsh, 1982; Yeager *et al.*, 1983; Rousar, Micka and Kimla, 1986). The transport step in the overall electrode process may limit the reaction rate when the former becomes slow compared with the inherent electron transfer rate at the interface; hence the importance of being able to predict that transport rate, so that the maximum (transport-limited) rate of electrode reaction can be calculated.

The flux ($\vec{N}_i$) of charged electro-active solute species *i* with mobility u_i ($u_i = 0$ for uncharged species) to an electrode surface, is due to the vectorial sum of:

- (first term) their migration in an electric field ($\nabla\phi$);
- (second term) diffusion in a concentration (*c*)/activity gradient;
- (third term) convection due to the local fluid velocity ($\vec{v}$):

$$\vec{N}_i = -u_i c_i \nabla\phi - D_i \nabla c_i + c_i \vec{v} \tag{7.33}$$

The current density in solution is a vector ($\vec{i}$) perpendicular to the equipotential surfaces in the solution, and according to Faraday's law it is given by

$$\vec{i} = F\sum_i z_i \vec{N}_i \tag{7.34}$$

When concentration gradients in the solution can be ignored (i.e. at current densities well below the mass transport limited rate), substitution of equation (7.34) into equation (7.33) gives

$$\vec{i} = -\kappa \nabla \phi \tag{7.35}$$

where

$$\kappa = F \sum_i z_i u_i c_i \tag{7.36}$$

is the specific conductivity of the solution, and where the convective terms sum to zero by the electro-neutrality requirement

$$\sum_i z_i c_i = 0 \tag{7.37}$$

The electro-neutrality condition holds in the bulk solution, but is violated in the electric double layer, since the intense electric field across the electrode–electrolyte interface causes a separation of charge between the two phases, though the system as a whole is electro-neutral.

The differential conservation law, equation (7.38), arises from a material balance on a small electrolyte volume element:

$$\frac{\partial c_i}{\partial t} = -\nabla \vec{N}_i \pm R_i \tag{7.38}$$

The last term R_i is the rate of homogeneous chemical reaction involving species i, occurring simultaneously with its electrochemical reaction; this is often zero, so that substitution of equation (7.33) into equation (7.38) gives

$$\frac{\partial c_i}{\partial t} = D_i \nabla^2 c_i - u_i \nabla (c_i \nabla \phi) - \vec{v} \nabla c_i \tag{7.39}$$

When excess supporting electrolyte is present, the migrational term can be neglected, as can the convection term at short times; the resulting simplified form of equation (7.39) is then Fick's second law, which in one dimension is

$$\frac{\partial c_i}{\partial t} = D_i \frac{\partial^2 c_i}{\partial x^2} \tag{7.40}$$

With appropriate boundary conditions, this can be used, e.g. to model the current–time transient in response to a step to a potential at which a reaction is transport controlled, the current then decaying linearly with time$^{0.5}$ at short times, prior to convective processes contributing to the overall transport rate (Bard and Faulkner, 1980; Macdonald, 1981).

When equation (7.38) is multiplied by the ionic charge z_i and summed over all ionic species i in solution,

$$\sum_i z_i \frac{\partial c_i}{\partial t} = \sum_i z_i D_i \nabla^2 c_i - \sum_i z_i u_i \nabla(c_i \nabla \phi) - \sum_i z_i \vec{v} \nabla c_i + \sum_i z_i R_i$$

the terms

$$\sum_i z_i \frac{\partial c_i}{\partial t} = 0 \qquad \text{and} \qquad \sum_i z_i R_i = 0$$

according to equation (7.37). In the absence of concentration gradients, as in a homogeneous electrolyte,

$$\sum_i z_i D_i \nabla^2 c_i = 0 \qquad \text{and} \qquad \sum_i z_i \vec{v} \nabla c_i = 0$$

Hence

$$-\sum_i z_i u_i \nabla(c_i \nabla \phi) = -\sum_i z_i u_i c_i \nabla^2 \phi = 0$$

so that

$$\nabla^2 \phi = \frac{\partial^2 \phi}{\partial x^2} + \frac{\partial^2 \phi}{\partial y^2} + \frac{\partial^2 \phi}{\partial z^2} = 0 \qquad (7.41)$$

which is Laplace's equation; ϕ is the potential at coordinates x, y, z and the electric field ($\vec{E}$) is:

$$\vec{E} = \nabla \phi = \frac{\partial \phi}{\partial x} + \frac{\partial \phi}{\partial y} + \frac{\partial \phi}{\partial z} \qquad (7.42)$$

Laplace's equation can be integrated with appropriate boundary conditions to give the potential field in solution ($\phi = f(x, y, z)$). This, together with a knowledge of the relationship between the electrode potential and current enables the spatial distribution of current in electrochemical cells/reactors, and enables calculation of the current flowing normal to the equipotential surfaces (Rousar, Micka and Kimla, 1986).

7.5 Electrochemical kinetics

7.5.1 *Charge transfer controlled electrochemical reactions*

For simple electron transfer reactions such as

$$O + ze^- \leftrightarrow R \tag{7.1}$$

the equilibrium electrode potential is given by the Nernst equation:

$$E_{\mathrm{O/R}}\ (\mathrm{V}) = E^0_{\mathrm{O/R}} + \frac{RT}{zF} \ln\left\{\frac{(O)}{(R)}\right\} \tag{7.43}$$

At electrode potentials more positive than that given by equation (7.43), the electro-active species R, if present, will be oxidized; conversely, at potentials more negative than $E_{\mathrm{O/R}}$, species O, if present, will be reduced. Hence, at electrode potentials other than that corresponding to the equilibrium value, an oxidation or reduction current, a direct measure of the reaction rate, would be detected in a potentiostatic circuit. A simple model of the kinetics (i.e. the steady state current–potential relationship) of such reactions can be constructed (Bockris and Reddy, 1973; Bard and Faulkner, 1980; Pletcher and Walsh, 1982; Southampton Electrochemistry Group, 1985) using the following simplifying assumptions:

- no adsorption of the reactant or product occurs on the electrode surface;
- the reaction does not involve phase transitions;
- the electron transfer step is the rate-determining step.

At $E_{\mathrm{O/R}}$, the net current density (i) due to the partial oxidation ($\overleftarrow{i}$) and reduction ($\overrightarrow{i}$) current densities is zero:

$$i = \overrightarrow{i} + \overleftarrow{i} = 0 \tag{7.44}$$

Current densities at an electrode of geometrical area A_{e} are related to the current (I) by

$$i = I/A_{\mathrm{e}} \tag{7.45}$$

At other than the reversible potential, net current will flow. The simplest mathematical description of the net current densities (i) due to the forward and reverse reactions is obtained by assuming first-order kinetics, i.e. the reaction rate is linearly dependent on the reactant concentration:

$$\text{Specific rate of reduction of } O = \vec{k}c_{\mathrm{O}}^{\infty}$$

$$-\vec{i} = zF\vec{k}c_{\mathrm{O}}^{\infty} \tag{7.46}$$

$$\text{Specific rate of oxidation of } R = \overleftarrow{k}c_{\mathrm{R}}^{\infty}$$

$$\vec{i} = zF\overleftarrow{k}c_{\mathrm{R}}^{\infty} \tag{7.47}$$

where $\vec{k}$ and $\overleftarrow{k}$ are potential-dependent rate constants and the superscript ∞ denotes a concentration in the bulk solution. Current densities (and overpotentials) are defined arbitrarily as negative for cathodic (C) and positive for anodic (A) electrode processes.

The observed current density (i) at any potential E vs. reference electrode is:

$$i = \vec{i} + \overleftarrow{i} \tag{7.48}$$

$$i = zF(\overleftarrow{k}c_{\mathrm{R}}^{\infty} - \vec{k}c_{\mathrm{O}}^{\infty}) \tag{7.49}$$

in which:

$$\vec{k} = \vec{k}_0 \exp\left\{-\frac{\alpha_{\mathrm{C}} zF}{RT}E\right\} \tag{7.50}$$

$$\overleftarrow{k} = \overleftarrow{k}_0 \exp\left\{\frac{\alpha_{\mathrm{A}} zF}{RT}E\right\} \tag{7.51}$$

Therefore,

$$i = zF\left[\overleftarrow{k}_0 c_{\mathrm{R}}^{\infty} \exp\left\{\frac{\alpha_{\mathrm{A}} zF}{RT}E\right\} - \vec{k}_0 c_{\mathrm{O}}^{\infty} \exp\left\{-\frac{\alpha_{\mathrm{C}} zF}{RT}E\right\}\right] \tag{7.52}$$

Hence, at the equilibrium potential ($E_{\mathrm{O/R}}$):

$$\vec{i} = \overleftarrow{i} = i_0 = zF\overleftarrow{k}_0 c_{\mathrm{R}}^{\infty} \exp\left\{\frac{\alpha_{\mathrm{A}} zF}{RT}E_{\mathrm{O/R}}\right\} = zF\vec{k}_0 c_{\mathrm{O}}^{\infty} \exp\left\{-\frac{\alpha_{\mathrm{C}} zF}{RT}E_{\mathrm{O/R}}\right\} \tag{7.53}$$

where i_0 is the **exchange current density**, a measure of the rate of the reaction under dynamic equilibrium conditions. The driving force to cause the reaction to depart from that condition is defined as the **overpotential**:

$$\eta = E - E_{\mathrm{O/R}} \tag{7.54}$$

Substitution of equations (7.53) and (7.54) into (7.52) gives the **Butler-Volmer equation**

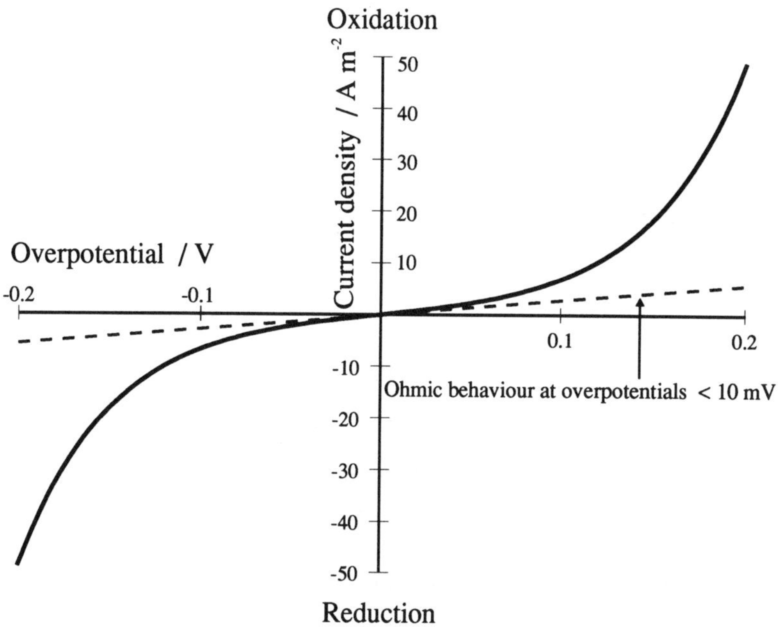

Figure 7.12 Example of the Butler-Volmer equation (7.55) plotted for a reaction with $i_0 = 1\,\mathrm{A\,m^{-2}}$ and transfer coefficients of 0.5, indicating the deviation from ohmic behaviour at overpotentials $> \sim 10\,\mathrm{mV}$.

$$i = i_0\left[\exp\left\{\frac{\alpha_A zF}{RT}\eta\right\} - \exp\left\{\frac{-\alpha_C zF}{RT}\eta\right\}\right] \tag{7.55}$$

where α_C and α_A are the so-called **transfer coefficients** for the cathodic and anodic reactions, respectively; they represent the fraction of the applied overpotential utilized in providing a driving force for reaction. For simple electrode reactions, $\alpha_C + \alpha_A = 1$ and $\alpha_C \approx \alpha_A \approx 0.5$. The Butler-Volmer equation (7.55) is plotted in Figure 7.12.

The first exponential term in equation (7.55) represents the contribution of the (anodic) reverse reaction to the net current. At sufficiently high cathodic overpotentials ($\eta < -0.1\,\mathrm{V}$), the effect of the (anodic) reverse reaction may be neglected, so that equation (7.55) simplifies to:

$$-i_C = -\vec{i} = i_0 \exp\left\{\frac{-\alpha_C zF}{RT}\eta\right\} \tag{7.56}$$

$$\log(-i_C) = \log(i_0) - \frac{\alpha_C zF}{2.303RT}\eta \tag{7.57}$$

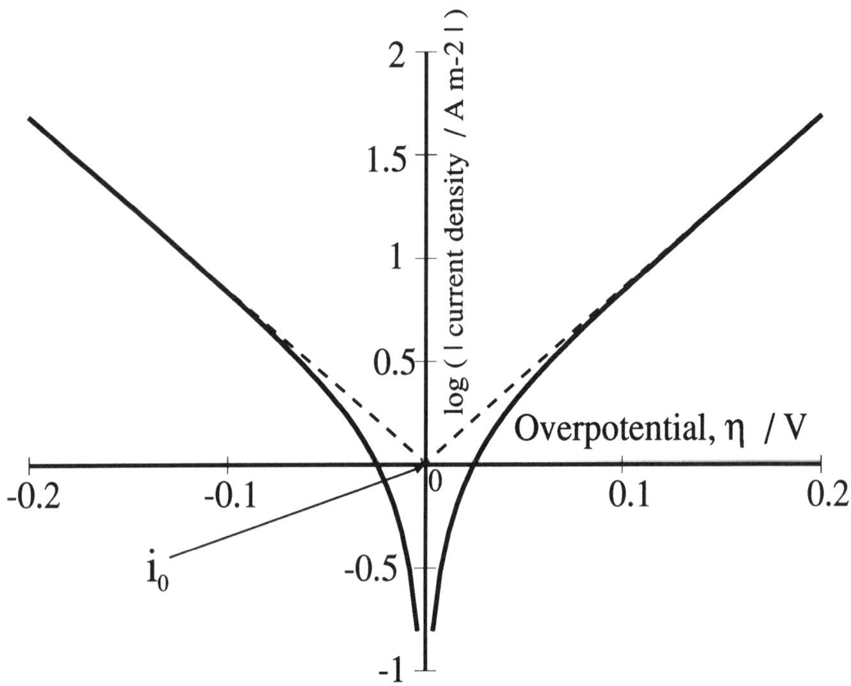

Figure 7.13 Semi-logarithmic plot of the data in Figure 7.12, though with the ordinate as the logarithm of the modulus of the current density. Extrapolation of the linear regions to zero overpotential enables the exchange current density (i_0) to be determined (equations (7.57) and 7.59)).

Similarly, at high anodic overpotentials ($\eta > 0.1\,\text{V}$), the effect of the (cathodic) forward reaction may be neglected and equation (7.55) simplifies to:

$$i_A = i_0 \exp\left\{\frac{\alpha_A zF}{RT}\eta\right\} \tag{7.58}$$

$$\log(i_A) = \log(i_0) + \frac{\alpha_A zF}{2.303RT}\eta \tag{7.59}$$

Figure 7.13 shows semi-logarithmic plots of the data in Figure 7.12, indicating how the linear regions, corresponding to equations (7.57) and (7.59) can be used to obtain the exchange current density i_0 by extrapolation to zero overpotential. The slopes of the linear portions enable the transfer coefficients (α_A and α_C) to be derived, according to equations (7.57) and (7.59).

The so-called Tafel equation is obtained by rearranging equation (7.57) to give

$$\eta = \frac{2.303RT}{\alpha_C zF} \log i_0 - \frac{2.303RT}{\alpha_C zF} \log(-i_C) \qquad (7.60)$$

or by rearranging equation (7.59) to give

$$\eta = \frac{2.303RT}{\alpha_A zF} \log i_A - \frac{2.303RT}{\alpha_A zF} \log i_0 \qquad (7.61)$$

Equation (7.55) can also be simplified at very low cathodic or anodic overpotentials ($\alpha zF\eta/RT \ll 1$, say $\eta < 10\,\text{mV}$), when the exponential terms may be linearized to give

$$|i| = \frac{i_0 zF}{RT} |\eta| \qquad (7.62)$$

i.e. ohmic behaviour is observed close to the reversible potential; it is a special case of the more general activation controlled charge transfer processes observed at higher overpotentials.

The standard rate constant (k^0) is related to the exchange current density by:

$$i_0 = zFk^0 (c_O^\infty)^{\alpha_A} (c_R^\infty)^{\alpha_C} \qquad (7.63)$$

i.e. it is the exchange current density that contains the dependence on reactant concentrations.

7.6 Mass transport controlled electrochemical reactions

For reaction at an electrode–electrolyte interface, electro-active species have to be transported to the interface (Bockris and Reddy, 1973; Newman, 1973; Bard and Faulkner, 1980; Pletcher and Walsh, 1982; Yeager *et al.*, 1983; Southampton Electrochemistry Group, 1985; Rousar, Micka and Kimla, 1986), and, depending on the reaction, the products may have to be transported away from the interface, i.e. the electron transfer and mass transport processes are linked in **series**, as shown in Figure 7.14.

At low overpotentials, transport rates are normally fast compared with the electrochemical reaction kinetics, i.e. electron transfer is the rate-determining step. However, because of the exponential increase of current density/reaction rates with increasing overpotential, according to equations (7.57) or (7.59), this cannot be sustained indefinitely, since the interfacial

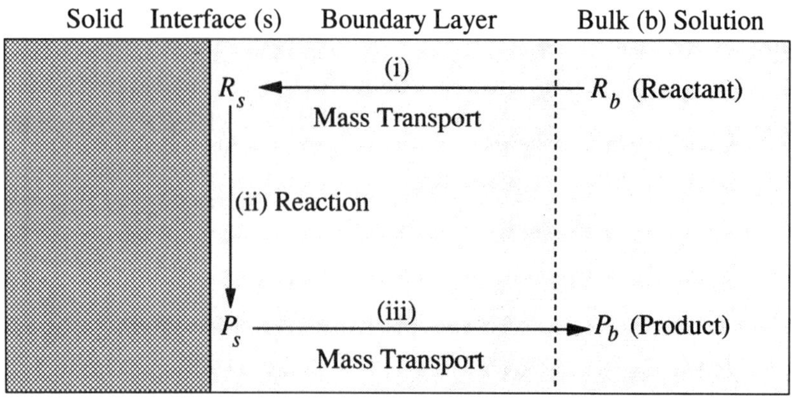

Figure 7.14 Mass transport processes coupled to reaction at a solid–solution interface.

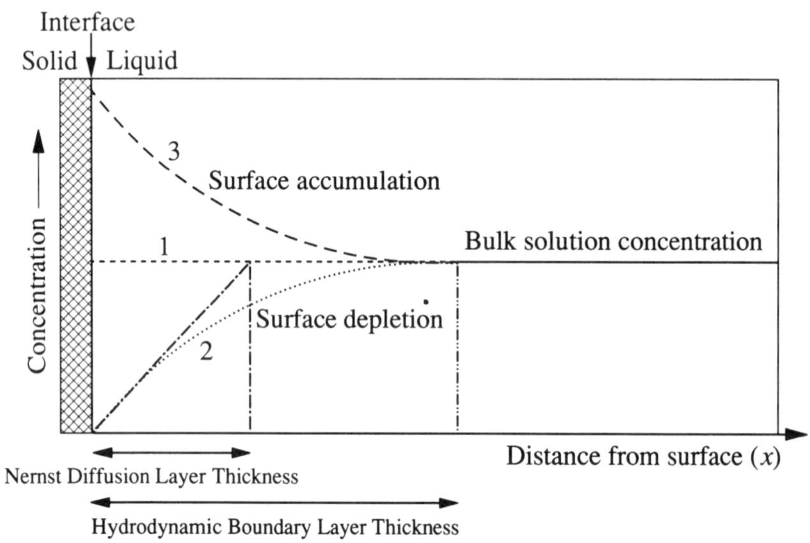

Figure 7.15 Schematic of interfacial concentration gradients due to heterogeneous reaction, in this example, of a solid–liquid interface. Case 1 corresponds to the equilibrium condition in the absence of adsorption, case 2 to a reactant from bulk solution being depleted by rapid reaction at the interface, and case 3 to the fast dissolution of the solid, resulting in accumulation of the product owing to its restricted rate of transport away from the surface.

reaction kinetics become faster than the rate of supply of reactant from the bulk electrolyte to an electrode surface. This leads to the reactant being progressively depleted at the interface and the development of a concentration gradient in the boundary layer (Figure 7.15); the rates of electron transfer and mass transport are comparable and both processes control the rate of the overall reaction, i.e. there is mixed control. As the overpotential is increased further, the surface concentration of electro-active species tends to zero, and in that limit the current density becomes independent of the potential, the reaction rate being determined wholly by the mass transport rate. Similar physicochemical models may be used to describe processes that involve:

- dissolution, with slow transport away from the interface, resulting in surface accumulation and possible precipitation (Figure 7.15);
- solution phase transport limitations in a porous product layer;
- solid state transport limited processes, e.g. the dissolution of iron from a chalcopyrite ($CuFeS_2$) lattice being limited by diffusion (and possibly migration if electric fields are present in a space charge layer in the solid surface) (Kelsall and Page, 1984; Dean and Kelsall, 1992; Kelsall *et al.*, 1992).

In aqueous electrochemical systems, supporting electrolytes are added to increase the solution conductivity, thereby decreasing the ohmic potential drop between electrodes (Figure 7.9), to make the current density distribution more uniform in reactors with larger electrodes and possibly to adjust the pH. For charged electro-active species, the mass transport limited current will be lower in the presence than in the absence of supporting electrolyte, since increasing the electrode potential does not then enhance the migrational component of the transport rate of electro-active species, most of the current being carried in the bulk solution by the supporting electrolyte. At the electrode, the net flux of all species is zero except for that of the reactant, so in the diffusion layer the supporting electrolyte ions and the reactant counter ions have zero net velocity, adopting concentration distributions such that the forces of migration and diffusion cancel. Beyond the boundary layer, Ohm's law (equation 7.35) applies in a well-stirred bulk electrolyte; convection cancels the migrational flux of the supporting electrolyte and augments that of the reactant.

In the presence of excess supporting electrolyte, the migrational term in equation (7.33) may be neglected; convection has the effect of thinning the hydrodynamic boundary layer at the electrode–electrolyte interface, thereby increasing the concentration gradient, though diffusion is the only transport process able to transport electro-active species through the stagnant boundary layer (Figures 7.14 and 7.15). According to Fick's first law in one dimension, the flux ($\vec{N}_i$) of species i to the electrode surface is proportional to the concentration/activity gradient, the proportionality constant being the diffusion coefficient, D_i:

$$\vec{N}_i = D_i \frac{(c_i^{\infty} - c_i^{S})}{\delta} \tag{7.64}$$

The term δ is the so-called Nernst diffusion layer thickness, a mathematical convenience having no physical reality but enabling a complex concentration–distance profile (Figure 7.14) to be linearized.

In the limit, the concentration c^S of electro-active species at the electrode surface tends to zero, so the flux becomes linearly dependent on the concentration (c_i^∞) of electro-active species (i) in bulk solution:

$$\vec{N}_i = \frac{D_i}{\delta} c_i^\infty = k_m c_i^\infty \tag{7.65}$$

The quotient (D_i/δ) is known as the mass transport rate constant, k_m. Hence, from equation (7.65) the mass transport limited current density (i_L) is also linearly related to the (c_i^∞):

$$i_L = zF\vec{N}_i = zF\frac{D_i}{\delta} c_i^\infty = zFk_\mathrm{m} c_i^\infty \tag{7.66}$$

7.7 Polarization curves

The simplest experimental means of investigating the kinetics of electrode reactions involves varying the electrode potential of a suitable working electrode, either manually point by point or with a slow linear sweep from a

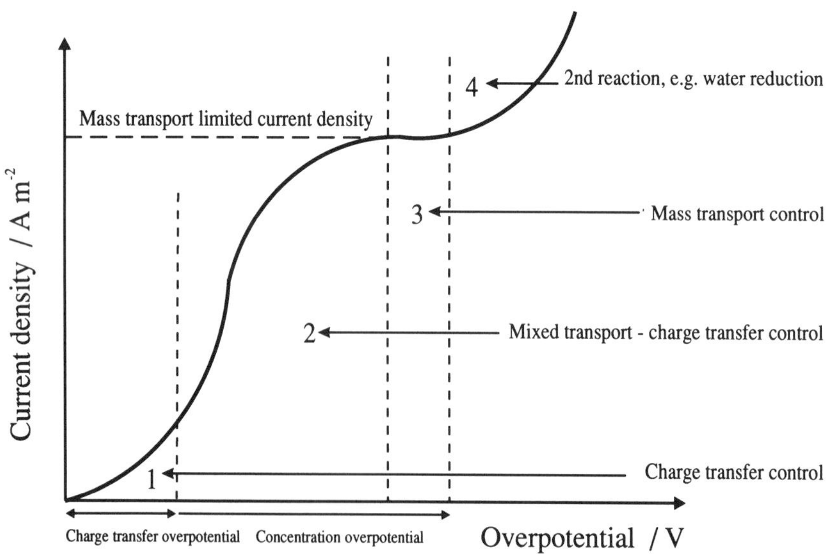

Figure 7.16 Schematic steady state polarization behaviour of the oxidation or reduction of electro-active species at an electrode in the presence of excess supporting electrolyte, showing different types of rate control with increasing overpotential.

wave-form generator, and measuring the steady state current response. Such a current–potential relationship, known as a polarization curve and shown schematically in Figure 7.16, often exhibits different regions of behaviour, corresponding to different determining steps (Bockris and Reddy, 1973; Newman, 1973; Bard and Faulkner, 1980; Pletcher and Walsh, 1982; Yeager *et al.*, 1983; Southampton Electrochemistry Group, 1985; Rousar, Micka and Kimla, 1986).

7.7.1 *Region 1: charge transfer cöntrol*

Mass transfer is fast compared with the charge transfer rate, so that, in the absence of adsorption, the bulk and surface concentrations are almost equal (subject to equation (7.67)) and charge transfer is the rate-determining step, leading to so-called charge transfer polarization; for a cathodic reduction

$$i_c = -i_0 \exp\left\{-\frac{\alpha_C zF}{RT}\eta\right\} \tag{7.56}$$

The exponential increase of current density with overpotential eventually causes the surface concentration of reactant to decrease significantly below its bulk value, leading to Region 2.

7.7.2 *Region 2: mixed charge transfer–mass transport control*

The depletion of the reactant at the surface causes so-called concentration polarization, which is additive with the charge transfer polarization. The surface concentrations c^s are related to the ratio of the respective mass transport current densities (i_L) of O and R and the actual current density (i):

$$\frac{c_R^s}{c_R^b} = \frac{i_L^R - i}{i_L^R} \quad \text{and} \quad \frac{c_O^s}{c_O^b} = \frac{i_L^O - i}{i_L^O} \tag{7.67}$$

A combination of charge transfer and mass transport control the overall rate of reaction; the current density is then given by

$$i = i_0\left[\left(\frac{c_R^s}{c_R^b}\right)\exp\left\{-\frac{\alpha_C zF}{RT}\eta\right\} - \left(\frac{c_O^s}{c_O^b}\right)\exp\left\{\frac{\alpha_A zF}{RT}\eta\right\}\right] \tag{7.68}$$

In the limit, the surface concentration of reactant falls to zero, causing Region 3.

7.7.3 Region 3: mass transport control

At sufficiently high overpotentials, the mass transport step becomes slower than the charge transfer rate and the former becomes the sole rate-determining step. The current density is then

$$i_L = zF\vec{N}_i = zFk_m c_i^\infty \tag{7.66}$$

where k_m ($m\,s^{-1}$) is the mass transport rate constant and c_i^∞ is the concentration of reactant in the bulk solution.

7.7.4 Region 4

At even higher overpotentials, a second reaction, e.g. in the case of cathodic reactions in aqueous electrolytes,

$$2H^+ + 2e^- \rightarrow H_2 \tag{7.5}$$

occurs in parallel with the primary reaction, resulting in a decrease in the latter's current efficiency (Φ), which is the proportion of the overall current (I_{total}) used in the primary reaction 1:

$$\Phi = \frac{I_{O/R}}{I_{total}} = \frac{I_{O/R}}{I_{O/R} + I_{H_2}} \tag{7.69}$$

Deviations from these steady state current–potential relationships may arise as a result of:

- the surface composition changing with time, resulting in the depletion of electro-active species and/or the formation of adsorbed product layers such as oxides, which passivate the surface to further reaction (see below);
- the potential changing rapidly with time, and the concentration gradients, in either the solid or the solution phase, taking longer to relax to their steady state profiles. Hence, in cyclic voltammetry, in which the potential is swept linearly between two limits, through one or more cycles, a current peak is observed at the onset on mass transport control, rather than the wave shown in Figure 7.16; at higher overpotentials, even though the thermodynamic driving force for reaction is increasing, the current decays as the reactant concentration gradients relax (decrease) at the electrode–electrolyte interface. In experiments in which the potential is stepped to a value corresponding to transport control, the current decays linearly with $(\text{time})^{0.5}$, which is characteristic of a diffusional process, the diffusion coefficient (D_i) of the reacting species i being derived from the slope of such a plot (Bard and Faulkner, 1980; Macdonald, 1981).

7.8 Mixed potential processes

Metal oxidation is required for dissolution by coupled electrochemical reactions which occur at a mixed potential (Figure 7.17), lying between the reversible potentials (E_r) of the metal (M^{z+}/M) and the oxidant couple (O/R), such as O_2/H_2O and/or H^+/H_2 (Pourbaix, 1966; Bockris and Reddy, 1973; Bard and Faulkner, 1980; Woods, 1981; Pletcher and Walsh, 1982; Southampton Electrochemistry Group, 1985). The oxidant anodically polarizes the metal (M), causing its dissolution by the reduction of the oxidized state (O) to the reduced state (R) of the oxidant, so that overall there is an electronic balance in the chemical reaction:

$$nM + zO \rightarrow nM^{z+} + zR \tag{7.70}$$

The thermodynamic requirement for such corrosion reactions is that:

$$E_{O/R} > E_{M^{z+}/M} \tag{7.71}$$

The ('mixed') potential measured in such a reacting system corresponds to the potential at which the currents due to oxidation and reduction processes are exactly equal. That condition is sensitive *inter alia*, to the mass transport

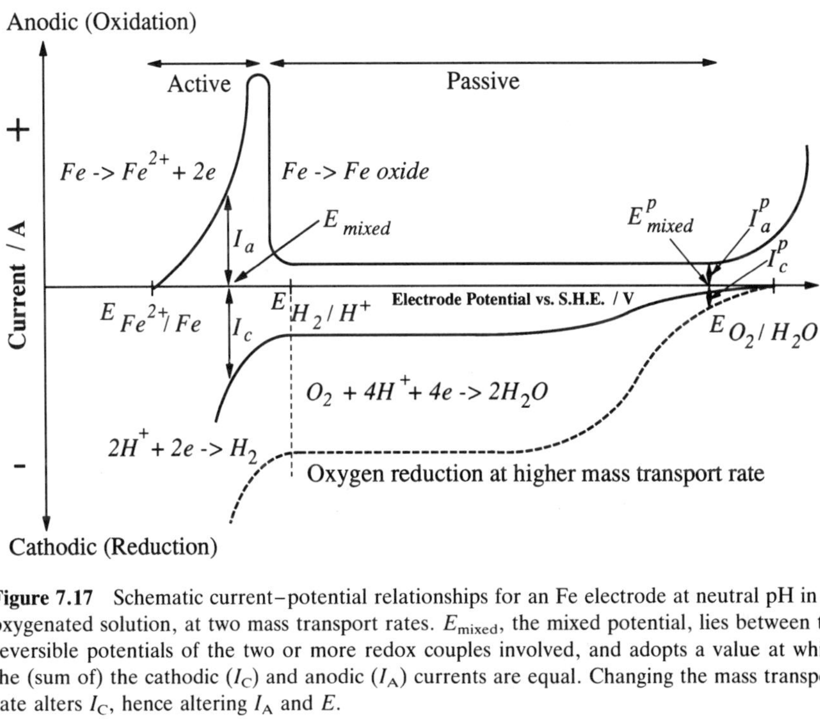

Figure 7.17 Schematic current–potential relationships for an Fe electrode at neutral pH in an oxygenated solution, at two mass transport rates. E_{mixed}, the mixed potential, lies between the reversible potentials of the two or more redox couples involved, and adopts a value at which the (sum of) the cathodic (I_C) and anodic (I_A) currents are equal. Changing the mass transport rate alters I_C, hence altering I_A and E.

rates, if either or both couples are transport-controlled. Figure 7.17 shows the effect of enhancing the oxygen transport rate to an Fe electrode; at neutral/alkaline solutions in which Fe oxides are formed, the effect may be to decrease the Fe dissolution rate, owing to the critical current density for active dissolution being exceeded, resulting in a so-called active–passive transition. The mixed potential then adopts a much more positive potential (Figure 7.17).

Similar arguments apply to the decomposition of metal sulfides with redox couples, though in addition reductive decomposition is possible, which is precluded with metals, except those few that form hydrides (SnH_4, PbH_4, etc.). The spontaneous oxidation of metal sulfides by reduction of dissolved oxygen occurs in the weathering of sulfide ore bodies in froth flotation processes (Woods, 1981).

7.9 Examples of the electrochemical behaviour of sulfide minerals

7.9.1 Cyclic voltammetric study of chalcocite (Cu_2S) oxidation

Figure 7.18 shows a cyclic voltammogram of a Cu_2S electrode in aqueous electrolyte at pH 9.2 for a single cyclic linear potential sweep, starting at a negative potential limit of −0.52 V vs. SCE (−0.278 V vs. SHE), in the predicted range of thermodynamic stability of Cu_2S (Figures 7.6 and 7.7), and scanning to 0.25 V vs. SCE (0.492 V vs. SHE) and back (Bard and

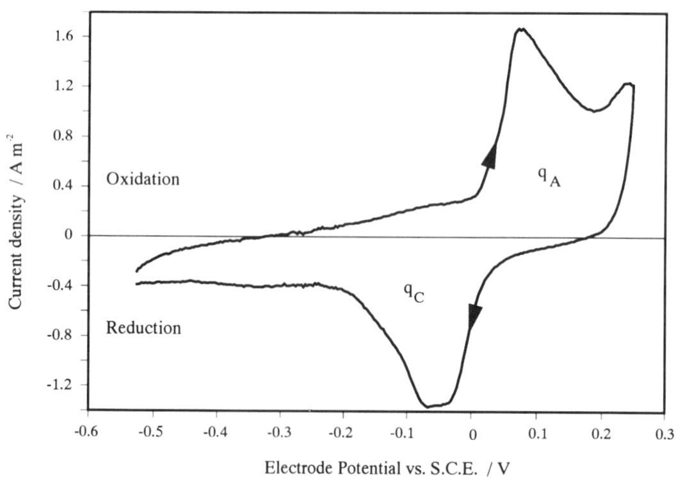

Figure 7.18 Cyclic voltammogram of a Cu_2S electrode in 0.1 M $Na_2B_4O_7$ (pH 9.2) subject to a single cycle, linear potential sweep at 10 mV s^{-1}, starting at −0.55 V, and scanning to 0.25 V and back. Integration of the resulting anodic and cathodic currents enables the anodic (q_A) and cathodic (q_C) charges to be determined (equation (7.3)). (After Liang, 1988.)

Faulkner, 1980; Southampton Electrochemistry Group, 1985; Goodenough, Kelsall and Liang, 1988; Liang, 1988). The total oxidation charge passed corresponds to only a few monolayers, depending on which reactants were formed and what the initial surface roughness factor was (ratio of real to geometrical surface area, estimated as four). Altering the mass transport rate had no effect on the voltammograms, i.e. the surface transformation reactions were not transport controlled.

Comparison of the voltammogram with the thermodynamic predictions in Figures 7.6 and 7.7 precludes the oxidation reaction being responsible for the oxidation currents at potentials $> \sim -0.3$ V vs. SCE:

$$Cu_2S + 4H_2O \rightarrow 2Cu + SO_4^{2-} + 8H^+ + 6e^- \qquad (7.31)$$

$$E \text{ vs. SHE (V)} = 0.5003 - 0.0788\,\text{pH} + 0.0099\log(SO_4^{2-})$$

as no current onset was observed at a potential just above the reversible potential for that reaction. For positive potential limits up to 0.1 V vs. SCE, the anodic (q_A) and cathodic (q_C) charges (areas under the oxidation and reduction current responses) were essentially equal, implying a facile reversible reaction, probably precluding phase transitions, which generally require large driving forces. This implies that the oxidation reaction was not

$$Cu_2S + 4H_2O \leftrightarrow 2CuO + SO_4^{2-} + 8H^+ + 6e^- \qquad (7.72)$$

being followed by the reduction reaction

$$2CuO + 4H^+ + 4e^- \rightarrow 2Cu + 2H_2O \qquad (7.73)$$

for which the ratio of the oxidation and reduction charges would be 6 : 4, the ratio of the numbers of electrons involved in reactions (7.72) and (7.73), sulfate ions being irreducible. The most probable reactions were the series summarized as

$$Cu_2S + xH_2O \leftrightarrow Cu_{2-x}S + xCuO + 2xH^+ + 2xe^- \qquad (7.74)$$

where $Cu_{2-x}S$ refers to the range of stable and metastable copper sulfides, with Cu_2S and CuS as their end members (Vaughan and Craig, 1977; Goodenough, Kelsall and Liang, 1988; Liang, 1988). Though in general, oxides are more stable than hydroxides, the latter tend to form in kinetic experiments in aqueous solutions. At electrodes, duplex structures with an inner oxide and an outer hydroxide layer have been reported; hence, reactions forming $Cu(OH)_2$ presumably occurred in parallel to those in which CuO was formed. Extending the positive potential limit to 0.25 V and 0.35 V, respectively, resulted in 0.91 and 0.75 of the oxidation charge being recovered during the negative-going potential sweep. This implies that at potentials > 0.1 V, the reaction

$$CuS + H_2O \leftrightarrow S + CuO + 2H^+ + 2e^- \qquad (7.75)$$

$$E \text{ vs. SHE (V)} = 0.8141 - 0.0591\,\text{pH}$$

began to occur, resulting in considerably less chemical reversibility of the oxidation processes. These hypotheses were confirmed by using X-ray photoelectron spectroscopy of Cu_2S electrodes electrochemically pre-treated at various potentials (Goodenough, Kelsall and Liang, 1988; Liang, 1988).

7.9.2 Rotating (ring-) disc electrodes

Levich (1962) solved analytically the equations describing the hydrodynamics of a downward-facing horizontal disc, immersed in a liquid and rotating about a vertical axis. The resulting flow is axially up towards the disc and radially out across its surface. The mass transport rate to the disc was predicted to increase with the square root of the rotation rate; for a mass transport controlled electrochemical reaction at a rotating disc electrode (RDE)

$$I_L = zF\vec{N}_i = 1.554 zFD_i^{2/3}\nu^{-1/6}c_i^{\infty}\omega^{1/2} \qquad (7.76)$$

where I_L is the limiting current, z is the number of moles of electrons per mole of reactant, D_i is the diffusion coefficient, ν is the kinematic viscosity, c_i^{∞} is the bulk concentration and ω is the rotation rate. For a known c_i^{∞}, linear plots of I_L vs. $\omega^{0.5}$ passing through the origin provide an important diagnostic test for processes the rate of which is controlled by transport to the electrode in solution (Albery and Hitchman, 1971; Bard and Faulkner, 1980; Pletcher and Walsh, 1982; Filinovsky and Pleskov, 1984; Southampton Electrochemistry Group, 1985).

A further sophistication of the RDE incorporates a ring electrode (Figure 7.19(a) and (b)) concentric with the disc but electronically isolated from it and controlled independently with a second potentiostatic circuit, enabling the disc and ring electrode currents to be measured independently. Such rotating ring–disc electrodes (RRDEs) are powerful tools for the investigation of the kinetics and mechanisms of electrode reactions involving soluble products and/or intermediates. The ring can be used to detect solution species from the disc electrode as they are driven radially out towards the ring electrode by the convective force of rotation; hence, the RRDE effectively works as an upstream generator and downstream detector.

Even if the species formed at the disc electrode are completely stable and totally soluble, only a fraction, the so-called collection efficiency (N_0), of the

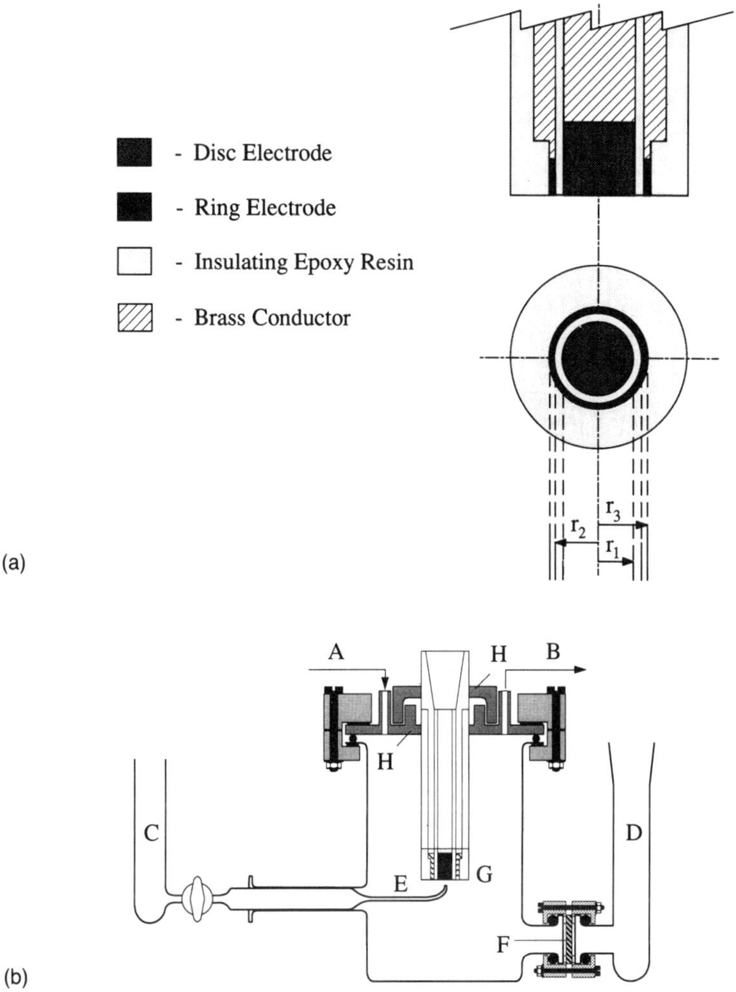

Figure 7.19 (a) Rotating ring–disc electrode assembly and electrochemical cell. (b) End sections of a rotating ring–disc electrode (RRDE). A, nitrogen in; B, nitrogen out; C, reference electrode compartment; D, counter electrode compartment; E, Luggin probe; F, cation exchange membrane to prevent counter-electrode products, such as oxygen, reaching the working electrode; G, rotating ring–disc electrode; H, lid with water seal to prevent oxygen ingress during rotation. (After Marsland, 1992.)

species formed at the disc electrode reach the surface of the ring and are detected as an oxidation or reduction current, depending on the ring electrode potential; the remaining species are dispersed into the bulk solution (Albery and Hitchman, 1971; Bard and Faulkner, 1980; Filinovsky and Pleskov, 1984).

For the system

$$O + ze^- \overset{disc}{\rightarrow} R \tag{7.77}$$

$$R \overset{ring}{\rightarrow} O + ze^- \tag{7.78}$$

in which reaction (7.78) is mass transport controlled, the collection efficiency, N_0, is defined by

$$N_0 = \frac{I_r}{I_d} \tag{7.79}$$

and was shown mathematically (Albery and Hitchman, 1971; Filinovsky and Pleskov, 1984) to depend only on the radii of the ring (r) and the disc (d), being independent of the electrode rotation rate.

7.9.3 Ring electrode detection of H_2S from reduction of a pyrite disc electrode

Figure 7.4 predicts that the reduction of pyrite in acid solutions occurs by the reaction:

$$FeS_2 + 4H^+ + 2e^- \leftrightarrow Fe^{2+} + 2H_2S \tag{7.29}$$

so that, in principle, measurement of the resulting current would enable the rate of evolution of H_2S to be determined:

$$\text{Rate of } H_2S \text{ generation (mol s}^{-1}) = \frac{I_{d,\,H_2S}}{F} \tag{7.80}$$

However, at potentials at which the dissolution reaction (7.29) occurs, H_2 (g) is discharged simultaneously:

$$H_2 + 2e^- \rightarrow H_2\ (g) \tag{7.5}$$

$$\text{Rate of } H_2 \text{ generation (mol s}^{-1}) = \frac{I_{d,\,H_2}}{2F} \tag{7.81}$$

as pyrite is a mineral with a low overpotential for hydrogen evolution; the Nernst equation (7.6) was not plotted in Figure 7.4 for the sake of clarity. This means that the fractional current efficiency (Φ), the proportion of the total current (I_t) used in the dissolution reaction (7.29) will be less than unity:

$$\Phi = \frac{I_{d,\,H_2S}}{I_{total}} = \frac{I_{d,\,H_2S}}{I_{d,\,H_2S} + I_{d,\,H_2}} \tag{7.82}$$

Slow potential sweeps of the pyrite disc electrode showed (Figure 7.20) that the reactions were at least partially (proton) transport controlled at the negative potentials required to drive reaction (7.29), as the resulting current depended on the electrode rotation rate, though not as simply as predicted by the Levich equation (7.76).

During the sweep, reaction (7.29) occurred, possibly involving metastable FeS_{2-x} phase intermediates:

$$FeS_2 + 2xH^+ + 2xe^- \leftrightarrow FeS_{2-x} + xH_2S \tag{7.83}$$

At potentials at which these reactions are thermodynamically feasible, the hydrogen evolution reaction (7.5) occurred in parallel, at potentials more negative than those given by equation (7.5). At very negative overpotentials (Figure 7.4), the products of reactions (7.83) and (7.29) can be reduced further to elemental iron:

$$Fe^{2+} + 2e^- \leftrightarrow Fe^0 \tag{7.84}$$

$$E_{Fe^{2+}/Fe}\ (\mathrm{V}) = -0.474 + 0.0296\log(Fe^{2+})$$

and/or

$$FeS_{2-x} + (4-2x)H^+ + (4-2x)e^- \leftrightarrow Fe^0 + (2-x)H_2S \tag{7.85}$$

Hence, a rotating (Ag/Ag_2S) ring–disc (FeS_2) electrode was used to enable discrimination of those contributions to the measured disc electrode current that produced H_2S, by using a ring electrode reaction that responded only to the H_2S generated at the disc and not to the co-evolved hydrogen, the dissolved concentration of which is limited by its low solubility. At suitably high anodic overpotentials, the oxidation of silver by the reaction, at pH < 7

$$Ag_2S + 2H^+ + 2e^- \leftrightarrow 2Ag + H_2S \tag{7.86}$$

$$E \text{ vs. SHE (V)} = -0.0664 - 0.0591\,\mathrm{pH} - 0.0296\log(H_2S)$$

is known (Marsland, Dawe and Kelsall, 1990; Marsland, 1992) to be transport controlled with respect to H_2S and so can act as a linear sensor for H_2S. Hence, by measuring the ring electrode current due to the oxidation reaction (7.86) and knowing the collection efficiency (N_0) of the RRDE, both the rate of the H_2S generation reaction and the rate of the H_2 evolution reaction on the disc can be determined:

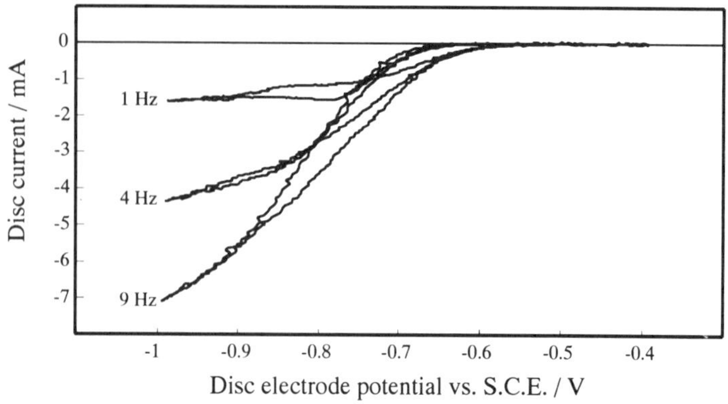

Figure 7.20 $5\,mV\,s^{-1}$ linear potential sweep voltammograms of a pyrite (FeS_2) rotating-disc electrode at pH 2, indicating the effect of increasing rotation rate in enhancing the reduction rate at low potentials. (Courtesy of I. Cservanyak, Imperial College.)

$$I_d = I_{d,\,H_2S} + I_{d,\,H_2} \tag{7.87}$$

and

$$I_r = 2N_0 \cdot I_{d,\,H_2S} \tag{7.88}$$

The 2 in equation (7.88) arises from reaction (7.29) involving one electron per H_2S molecule generated, but reaction (7.86) producing two electrons per H_2S molecule detected. Hence, the rates of H_2S generation and H_2 evolution are given by:

$$I_{d,\,H_2S} = \frac{I_r}{2N_0}; \qquad I_{d,\,H_2} = I_d - \frac{I_r}{2N_0} \tag{7.89}$$

Figure 7.21(a) shows the potential dependence of the total FeS_2 disc current, Figure 7.21(b) shows the ring electrode current response at a potential at which reaction (7.86) is mass transport controlled, and Figure 7.21(c) gives the partial disc current due to reactions other than those producing H_2S, calculated using the theoretical collection efficiency, $N_0 = 0.2045$ (Albery and Hitchman, 1971; Filinovsky and Pleskov, 1984) for the ring–disc electrode used.

The inflexion in the voltammogram at -0.8 to $-1.0\,V$ vs. SCE, shown in Figure 7.21(a), is characteristic of a nucleation and growth process; the current on the negative-going potential sweep was greater than that on the positive going sweep, indicating an additional reaction, a more catalytically active surface and/or an increased area. The response, shown in Figure 7.21(b), of the ring electrode to the potential sweep at the disc electrode,

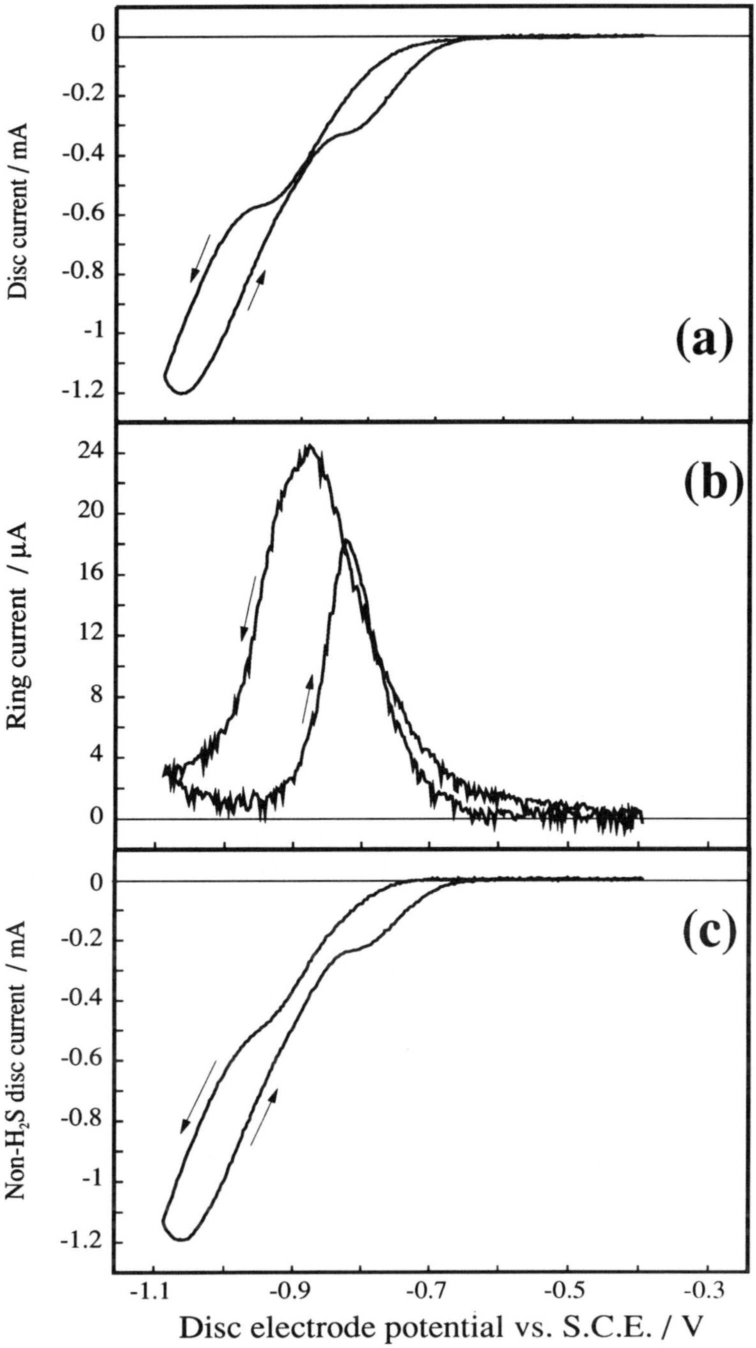

Figure 7.21 Disc and ring electrode current responses to a $5\,mV\,s^{-1}$ linear potential sweep at the disc of a rotating (Ag/Ag_2S) ring–(FeS_2) disc electrode (RRDE); pH = 4, 4 Hz rotation rate. (a) Total disc electrode current due to FeS_2 reduction; (b) Ring electrode current due to H_2S generated at the disc electrode detected by the Ag ring electrode oxidizing to Ag_2S; (c) Calculated non-H_2S generating disc electrode current. (Courtesy of I. Cservanyak, Imperial College.)

exhibited a sharp current peak, probably reflecting the depletion of sulfur in the pyrite disc surface as H_2S was evolved and, at lower potential, the formation of elemental iron. As Fe has a relatively low hydrogen overpotential, its formation by reaction (7.85) catalysed hydrogen evolution.

From the ring currents I_r, the two contributions to the disc current I_d were calculated, using equation (7.89) and:

$$I_d = I_{d,\,H_2S} + I_{d,\,Non\text{-}H_2S} \tag{7.90}$$

Hence, the data shown in Figure 7.21(c) were obtained, those contributions to the overall disc electrode currents resulting primarily in H_2 evolution, minor time-dependent components being due to reactions (7.84) and (7.85).

7.10 Conclusions

This brief summary of electrochemical thermodynamics and kinetics is intended to provide some indication of the power and utility of electrochemical techniques for investigating the chemical behaviour of sulfide minerals. More details can be found in three symposium proceedings volumes (Richardson, Srinivasan and Woods, 1984; Richardson and Woods, 1988; Richardson and Woods, 1992).

However, data from electrochemical techniques alone contain no direct information, as do data from spectroscopic techniques, about the chemical nature metal sulfide–solution interfaces, though they are clearly sensitive to their composition and properties. Hence, many authors have reported results of electrochemical experiments coupled to subsequent surface analysis, e.g. by X-ray photoelectron spectroscopy (XPS) and Auger electron spectroscopy (AES). In addition, a range of techniques has been developed in recent years for interrogating such interfaces at a molecular level in electrolytes and under potential control, e.g. Fourier transform infrared (FTIR) spectroscopy, scanning-tunnelling microscopy (STM) and spectroscopy (STS), and atomic force microscopy (AFM). These, coupled to electrochemical techniques, are only just beginning to be used to study metal sulfides; their power to provide more detailed information about the properties and behaviour of metal sulfide–solution interfaces will become evident in the near future.

Appendix 7.A Nomenclature

Symbol	Physical Meaning	Units
A_e	electrode area	m^2
c_i	electrolyte concentration of species i	$mol\,m^{-3}$
d	inter-electrode distance	m
D_i	diffusion coefficient of species i	$m^2\,s^{-1}$
e	electronic charge, 1.60219×10^{-19}	C
$\vec{E}$	electric field	$V\,m^{-1}$
E	electrode potential vs. reference electrode	V
E^0	equilibrium electrode potential under standard conditions vs. reference electrode	V
E_r	equilibrium electrode potential vs. reference electrode	V
F	Faraday constant (charge on 1 mole of electrons = $L_{Av}e$), 96485	$C\,mol^{-1}$
ΔG^0	standard Gibbs energy change for reaction	$kJ\,mol^{-1}$
i	current density	$A\,m^{-2}$
i_L	mass transport limited current density	$A\,m^{-2}$
i_0	exchange current density	$A\,m^{-2}$
I, I_t	current, total current	A
I_L	mass transport limited current	A
I	ionic strength, $0.5\sum_i c_i z_i^2$	$mol\,m^{-3}$
k_B	Boltzmann constant, 1.381×10^{-23}	$J\,K^{-1}$
k_m	mass transport rate constant	$m\,s^{-1}$
$k_{(0)}$	(standard) first-order rate constant	$m\,s^{-1}$
K	activity-based equilibrium constant	-
L_{Av}	Avogadro's constant, 6.02205×10^{23}	mol^{-1}
M	molar mass	$g\,mol^{-1}$
N_i	flux of species i	$mol\,m^{-2}\,s^{-1}$
N_0	rotating ring–disc electrode collection efficiency	-
q	charge density	$C\,m^{-2}$
Q	partial charge	C
R	gas constant, 8.31441	$J\,mol^{-1}\,K^{-1}$
R	electrolyte resistance	ohm
R_i	chemical reaction rate	$mol\,m^{-3}\,s^{-1}$
t	time	s
T	absolute temperature	K
u_i	ionic mobility	$m^2\,V^{-1}\,s^{-1}$
U	electrochemical cell terminal voltage	V
v	electrolyte velocity	$m\,s^{-1}$
V	volume	m^3
z_i	number of elementary charges on species i or number of moles of electrons per mole of electro-active species	-
α	electrochemical kinetics transfer coefficient, often 0.5	-
δ	Nernst diffusion layer thickness	m
ε_0	vacuum permittivity, 8.854×10^{-12}	$C\,V^{-1}\,m^{-1}$
ε_r	relative permittivity	-
ϕ	potential	V
Φ	(fractional) current efficiency	-
κ	electrolyte specific conductivity, $F\sum_i z_i u_i c_i$	$ohm^{-1}\,m^{-1}$

Appendix 7.A *Continued*

Symbol	Physical Meaning	Units
η	overpotential ($E - E_r$)	V
ν	electrolyte kinematic viscosity	$m^2 s^{-1}$
μ	dynamic viscosity of water, 8.95×10^{-4}	$N m^{-2} s$
ρ	density	$g m^{-3}$
σ	electrolyte conductivity	$A V^{-1} m^{-1}$
ω	RDE rotation rate	s^{-1}
ζ	zeta potential	V
∇	$\frac{\partial}{\partial x} + \frac{\partial}{\partial y} + \frac{\partial}{\partial z}$	–
∇^2	$\frac{\partial^2}{\partial x^2} + \frac{\partial^2}{\partial y^2} + \frac{\partial^2}{\partial z^2}$	–

Subscripts and superscripts

A anode
∞ bulk solution value (of concentration)
C cathode
e electrolyte
m metal
s surface

References

Albery, W.J. and Hitchman, M.L. (1971) *Ring-Disc Electrodes*. Clarendon Press, Oxford.

Bard, A.J. and Faulkner, L.R. (1980) *Electrochemical Methods*. Wiley, New York.

Bard, A.J., Parsons, R. and Jordan, J. (eds) (1985) *Standard Potentials in Aqueous Solution*. Marcel Dekker, New York.

Biegler, T. and Swift, D.A. (1979) Anodic behaviour of pyrite in acid solutions. *Electrochim. Acta*, **24**, 415–20.

Biernat, R.J. and Robins, R.G. (1969) High temperature potential/pH diagrams for the sulfur–water system. *Electrochim. Acta*, **14**, 809–20.

Bockris, J.O'M. and Reddy, A.K.N. (1973) *Modern Electrochemistry*, Vol. 2. Plenum Press, New York.

Dean, F.W.H. and Kelsall, G.H. (1992) Photoelectrochemical behaviour of chalcopyrite ($CuFeS_2$) in alkaline solutions, in *3rd International Symposium on Electrochemistry in Mineral and Metal Processing* (eds P.E. Richardson and R. Woods), Electrochemical Society, New Jersey, pp. 297–317.

El-Raghy, S.M. and El-Demerdash, M.F. (1989) Computation of E_h-pH diagrams for M–S–H_2O systems: a new approach. *J. Electrochem. Soc.*, **136**, 3647–54.

Filinovsky, V.Yu. and Pleskov, Yu.V. (1984) Rotating disk and rotating ring-disk electrodes, in *Comprehensive Treatise of Electrochemistry*, Vol. 9 (eds E. Yeager *et al.*), Plenum Press, New York, pp. 293–350.

REFERENCES

Goodenough, M.R., Kelsall, G.H. and Liang, Z. (1988) Electrochemical decomposition of covellite (CuS) particles in a bipolar packed bed electrode, in *2nd International Symposium on Electrochemistry in Mineral and Metal Processing* (eds P.E. Richardson and R. Woods), Electrochemical Society, New Jersey, pp. 339–59.

Kelsall, G.H. and Page, P.W. (1984) Aspects of chalcopyrite electrochemistry, in *Symposium on Electrochemistry in Mineral and Metal Processing* (eds P.E. Richardson, S. Srinivasan and R. Woods), Electrochemical Society, New Jersey, pp. 303–16.

Kelsall, G.H. and Thompson, I. (1993a) The redox chemistry of H_2S oxidation by the British Gas Stretford Process. I. Thermodynamics of sulphur–water systems. *J. Appl. Electrochem.*, **23**, 279–86.

Kelsall, G.H. and Thompson, I. (1993b) The redox chemistry of H_2S oxidation by the British Gas Stretford Process. II. Electrochemical kinetics of HS^-/S redox systems. *J. Appl. Electrochem.*, **23**, 287–95.

Kelsall, G.H., England, K., Yin, Q. and Vaughan, D.J. (1992) Electrochemical oxidation of chalcopyrite ($CuFeS_2$) in alkaline solutions, in *3rd International Symposium on Electrochemistry in Mineral and Metal Processing* (eds P.E. Richardson and R. Woods), Electrochemical Society, New Jersey, pp. 318–41.

Kuhn, A.T., Charna, M.S. and Kelsall, G.H. (1983) A review of the air oxidation of aqueous sulphide solutions. *J. Chem. Biotechnol.*, **33A**, 406–14.

Levich, V.G. (1962) *Physicochemical Hydrodynamics*. Prentice-Hall, Englewood Cliffs, New Jersey.

Liang, Z. (1988) *Electrochemical behaviour of cuprous sulfide (Cu_2S) in alkaline solutions*. PhD thesis, University of London.

Macdonald, D.D. (1981) *Transient Techniques in Electrochemistry*. Plenum Press, New York.

Marsland, S.D. (1992) *H_2S Generation by Iron Sulfide Decomposition in Petroleum Reservoirs*. PhD thesis, University of London.

Marsland, S.D., Dawe, R.A. and Kelsall, G.H. (1990) An electrochemical approach to oil reservoir souring. *Trans. I. Chem. E.*, **68A**, 357–64.

Mishra, K.K. and Osseo-Asare, K. (1988a) Electrodeposition of H^+ on oxide layers at pyrite (FeS_2) surfaces. *J. Electrochem. Soc.*, **135**, 1898–901.

Mishra, K.K. and Osseo-Asare, K. (1988b) Aspects of the interfacial electrochemistry of semiconductor pyrite (FeS_2). *J. Electrochem. Soc.*, **135**, 2502–9.

Newman, J. (1973) *Electrochemical Systems*, Prentice-Hall, New Jersey.

Page, P.W. (1989) *The Electrochemistry of Pyrite (FeS_2), Pyrrhotite (FeS), Chalcopyrite ($CuFeS_2$) and Pentlandite ($(Fe,Ni)_9S_8$)*. PhD thesis, University of London.

Pletcher, D. and Walsh, F.C. (1982) *Industrial Electrochemistry*. Chapman & Hall, London.

Pourbaix, M. (1966) *Atlas of Electrochemical Equilibria in Aqueous Solutions*. Pergamon Press, Oxford.

Richardson, P.E. and Woods, R. (eds) (1988) *2nd International Symposium on Electrochemistry in Mineral and Metal Processing*. Electrochemical Society, New Jersey.

Richardson, P.E. and Woods, R. (eds) (1992) *3rd International Symposium on Electrochemistry in Mineral and Metal Processing*. Electrochemical Society, New Jersey.

Richardson, P.E., Srinivasan S. and Woods, R. (eds) (1984) *Symposium on Electrochemistry in Mineral and Metal Processing*. Electrochemical Society, New Jersey.

Rousar, I., Micka, K. and Kimla, A. (1986) *Electrochemical Engineering I & II*. Elsevier, Amsterdam.

Southampton Electrochemistry Group (1985) *Instrumental Methods in Electrode Kinetics*. Ellis Horwood, Chichester.

Vaughan, D.J. and Craig, J.R. (1977) *Mineral Chemistry of Metal Sulfides*. Cambridge University Press, Cambridge.

Wadsley, M. (1992) Stoichiometric considerations in multicomponent aqueous stability diagrams. *Hydrometallurgy*, **29**, 91–108.

Wagman, D.D., Evans, W.H., Parker, V.B. *et al.* (1982) The NBS tables of chemical thermodyamic properties. *J. Phys. Chem. Ref. Data*, **11**, (Suppl. 2).

Welham, N.J. (1992) *Selective Leaching of Gold from Pyrite (FeS_2) Coupled to Gold Electrowinning*. PhD thesis, University of London.

Woods, R. (1981) *Mineral Flotation*, in *Comprehensive Treatise of Electrochemistry*, Vol. 2 (eds J.O'M. Bockris *et al.*), Plenum Press, New York, pp. 571–95.

Woods, R., Yoon, R.H. and Young, C.A. (1987) E_h-pH diagrams for stable and metastable phases in the copper-sulfur-water system. *Int. J. Min. Proc.*, **20**, 109–20.

Yeager, E., Bockris, J.O'M., Conway, B.E. and Sarangapani, S. (eds) (1983) *Comprehensive Treatise on Electrochemistry, Vol. 6, Electrodics: Transport*. Plenum Press, New York.

Yokokawa, H., Sakai, N., Kawada, T. and Dokiya, M. (1990) Generalized electrochemical potential diagrams for complex aqueous (M-X-H-O-e^-) systems. *J. Electrochem. Soc.*, **137**, 388–98.

CHAPTER EIGHT

Surface chemistry of sulfide flotation

P.E. Richardson

8.1 Introduction

Mineral flotation is the technology of attaching air bubbles to the surfaces of particles in a highly selective manner to separate and concentrate individual mineral components. With its introduction in Australia and the USA at the turn of this century, it quickly grew and remains the most important unit process for concentrating minerals in terms of tonnages of materials processed and diversity of applications. Its applications encompass the entire mineral spectrum: sulfide ores, non-sulfide ores (oxides, carbonates and sulfates), non-metallic minerals (silicates and insoluble salts), sulfur, coal, phosphates, precious metals and bitumen from tar sands (Leja, 1982).

Flotation is an early and highly successful application of surface chemistry in which the wettability of solid surfaces is controlled by chemical means. The fundamental thermodynamic requirement in flotation is the attachment of an air bubble to a mineral surface in an aqueous medium. This attachment requires that the work of adhesion of water to the surface of the solid, W_A, be less than the work of cohesion of water, W_C, that is

$$W_A < W_C \quad (8.1)$$

According to Fowkes (1967), the work of adhesion of water to a solid surface depends on three independent short range components comprising dispersion forces, W_d, hydrogen bonding, W_h, and ionized groups on the surface, W_i, where,

$$W_A = W_d + W_h + W_i \quad (8.2)$$

From considerations of W_A and W_C, Laskowski and Kitchener (1969) have shown that all solids would be hydrophobic if they did not carry polar or ionic groups (the work of cohesion of pure water is 146 erg cm^{-2}; the dispersion force component, W_d, is less than 146 erg cm^{-2} for all solids). The clean (non-oxidized) surfaces of only a few minerals meet the criteria $W_A < W_C$ (e.g. graphite, high rank coals, bitumen and other hydrocarbons, sulfur, talc, molybdenite and stibnite).

Mineral Surfaces. Edited by D.J. Vaughan and R.A.D. Pattrick. Published in 1995 by Chapman & Hall, London. ISBN 0 412 56340 1

Naturally hydrophobic minerals can be separated from associated hydrophilic minerals by exploiting their inherent floatability. Alternatively, their flotation may be depressed through oxidation or the incorporation of hydrophilic polar or ionic groups on the surface.

Most minerals have polar surfaces and may also have ionic groups on the surface. Such surfaces are inherently hydrophilic and non-floatable. These may be made floatable by using surfactants, which adsorb or react with solid surfaces to decrease the work of adhesion of water.

The surface chemistry of flotation selectively controls the wettability of one or a few minerals, in a complex mixture of minerals, to obtain differential separations. This is accomplished using a wide variety of inorganic or organic chemicals, classified as flotation modifiers and flotation collectors. Modifiers are often, but not always, inorganic chemicals that promote collector adsorption or depressants that increase the polar and hence hydrophilicity of surfaces (or otherwise inhibit collector adsorption). Collectors are organic surfactants with a hydrophilic polar end, which may or may not be ionized, and a hydrophobic or non-polar group, which is usually a hydrocarbon but may be a fluorocarbon or a siloxane. The polar end generally interacts specifically with the mineral surface or adsorbs in the double layer just outside the surface; the hydrocarbon end has a higher chemical affinity for the gas phase than for water.

Flotation is a complex sequence of events and the overall process depends on many factors other than the surface chemistry of mineral–collector and mineral–modifier interactions. Other factors include: bubble–particle collisions; the thinning of liquid films during solid–bubble attachment; the use of surfactants other than collectors that act primarily at the water–gas interface and effect the kinetics of bubble attachment; the hydrodynamics of particle–water slurries; and a host of physicochemical processes involving the solution, as well as the multiple component solid phases that are present.

This chapter is directed at the reactions that occur at the surfaces of minerals and that are exploited to control their floatability. A single chapter cannot comprehensively treat all aspects of the surface chemistry of flotation and therefore the choice was made to discuss, in some detail, the surface chemistry of sulfide flotation. This choice also reflects the author's particular interest in the relationship between the flotation of sulfides and a mineralogical property (electronic structure) of the sulfide minerals.

8.2 Sulfide flotation: general

Most sulfide minerals important in flotation are either semiconducting or metallic, with electrical conductivities sufficient to support charge-transfer electrochemical reactions when they are in contact with an aqueous phase. Current theories of sulfide flotation are based on mixed potential electrochemical reactions. Mixed potential processes involve two simultaneous reactions, i.e. an anodic and a cathodic reaction such that, at equilibrium, overall charge neutrality is maintained at the surface. Hydrophobic surfaces are produced by an anodic reaction between sulfides and thiol-type collectors, while dissolved oxygen is simultaneously reduced on the sulfides.

Most sulfides are also unstable in the presence of oxygen and water, and when exposed by mining they tend to oxidize by similar mixed potential electrochemical reactions. Because the flotability of sulfides involves a balance between small (submonolayer) quantities of hydrophobic and hydrophilic groups on the surface, and because the chemical composition of the surface influences subsequent reactions with thiol collectors, the oxidation of sulfides are of fundamental importance in flotation. Several distinct effects of oxidation can be distinguished.

1. A number of sulfides exhibit 'self-induced' flotation, i.e. flotation in the absence of a collector, which is related to an oxidation product on the surface (Trahar, 1984).
2. Oxidation products involving the cation of sulfides are highly polar; when not soluble or displaced by the collector the resulting hydrophobicity represents a balance between coverage with the hydrophobic collector and coverage with the hydrophilic oxide groups (Shannon and Trahar, 1986; Senior and Trahar, 1991).
3. Oxidation products are often soluble, producing metal cations in solution, which then may adsorb or react with other minerals. This may cause the surfaces of different minerals to be similar with regard to interactions with thiol collectors (Guy and Trahar, 1985; Gebhardt and Richardson, 1987).
4. Thiol collectors are known to undergo ion exchange reactions with oxidation products on the surfaces of sulfides, providing an alternative reaction mechanism (Sutherland and Wark, 1955).

8.3 Self-induced flotation

For years the natural or inherent hydrophobicity or hydrophilicity of 'clean' surfaces of sulfides was a controversial and debated topic. With the exception of sulfides whose predominant cleavage surfaces are created by the rupture of van der Waals bonds (e.g. molybdenite, realgar and orpiment, which are highly floatable), most sulfides were thought to be hydrophilic. However, reports of flotation and finite contact angles in the absence of conventional collectors led a number of workers to challenge this view. Sutherland and Wark (1955) discussed the difficulties in reaching definitive conclusions on this topic; they concluded 'the authors . . . cannot agree whether a galena surface prepared in the presence of water and absence of oxygen, is inherently floatable.' Using arguments based on the ability of the constituents of water to hydrogen bond with sulfide minerals, Finkelstein *et al.* (1975)

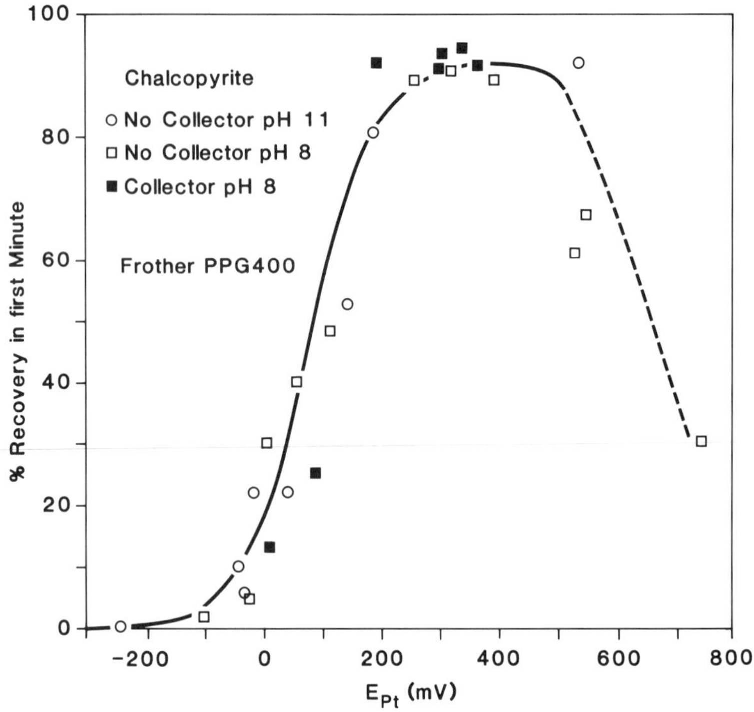

Figure 8.1 The flotation of chalcopyrite in the presence and absence of a collector as a function of potential. (Adapted from Trahar, 1984, with permission of the Australasian Institute of Mining and Metallurgy.)

suggested that there were no strong wetting forces at a clean (non-oxidized) sulfide surface. As concluded by these researchers, 'sulfide minerals are neither strongly hydrophobic nor strongly hydrophilic and whether they float or not can be markedly influenced by relatively minor changes at their surfaces.'

The topic of the inherent flotability of sulfides was advanced considerably by Heyes and Trahar (1977), who demonstrated that chalcopyrite could be floated in the absence of a collector under slightly oxidizing conditions, but could not be floated under reducing conditions. Figure 8.1 shows their results for the flotation of chalcopyrite, in the absence and presence of a collector, as a function of the potential of a platinum wire inserted in the flotation pulp. Guy and Trahar (1984), Heyes and Trahar (1984) and Trahar (1984) subsequently determined that a number of sulfides exhibit some degree of flotation under oxidizing conditions. They established that chalcopyrite, galena, pyrrhotite and pentlandite are strongly floatable in the absence of a collector, whereas bornite, chalcocite, pyrite and sphalerite show lesser degrees of flotation. As stated by Trahar (1984), there is

'. . . little room for doubt that some sulfides are capable of rapid and complete flotation in the absence of conventional collectors . . .' Trahar referred to flotation resulting from moderate oxidation as 'self-induced' flotation to distinguish it from collector-induced flotation and from the natural or inherent flotation exhibited by sulfides whose predominant cleavage surfaces are produced by rupture of van der Waals bonds.

In recent years, numerous studies have been directed at identifying the hydrophobic surface entity responsible for self-induced flotation. These studies have been carried out on surfaces prepared by fracture, cleavage and abrasion, and on particulate bed electrodes. *In situ* techniques have included, for the most part, linear sweep voltammetry, cyclic voltammetry and intermittent galvanostatic polarization. *In situ* electrochemical techniques have been supplemented with *ex situ* X-ray photoelectron spectroscopy (XPS).

It is now known that sulfides oxidize through several different pathways, indicated by the following reactions:

$$MS + 2H_2O = M(OH)_2 + S + 2H^+ + 2e^- \qquad (8.3)$$

$$MS + 2\lambda H_2O = \lambda M(OH)_2 + M_{1-\lambda}S + 2\lambda H^+ + 2\lambda e^- \qquad (8.4)$$

$$2nMS + 2nH_2O = MS_{2n} + nM(OH)_2 + 2nH^+ + 2ne^- \qquad (8.5)$$

$$MS + 6H_2O = M(OH)_2 + SO_4^{2-} + 10H^+ + 8e^- \qquad (8.6)$$

$$2MS + 7H_2O = 2M(OH)_2 + S_2O_3^{2-} + 10H^+ + 8e^- \qquad (8.7)$$

Because most sulfide flotation circuits operate above pH 7, the reactions have been written for basic solutions, assuming the metal hydroxide is the stable cation product. The metal hydroxide may be soluble under some conditions and metal oxides or metal-hydroxide complexes may also be formed.

Reactions (8.3)–(8.5) each produce two products: one an oxidation product of the cation that is expected to be polar and hydrophilic; the other corresponding to an increase in sulfur in one of three different forms – elemental sulfur (reaction 8.3), excess sulfur in the surface region due to a metal-deficiency (reaction 8.4), and polysulfide (reaction 8.5). It is known that sulfur (Wark and Cox, 1935) and sulfur-rich surfaces (Gaudin, Miaw and Spedden, 1957) are highly hydrophobic. Both of these have been proposed as the hydrophobic entity responsible for self-induced flotation.

Reaction (8.3) was proposed by Richardson and Maust (1976) for galena oxidation in alkaline solutions based on voltammetry and photovoltaic studies. Figure 8.2 shows cyclic voltammetry curves for galena as a function of pH in quiescent and stirred solutions. The reactions were identified from the potentials at which they occurred and the pH dependence of these potentials. Soluble vs. insoluble products were identified by comparing the currents in quiescent and agitated solutions. Agitation disperses soluble

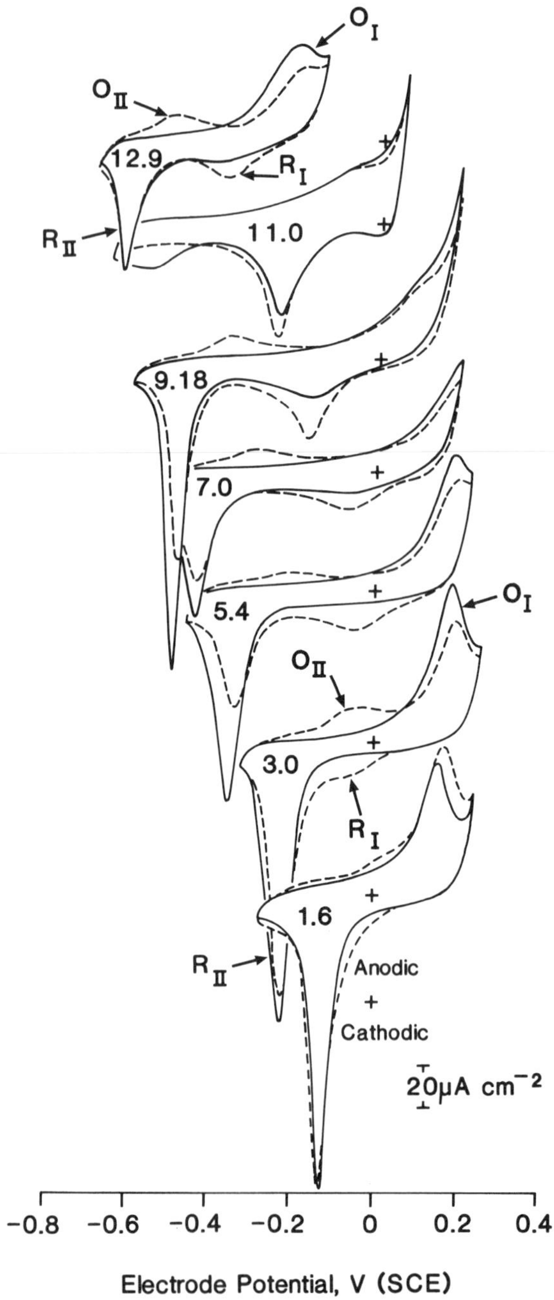

Figure 8.2 Cyclic voltammetry curves for galena as a function of pH in quiescent (dashed curves) and stirred (solid curves) solutions. Sweep rate: $25\,mV\,s^{-1}$. (From Richardson and Maust, 1976.)

products into the bulk solution; under quiescent conditions the soluble products remain in the vicinity of the electrode where they can undergo further reactions.

The current peak, O_I, represents the anodic oxidation of galena. All of the other peaks on the voltammograms were shown to be related to this peak. R_I was attributed to the reduction of Pb(II) species produced by oxidation, R_{II} to the reduction of sulfur produced during oxidation, and O_{II} to the re-oxidation of the sulfur reduction product. The sulfur reduction product (HS^- in alkaline, H_2S in basic solution) is soluble and dispersed in agitated solutions, thus O_{II} disappears with agitation of the solution. The

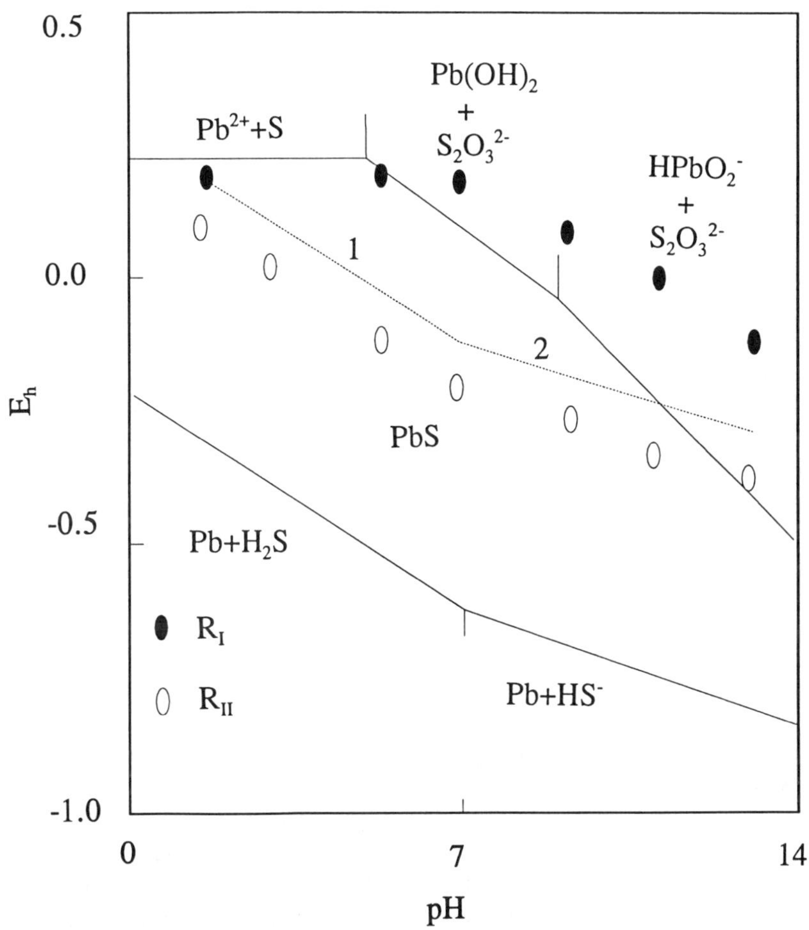

Figure 8.3 Pourbaix diagram for galena and for the sulfur/H_2S (line 1) and sulfur/HS^- (line 2) reactions. Activities of soluble Pb and sulfur species = 10^{-4} M. Experimental points are the peak potentials of R_I and R_{II} obtained from the voltammograms in Figure 8.2.

Table 8.1 Products of anodic oxidation of sulfide minerals in basic solutions from electrochemical investigations

Mineral	Oxide	Sulfide	Major S	Minor S
$Fe_{1-x}S$	$Fe(OH)_3$	–	S	SO_4^{2-}
FeS_2	$Fe(OH)_3$	–	SO_4^{2-}	S
$CuFeS_2$	$Fe(OH)_3$	CuS	S	–
Cu_5FeS_4	$Fe(OH)_3$	Cu_5S_4	–	–
Cu_2S	$Cu(OH)_2$	$Cu_{2-x}S$	–	–
CuS	$Cu(OH)_2$	–	S	$S_2O_3^-$
PbS	$Pb(OH)_2$	–	S	$S_2O_3^{2-}$
$(Fe,Ni)_9S_8$	$Fe(OH)_3$	n.i.*	S	SO_4^{2-}

* n.i. = not identified.
From Woods (1984), with permission of The Australasian Institute of Mining and Metallurgy.

peak potentials are plotted on the Pourbaix diagram of galena in Figure 8.3. There is good agreement between the experimental peak potentials and their pH dependence and the thermodynamic potentials.

Subsequent studies (Lamache, Bauer and Pegouret, 1981; Hamilton and Woods, 1984; Pritzker and Yoon, 1984) have concluded that elemental sulfur is a major oxidation product of galena at all pHs, with divalent lead ions being formed in acid, and lead oxide, hydroxide and plumbite ions being formed in alkaline solution.

The minor and major oxidation products for several sulfides have been listed by Woods (1984) as shown in Table 8.1. Sulfur is a major oxidation product on pyrrhotite, chalcopyrite, covellite and pentlandite, and a minor product on pyrite.

Trahar (1984) has presented several arguments for and against sulfur as the hydrophobic entity. First, in alkaline solutions, sulfur is unstable with respect to sulfate; however, the production of sulfate through oxidation of sulfides is known to be kinetically hindered. Second, attempts to leach sulfur from mineral surfaces has not produced positive results (Heyes and Trahar, 1977; Luttrell and Yoon, 1983). Some evidence has been provided for the existence of sulfur as an oxidation product on sulfides by mass spectroscopy (Ralston and Healy, 1980a, b) and XPS (Buckley and Woods, 1981; Luttrell and Yoon, 1983).

Sulfur has been electro-deposited on pyrite (Walker, Walters and Richardson, 1986) and pyrite and chalcopyrite (Buckley, Woods and Wouterlood, 1988) in alkaline solutions containing sodium sulfide and has a pronounced effect on the flotability and contact angle. Figure 8.4 shows a cyclic voltammetry curve for the oxidation of sulfide ions to sulfur on pyrite and the corresponding contact angle and flotation response that accompanies sulfur deposition. Elemental sulfur was identified by XPS. The data provide

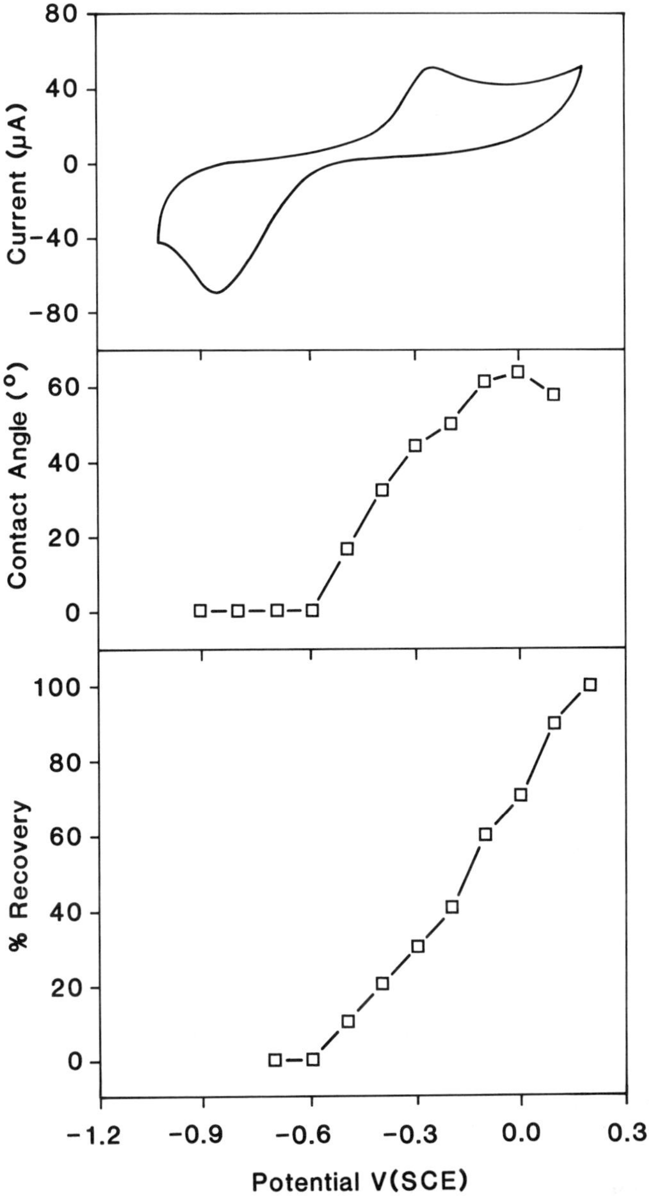

Figure 8.4 Flotation recovery (bottom), contact angle (middle) and current–voltage curve (top) of pyrite in 0.01 M HS^- dissolved in 0.05 M $Na_2B_4O_7$. (From Walker, Walters and Richardson, 1986.)

convincing evidence that a hydrophobic sulfur entity can be produced on the surfaces of sulfides in alkaline solutions and that it can exist for sufficiently long periods to influence flotation. Yoon (1981), Luttrell and Yoon (1983) and Trahar (1983) found that sodium sulfide could be used to obtain consistently high recoveries in the flotation of chalcopyrite, providing additional support that sulfur in one form or another provides sufficient hydrophobicity for flotation.

Reaction (8.4), often referred to as the metal-deficient model, was proposed on the basis of XPS studies on the oxidation of galena (Gardner and Woods, 1979a; Buckley and Woods, 1984) and the oxidation of pyrrhotite and pyrite (Hamilton and Woods, 1981; Buckley and Woods, 1987) and chalcopyrite (Buckley and Woods, 1981; Buckley, Hamilton and Woods, 1985). This reaction was suggested by observations that the binding energies of the cations of these sulfides, with slight to moderate oxidation, shift to known binding energies of the metal oxidation products, whereas there appears to be little or no shift in the binding energy of sulfur. Figure 8.5 shows XPS spectra obtained by Buckley, Hamilton and Woods (1985) on a galena sample after several stages of oxidation. Exposure of a fresh galena surface to air for one day does not change the S(2p) spectrum (curves A and B), but new intensity on the high binding energy side of the Pb(4f) peaks show that an oxidized lead species has formed. Immersion in dilute acetic acid for 35 days causes a significant shift in the S(2p) spectrum (curve C) from the original PbS position (curve A), but still does not correspond to the binding energy of elemental sulfur. Oxidation with dilute peroxide and washing to remove sulphate yields two S(2p) peaks, one corresponding to sulfur in PbS, the other to elemental sulfur with a binding energy of 163.5 eV. To explain this behaviour, Buckley, Hamilton and Woods proposed that with oxidation, the more highly mobile cations move to the surface, where a metal oxide layer is produced. The less mobile sulfur remains in normal lattice positions. Thus a number of atomic layers beneath the oxide film may be depleted of cations. The oxide (hydroxide) layer may be soluble or may be removed in the turbulence of a flotation cell, leaving a surface that is predominately sulfur-rich and hydrophobic, accounting for self-induced flotation.

Richardson and Maust (1976) discussed a cation vacancy model for the oxidation of galena; however, they proposed that a sulfur-rich region formed by oxidation, and a metal-rich region formed by reduction, is appropriate only within the range that sulfur and lead are soluble in lead sulfide, i.e. about 0.1% (Scanlon, 1959). Buckley and Woods (1984) suggest that reaction (8.4) is not limited by bulk solubility. This is one uncertainty concerning the metal-deficient sulfide model. Another is that voltammetry curves frequently show a well-defined peak associated with the reduction of the sulfur-containing oxidation product (Figure 8.2). This behaviour is consistent with a second solid phase such as sulfur on the surface. Sulfur

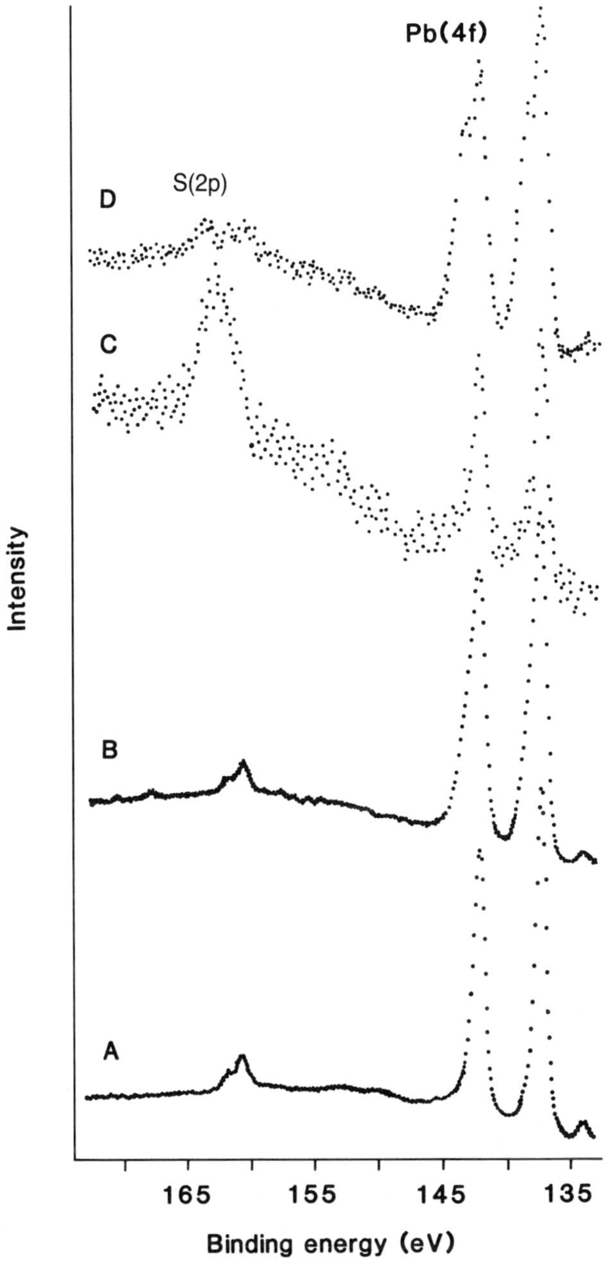

Figure 8.5 S(2p) and Pb(4f) photoelectron spectra at 150 K of galena surfaces: (A) exposed to air for 1 min; (B) exposed to air for 1 day; (C) washed after immersion in dilute acetic acid for 35 days; and (D) washed after oxdiation with hydrogen peroxide solution. (From Buckley, Hamilton and Woods, 1985, with permission of *J. Appl. Surf. Sci.*)

in solid solution, as implied by the metal-deficient model, would have a continuous range of activities and would not, therefore, be expected to give a well-defined reduction peak.

Reaction (8.5) was suggested by Luttrell and Yoon (1983) from studies of the oxidation of chalcopyrite. They found XPS spectra for sulfur similar to those found by Buckley and Woods (1981), i.e. three sulfur 2p components that could be fitted to $CuFeS_2$, CuS, and S^0 (Table 8.1). They suggested the non-sulfide component could be assigned to the formation of a copper polysulfide, which they assumed to be hydrophobic. One of the controversial topics in current research concerns distinguishing the oxidation product of reactions (8.4) and (8.5) with regard to the form of sulfur. In a metal-polysulfide, it would be expected that the sulfur adjacent to the metal would exhibit a different bonding energy to sulfur sharing other sulfurs as nearest neighbours. This has been the case for the $[Pt(S_5)_3]^-$ complex, where two and possibly three distinct electronic environments for the sulfur atoms were present (Termes, Buckley and Gillard, 1987). No definitive experiment has been designed to distinguish metal-deficient sulfur from polysulfide as an oxidation product on sulfide minerals. To distinguish different binding energies for sulfur in a polysulfide, the number of sulfurs would presumably have to be greater than two.

Reactions (8.6) and (8.7) leading to sulfate or thiosulfate are kinetically slow and occur under stronger oxidizing conditions. Reactions (8.3) and (8.4) may be precursor reactions to sulfate and thiosulfate formation. These reactions remove sulfur from the surface and do not contribute to self-induced flotation, but may be responsible for the depression of sulfides under stronger oxidizing conditions (Figure 8.1).

The oxidation of chalcocite highlights one of the difficulties in studying the adsorption process on sulfide surfaces when their composition changes due to oxidation. Koch and McIntyre (1976) used rest potential, reflectance and X-ray diffraction measurements to study the electro-oxidation of thin films of chalcocite. The oxidation rates and film thicknesses were such that equilibrium could be established throughout the samples. From the dependence of the potential of the electrode during oxidation (Figure 8.6), they determined the step-wise oxidation sequence

$$Cu_2S \rightarrow Cu_{1.93}S \rightarrow Cu_{1.83}S \rightarrow Cu_{1.67}S \rightarrow Cu_{1.4}S \rightarrow CuS \tag{8.8}$$

corresponding to the progressive formation of five distinct copper-sulfide phases. In natural samples used in flotation research, some of these phases may co-exist, in particular chalcocite and djurleite (Young, Woods and Yoon, 1988). In electrochemical experiments, the surface region may convert to one or more of these phases.

Tertiary sulfides, such as chalcopyrite and bornite, have a more complex oxidation sequence than binary sulfides. Gardner and Woods (1979b) and

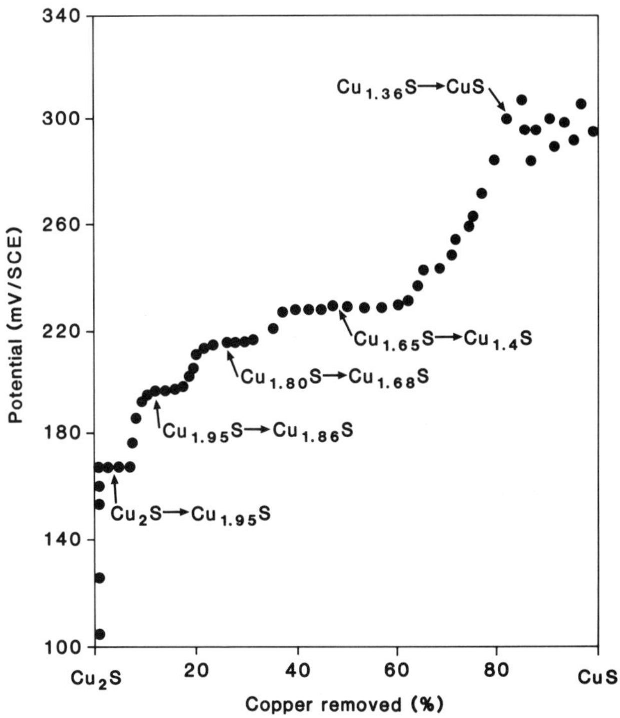

Figure 8.6 Rest potential during the electro-oxidation of Cu_2S in acid solution. The vertical steps represent two phase transistion regions, the plateaux single phase regions indicated on the diagram. (Adapted from Koch and McIntyre, 1976, with permission of Elsevier Sequoia SA, Lausanne.)

Buckley, Hamilton and Woods (1985) have suggested chalcopyrite oxidizes by the reaction

$$CuFeS_2 + 3H_2O = CuS_2 + Fe(OH)_3 + 3H^+ + 3e^- \qquad (8.9)$$

and attributed its self-induced flotation to CuS_2. Buckley, Hamilton and Woods (1984) have found that bornite oxidizes by the reaction

$$Cu_5FeS_4 + 3H_2O = Cu_5S_4 + Fe(OH)_3 + 3H^+ + 3e^- \qquad (8.10)$$

8.4 Collector-induced flotation

The flotation of sulfide minerals is carried out almost exclusively with thiol-type collectors. Thiol collectors have a high chemical reactivity for heavy ions in solution or at mineral surfaces. These include O-alkyl dithiocarbonates (xanthates), dialkyl dithiophosphates (aeroflots), dialkyl dithiocarbamate, and mercaptobenzothiazole. Because of their short hydrocarbon chains and high solubilities as anions, thiol collectors only weakly adsorb at the gas–liquid interface and thus act mainly as surfactants with a high chemical affinity for reacting with heavy metal sulfides. Xanthates ($ROCS_2^-$, R = CH_2, with R = 2–16) were first used in flotation in 1923 (Leja, 1982) and are the most widely used collectors for sulfides. The xanthates are also the most studied and best understood collectors with regard to their interactions with sulfide minerals. As a result xanthate–sulfide systems have assumed the role of model systems for sulfide flotation research and will be the subject of much of the remainder of this chapter.

Early theories of the collecting action of thiol collectors were based on chemical reactions but were inconsistent with established values of the solubility products of metal xanthates. This led to models based on adsorption processes rather than chemical reactions (Sutherland and Wark, 1955). An adsorption theory for collector adsorption was attractive because it qualitatively accounted for the empirical Barsky relationship (Barsky, 1934), in which the upper pH limit of flotation was known to shift to higher values with increasing collector concentration, suggesting competition between the adsorption of collector and hydroxide ions. Early adsorption theories assumed the adsorption of charged anions, which, from electrostatic considerations, could lead to only minimal surface coverage. The adsorption of a charged anion, without the adsorption of a charge-compensating cation, would not be expected to produce a hydrophobic surface because of charge–dipole or coulombic interactions with water or its constituents. To overcome these problems, models in which the adsorption of the undissociated acid (Cook and Nixon, 1950; Cook and Wadsworth, 1957) and oppositely charged ions (Sutherland and Wark, 1955) were proposed.

The currently accepted mixed potential model was suggested by Salamy and Nixon (1954) as a method to maintain electrical neutrality at the interface during collector adsorption. In this model, the reaction between the collector and the surface takes place by an anodic process, while dissolved oxygen is separately reduced at the surface to remove the electrons donated by the oxidation process. The mixed potential model of sulfide flotation with thiol collectors is analogous to the mixed potential model of metal corrosion, where electrons left at the metal during dissolution are consumed by oxygen reduction. This model also accounts for the known necessity for the presence of oxygen in the flotation of sulfide minerals with xanthate collectors (Plaksin and Bessonov, 1957).

8.5 Mixed potential theory

8.5.1 Anodic reactions

Several anodic reactions between thiol collectors and sulfides have now been identified. These are:

1. Charge transfer chemisorption of the thiol ion (X^-)

$$X^- = X_{ads} + e^- \tag{8.11}$$

2. Oxidation of the thiol ion to its disulfide (X_2)

$$2X^- = X_2 + 2e^- \tag{8.12}$$

3. Formation of a metal thiolate

$$MS + 2X^- = MX_2 + S + 2e^- \tag{8.13}$$

Reaction (8.13) may also proceed through the formation of sulfur-oxy anions such as sulfate or thiosulfate

$$MS + 2X^- + 4H_2O = MX_2 + SO_4^{2-} + 8H^+ + 8e^- \tag{8.14}$$

or take place via separate steps which occur at different times

$$MS + 2H_2O = M(OH)_2 + S^0 + 2H^+ + 2e^- \tag{8.15}$$

$$M(OH)_2 + 2X^- = MX_2 + 2OH^- \tag{8.16}$$

Although reaction (8.16) does not involve electron transfer, reaction (8.15) as its precursor does, and all of the reactions that lead to a hydrophobic surface are electrochemical in nature or involve an electrochemical oxidation process.

8.5.2 Cathodic reactions

Dissolved molecular oxygen is the most important oxidant in a number of metallurgical processes, including flotation, and the kinetics of its reduction on sulfides has been the subject of a number of studies (Tolun and Kitchener, 1964; Biegler, Rand and Woods, 1975, 1977; Biegler, 1976; Rand, 1977; Ahmed, 1978a, b; Richardson and O'Dell, 1985).

The most illuminating studies from the viewpoint of flotation are those of Rand (1977) and Biegler, Rand, and Woods (1977), which compared the reduction of oxygen on a number of sulfide minerals as shown in Figure 8.7. The current–potential curves in this figure represent the relative catalytic ability of the several sulfides to reduce oxygen. It is clear from the figure that there are significant differences in the ability of different sulfides to reduce oxygen. Pyrite was the most active mineral studied and it has an activity for oxygen reduction slightly less than noble metals. Plaksin and Bessonov (1957) reported that the requirement for oxygen in flotation

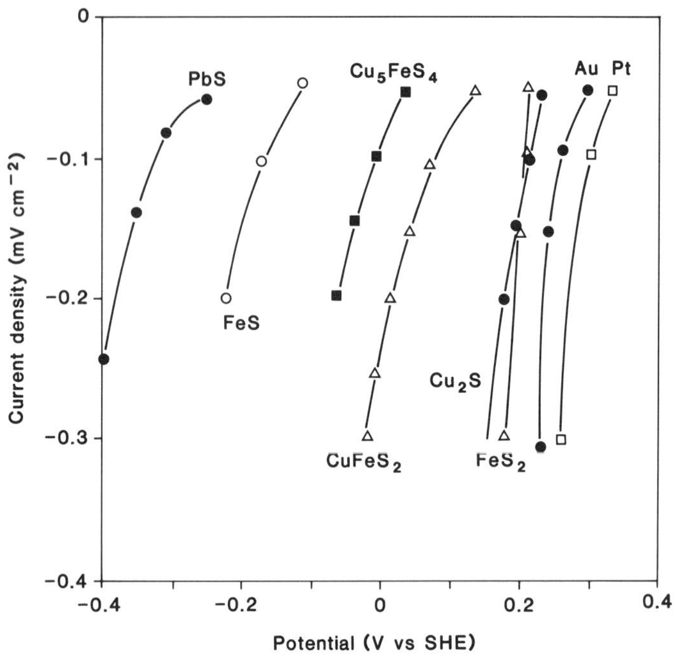

Figure 8.7 Current–potential curves obtained on rotating disc electrode of several sulfide minerals. The currents represent primarily the reduction of oxygen. (Adapted from Rand, 1977, with permission of Elsevier Sequoia SA, Lausanne.)

increased in the order galena < pyrite < chalcopyrite < pyrrhotite < arsenopyrite in alkaline solution. Woods (1976) has pointed out that this is the same order as the catalytic ability of the sulfides to reduce oxygen, with the exception that pyrrhotite and arsenopyrite are in reverse order and not including galena because it is known that the hydrophobic species for galena is different than that adsorbed on the other sulfides.

Beigler, Rand and Woods (1975) studied in detail the kinetics of oxygen reduction on pyrite. In acid solution, the overall reaction involved four electrons

$$O_2 + 4H^+ + 4e^- = 2H_2O \tag{8.17}$$

Hydrogen peroxide was detected in solution and analysis of the current–voltage curves indicated the reaction proceeded via

$$O_2 + 2H^+ + 2e^- \rightarrow H_2O_2, \qquad H_2O_2 + 2H^+ + 2e^- \rightarrow 2H_2O \tag{8.18}$$

The first electron step

$$O_2 + e^- \rightarrow O_2^- \qquad (8.19)$$

was rate determining.

The presence of hydrogen peroxide as an intermediate in oxygen reduction is consistent with reports by Jones and Woodcock (1978) that hydrogen peroxide reacts with xanthates to form perxanthate ($ROCS_2O^-$) and that perxanthate is present in industrial concentrators. It is also consistent with observations of perxanthate in microflotation cells when sulfide minerals are potentiostated at negative potentials (Richardson *et al.*, 1984).

8.6 Flotation and xanthate reactions: single sulfide minerals

8.6.1 Chalcocite

Cyclic voltammetry was used by Kowal and Pomianowski (1973) to study the reactions between ethyl xanthate and chalcocite (85% Cu_2S, 15% Cu_5FeS_4). Two closely spaced anodic current peaks were observed, which they assigned to the adsorption of a monolayer of ethyl xanthate. The two current peaks were believed to represent chemisorption at surface copper sites and the separation into two components may correspond to the copper at the surface having different crystallographic environments or to the existence of additional copper sulfide phases in the sample. A third anodic process was observed at more oxidizing potentials and attributed to the formation of a bulk-like copper xanthate.

The flotation of chalcocite with ethyl xanthate as a function of potential has been studied by Heyes and Trahar (1979), O'Dell *et al.* (1984), Richardson *et al.* (1984) and Basilio, Pritzker and Yoon, (1985). Heyes and Trahar and Basilio, Pritzker and Yoon used reducing and oxidizing reagents to control the potential; Richardson *et al.* controlled the potential of a bed of mineral particles with a potentiostat, packing the bed for electrochemical conditioning and releasing the pressure to determine the flotation response. Figure 8.8 shows a typical voltammetry curve obtained by O'Dell *et al.* on a packed chalcocite bed electrode. The two chemisorption processes are evident between -0.2 and $0\,V$ and the reaction to produce a bulk-like copper xanthate evident at potentials positive of $0\,V$. The flotation response after conditioning at each potential and the flotation response obtained by Heyes and Trahar using reducing and oxidizing reagents to control the potential are superimposed on the voltammetry curve and clearly establish that the flotation edge corresponds to the reactions attributed to chemisorption.

Figure 8.9 shows recovery–potential curves obtained by the three different research groups. The lower flotation edge obtained by the different techniques compare favourably, and establish that the lower flotation edge does not depend on pH.

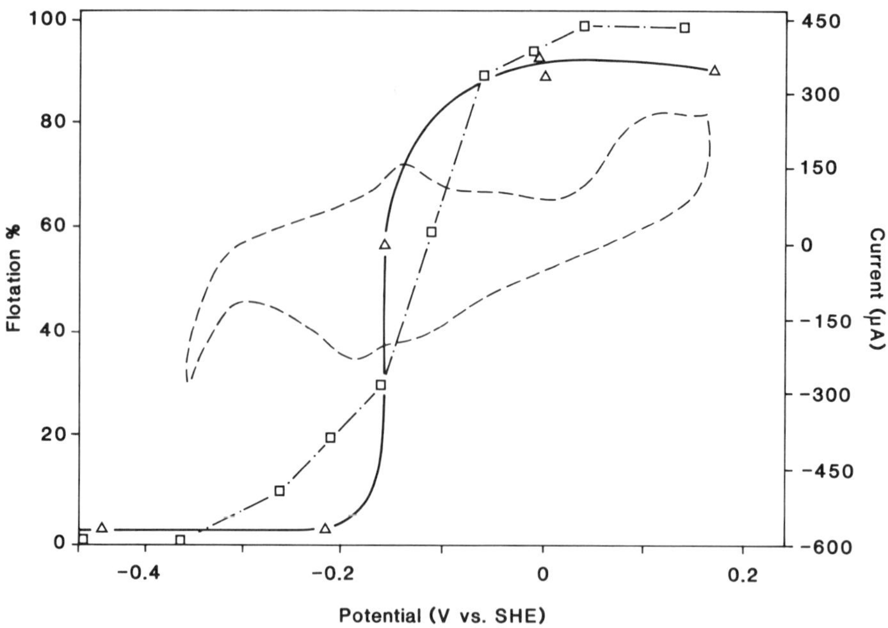

Figure 8.8 Flotation response of chalcocite in the presence of ethyl xanthate. Solid curve, data from Heyes and Trahar (1979); dotted curve, data from Richardson *et al.* (1984). Voltammetry curve, dashed line, is a cyclic voltammogram at $5\,mV\,s^{-1}$ for a chalcocite bed electrode, data from O'Dell *et al.* (1984). (From Woods and Richardson, 1986.)

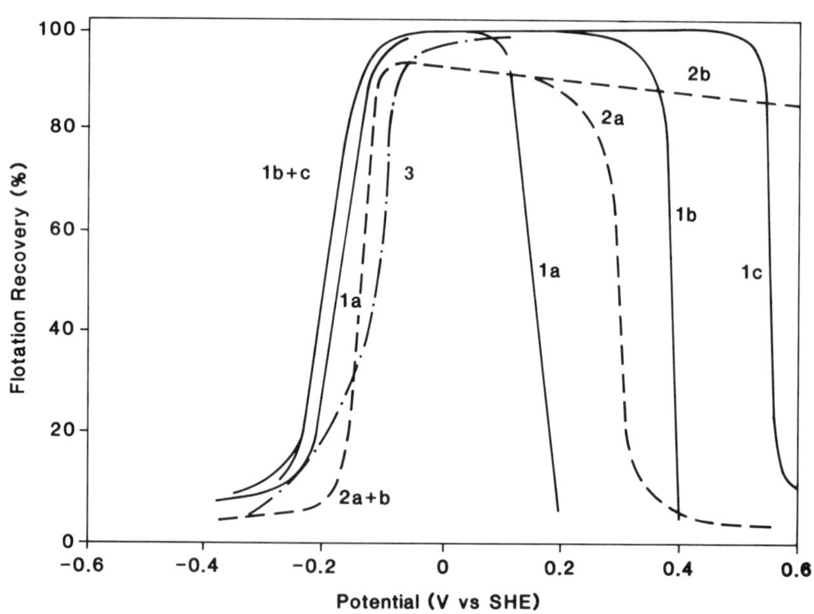

Figure 8.9 Recovery–potential curves of chalcocite with ethyl xanthate as a function of potential. Curve la, pH 11; curve 1b, pH 8; curve 1c, pH 5 (from Basilio, Pritzker and Yoon, 1985). Curve 2a, pH 11; curve 2b, pH 8 (from Heyes and Trahar, 1979). Curve 3, pH 9.2 (from Richardson *et al.*, 1984). (From Woods and Richardson, 1986.)

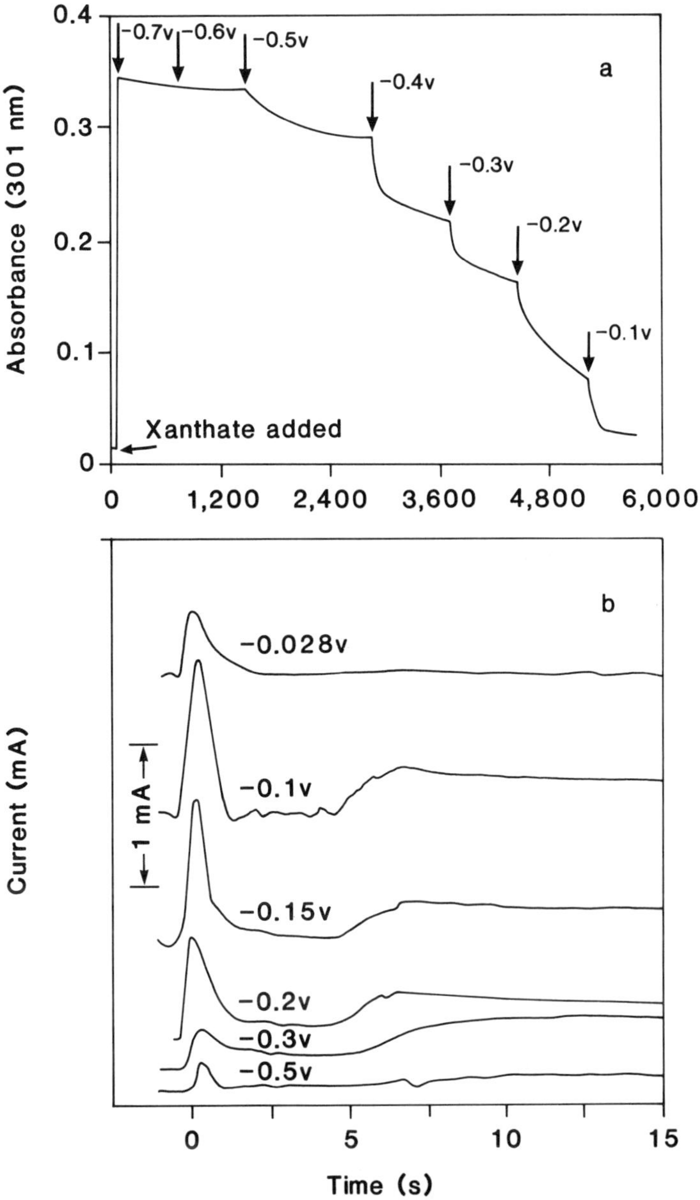

Figure 8.10 Chalcocite bed electrode in the presence of xanthate. (a) Xanthate (2×10^{-5} M) added at -0.46 V and the potential subsequently stepped to 0.14 V in 100 mV increments; (b) individual chalcocite bed electrodes potentiostated at several potentials between -0.46 and 0.22 V and xanthate (1.44×10^{-5} M) added while measuring the bed current. (From Richardson *et al.*, 1984.)

Xanthate anions have electronic adsorption spectra with well-defined peaks at 301 and 227 nm. Using these peaks, Richardson *et al.* (1984) continuously monitored the solution phase during their studies on mineral bed electrodes. Figure 8.10(a) shows the rate of uptake of xanthate as a function of bed potential and Figure 8.10(b) the current peaks associated with adding xanthate to beds held at different potentials. Flotation commenced at −0.26 V, and increased with increasing potential, establishing an excellent correlation between xanthate uptake and flotation. When beds are potentiated at different potentials before adding xanthate, the composition of the surface may change by reactions of the form of (8.3)–(8.7) and ion exchange reactions may become significant at oxidizing potentials. This is evident in Figure 8.10(b), where the current peak at 0.22 V is much less than that observed at lower potentials. The reduced charge passed at 0.22 V indicates that xanthate undergoes an ion exchange reaction with oxidation products of chalcocite. A soluble Cu(II) species was also observed by ultraviolet spectroscopy at 0.22 V and a part of the reduced current is attributed to a homogenous reaction of xanthate with the soluble copper species.

An upper flotation edge that depends on pH is evident in Figure 8.9. Heyes and Trahar (1979) attributed the decrease in flotation at high potentials to the oxidation of copper xanthate to dixanthogen by the reaction

$$2CuX + 2H_2O = 2CuO + X_2 + 4H^+ + 4e^- \qquad (8.20)$$

One problem with assigning the drop in potential to reaction (8.20) is that dixanthogen is highly hydrophobic, suggesting that the drop in flotation could be explained by oxidation of chalcocite (Woods and Richardson, 1986)

$$Cu_2S + \lambda H_2O \rightarrow Cu_{(2-\lambda)}S + \lambda CuO + 2\lambda H^+ + 2\lambda e^- \qquad (8.21)$$

Reactions (8.20) and (8.21) have the same pH dependence. The formation of significant quantities of CuO may be the main reason for depression at high potentials.

Chemical identification of a chemisorbed species followed by the formation of copper xanthate to multilayer coverage has been obtained by Leppinen, Basilio and Yoon (1988). These authors used FTIR spectroscopy with an attenuated total reflection (ATR) element to study xanthate adsorption on chalcocite contained in a carbon paste electrode.

Figure 8.11 shows the FTIR signal intensity of ethyl xanthate adsorbed on chalcocite as a function of the potential of the carbon paste electrode. The FTIR spectra (not shown) are consistent with three distinct reaction products of the collector: a chemisorption region below −7.0 mV, the formation of copper ethyl xanthate (CuX) at potentials between −7.0 and 200 mV, and the formation of multilayers of CuX at higher potentials. In the later poten-

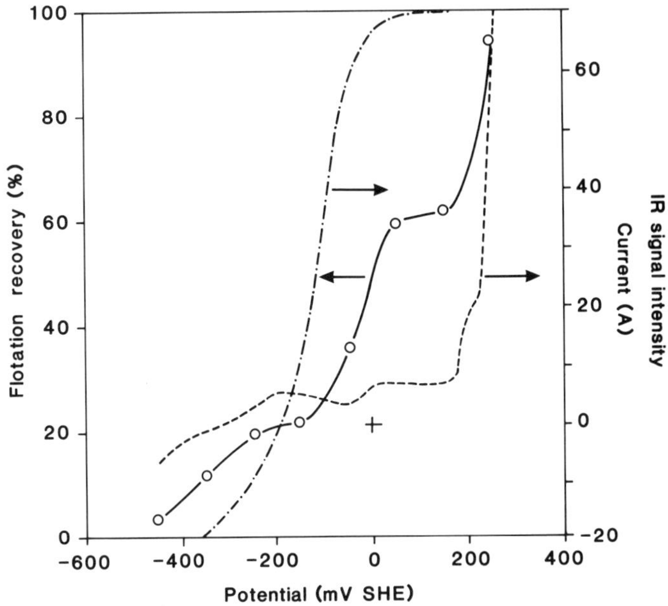

Figure 8.11 FTIR signal intensity of ethyl xanthate adsorbed on chalcocite at different potentials (solid line) (pH 9.2, 10^{-3} M ethyl xanthate) and comparison with the voltammogram (dashed line) of Cu_2S in a carbon paste electrode (10^{-4} M xanthate) and the flotation response obtained by Richardson and Walker (1985) (dot-dash line). (From Leppinen, Basilio and Yoon, 1988, with permission of The Electrochemical Society, Inc.)

tial range, Leppinen, Basilio and Yoon suggest that multilayers of CuX are formed by an EC-type mechanism in which the mineral is oxidized to release copper ions which then react with xanthate to form bulk CuX on the surface. They suggest that the multilayer product is different from the CuX formed between −0.7 and 0.2 V.

The current due to xanthate oxidation on the carbon paste electrode and flotation results of Richardson and Walker (1985) are superimposed on the FTIR signal intensity curves of Figure 8.11. There is a good correlation between flotation, the xanthate oxidation current, and the IR signal intensities of the surface xanthate species.

Woods, Young and Yoon (1990) have recently constructed chemisorption isotherms for the chalcocite–water–xanthate systems. They determined the free energy of formation of copper ethyl xanthate relative to xanthate from rest potential measurements and determined the equilibrium quantity of xanthate adsorbed at various potentials from cathodic stripping curves. The coverage with chemisorbed xanthate was found to follow the Frumkin isotherm

$$\theta/(1-\theta)e^{g\theta} = Ka_{\mathrm{A}}e^{\gamma FE/RT} \tag{8.22}$$

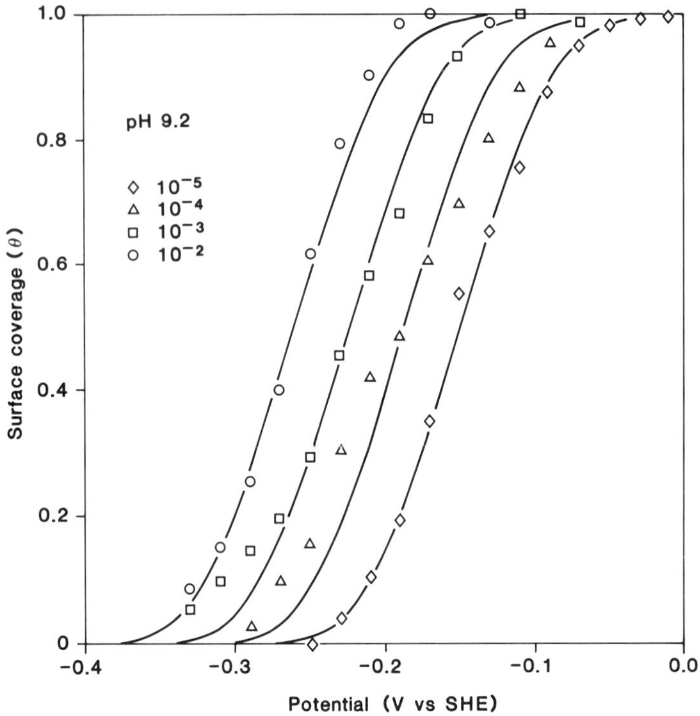

Figure 8.12 Comparison of experimentally determined coverages of chalcocite with ethyl xanthate at pH 9.2 as a function of potential with Frumkin isotherms calculated for $g = 4$, $\gamma = 1.6$ and $K = 8.4 \times 10^9$. (From Woods, Young and Yoon, 1990, with permission of Elsevier Science Publishers BV, Amsterdam.)

where θ is the fractional coverage, g and K are constants, a_A is the activity of the adsorbate in solution, γ is the electrosorption valency, and the other terms have their usual meaning. The term $\exp(g\theta)$ arises from variation in the free energy of adsorption with coverage because of either heterogeneity of the surface or interactions between adsorbed molecules. The value of the electrosorption valency reflects changes in the double layer associated with adsorption and the number of electrons involved in the charge transfer reaction.

Figure 8.12 compares the experimentally determined coverage of chalcocite with ethyl xanthate at pH 9.2 as a function of potential with the Frumkin isotherms calculated for $g = 4$, $\gamma = 1.6$ and $K = 8.4 \times 10^9$. Considering that the composition of the electrode, aside from ethyl xanthate adsorption, is a function of electrode potential (Figure 8.6), the fit of the experimental data to the Frumkin isotherm is quite good. Isotherms for xanthate adsorption on copper (Woods, Young and Yoon, 1990), and silver (Woods *et al.*, 1992) have also been determined.

Chander and Fuerstenau (1975a, b) investigated the reactions of dithiophosphate on chalcocite. They observed two anodic processes on voltammograms. The first was assigned to the chemisorption of dithiophosphate on the surface, the second to copper dithiophosphate formation. They also studied the flotation behaviour and contact angles for this system and found good correlations between the development of a finite contact angle, flotation and the anodic oxidation reactions. They also found that dithiophosphate could react with oxidized chalcocite by an ion-exchange mechanism.

8.6.2 Galena

For a number of years, reactions in the galena–water–xanthate system were the subject of extensive studies and this system assumed, for a while, the status of a model system for sulfide flotation. However, with modern electrochemical and flotation techniques, the flotation of galena has been shown to be more complex than the flotation of a number of other sulfides. This system is not as well understood as the chalcocite–xanthate system. Tolun and Kitchener (1964) concluded from steady state polarographic measurements that dixanthogen was the active collector. Lead xanthate (PbX_2) was also observed to form, but bubbles would not adhere to the surface when this compound was present. Toperi and Tolun (1969) used contact angle and rest potential measurements to conclude that the flotation of galena occurs when lead xanthate, xanthate ion and dixanthogen coexist. Woods (1971, 1972), using cyclic voltammetry, observed an anodic current peak that he attributed to the chemisorption of a monolayer of xanthate by reaction (8.6). At high potentials, he suggested that lead xanthate and dixanthogen were formed. Poling (1976) used infrared spectroscopy to study xanthate reactions on vapour-deposited lead sulfide films and concluded that the initial reaction layer consisted of chemisorbed lead sulfide in one-to-one coordination with surface lead, a result consistent with, and supporting, the chemisorption process observed by Woods. Leppinen, Basilio and Yoon (1988), using *in situ* FTIR, also found that the first adsorbed species on galena have spectra that are different from that of lead xanthate, whereas the spectra of multilayers correspond to bulk lead xanthate. The spectra showed no evidence of dixanthogen. This supports the conclusions of Allison and Finklestein (1971) that the major surface product is a lead xanthate with only 1% or less diethyl dixanthogen. Buckley and Woods (1991) used cathodic stripping voltammetry to study ethyl xanthate adsorption on freshly exposed galena surfaces. By comparing the results with flotation values reported in the literature, they concluded that flotation commences with very low coverages of xanthate and that maximum recovery is attained when only a fraction of a monolayer is adsorbed.

It is likely that chemisorbed xanthate, lead xanthate and dixanthogen may all be formed on galena, depending not only on the potentials employed but

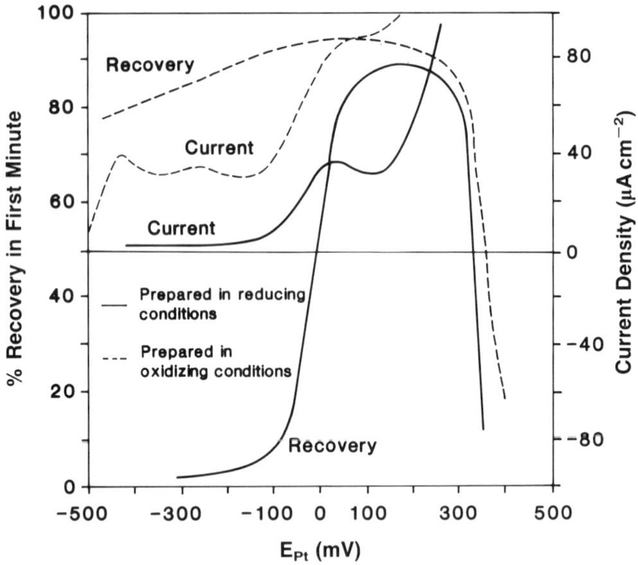

Figure 8.13 Anodic current and flotation response of galena as a function of potential in the presence of ethyl xanthate at pH 8 after preparation in oxidizing and reducing environments. (From Guy and Trahar, 1985, with permission of Elsevier Science Publishers BV, Amsterdam.)

also on the prior electrochemical history of the samples. For example, a lower flotation edge is observed, but the potential of this lower edge has widely different values. Guy and Trahar (1984, 1985) ground galena under reducing (steel mill in a nitrogen atmosphere) and oxidizing (ceramic mill in air) environments and found that the flotation edge shifted by several hundred millivolts. Figure 8.13 shows their results and the results of voltammetric studies by Woods (1971). The flotation of galena ground in a reducing environment commences at the same location as the anodic peak on the voltammetry curve, near $-0.1\,V$. Woods attributed this anodic peak to xanthate chemisorption. When galena is ground in an oxidizing environment, flotation commences at a much lower potential. An anodic current was also observed by Woods (1971) in this potential region on galena that had been previously oxidized. He attributed this anodic current peak to the formation of lead xanthate from excess lead in the surface produced by prior oxidation. An upper flotation edge is also observed for galena, which Guy and Trahar (1984) suggest is due to the oxidation of lead xanthate to dixanthogen, and Woods suggests is due to the oxidation of galena. The relevant reactions are similar to those for the upper flotation edge of chalcocite discussed in the previous section.

Figure 8.13 suggests that dixanthogen is not a prerequisite for galena flotation. Strong flotation is observed on samples prepared under reducing conditions at potentials corresponding to the chemisorption region. The relevant reactions may be

$$PbS + X^- \rightarrow PbS\vdots X_{ads} \tag{8.23}$$

$$PbS + \lambda X^- \rightarrow Pb_{(1-\lambda)}S\vdots \lambda PbX \tag{8.24}$$

For samples prepared under oxidizing conditions, flotation at potentials much more negative than both the chemisorption region and the reversible xanthate/dixanthogen couple suggest that the hydrophobic species is lead xanthate, PbX_2.

Plaksin and Shafeev (1960) proposed that the semiconducting properties of galena and other sulfides may have a significant influence on the chemisorption of xanthate collectors. They suggested that chemisorbed oxygen acts as an acceptor-like surface state, localizing electrons at the surface and increasing the free hole concentration in the valence band. If it is assumed that oxidation reactions involving xanthate occur via the valence band, the increase in free hole concentration would increase xanthate coverage.

It has been shown from photovoltage and capacitance measurements (Richardson and O'Dell, 1985) that fresh surfaces of galena produced by cleavage under an electrolyte instantaneously assume the flat band potential.

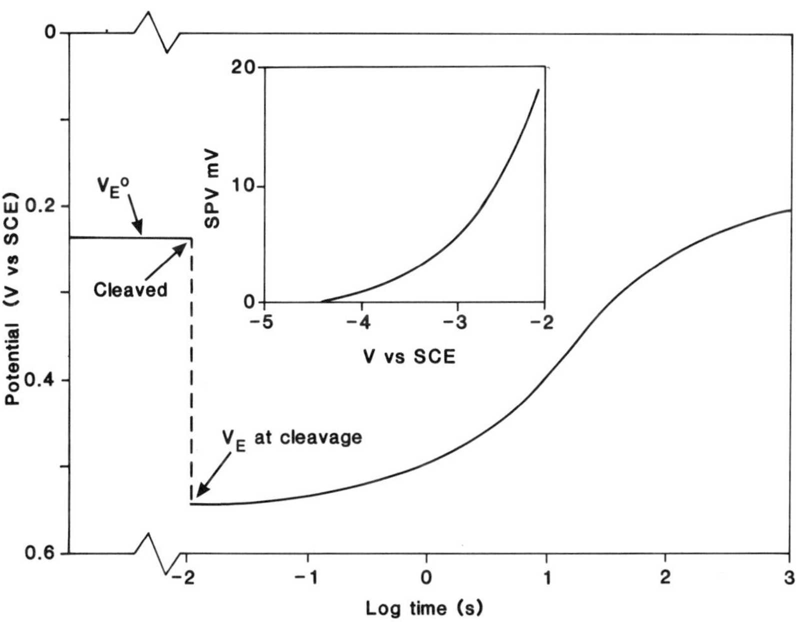

Figure 8.14 Potential–time curve for a fresh surface of galena produced by cleavage. Inset shows a plot of the surface photovoltage following cleavage. (From Richardson and O'Dell, 1985.)

A typical potential–time curve for fresh surfaces produced by cleavage is shown in Figure 8.14. The more n-type the galena, the more negative the instantaneous electrode potential (flat band potential) at cleavage. It can be seen from the Pourbaix diagram for galena (Figure 8.3) that fresh surfaces have a potential in the stability field of galena. Within seconds after cleavage, fresh surfaces begin to oxidize, as indicated by an increase in potential, reaching a more positive steady state potential (mixed potential) 2–4 min after cleavage.

The photovoltage the instant after cleavage is zero, indicating the electrodes assume the flat band condition. With subsequent oxidation, a finite photovoltage developed that increased with increasing oxidation, indicating upward band bending and establishing that oxidation produces a depletion layer on n-type and an accumulation layer on p-type galena electrodes. At the steady state rest potential, which is independent of semiconducting properties, the surface of all electrodes were highly p-type, owing to the upward band bending caused by oxidation.

Richardson and O'Dell (1985) also reported that the cleavage potentials of fresh surfaces of n- to slightly p-type electrodes were more negative than the potential for xanthate chemisorption (Woods, 1971); thus, oxidation is a necessary condition before xanthate can adsorb. No highly p-type electrodes were available, but extrapolation of the flat band potential indicates that highly p-type samples would have a cleavage potential in the range where xanthate chemisorbs.

It may be of interest to mineralogists that the space charge layer on galena can be readily and reversibly polarized and that the surface photovoltage can be used easily to determine carrier type. Figure 8.15 shows the potential dependence of the photovoltage (bottom), capacitance (middle) and interfacial current (top) for a synthetic PbS electrode. The shape and magnitude of the photovoltage and capacitance curves follow the classical behaviour expected for semiconductors without significant trapping or recombination of charge carriers at the surface (Frankl and Ulmer, 1966; Many, Goldstein and Grover, 1971). The positive branch corresponds to upward band bending, the negative to downward band bending, and the cusp to flat bands. The ratio of the saturation photovoltages under strong anodic and cathodic polarization gives the electron–hole ratio in the bulk. Natural galena crystals have been shown to have similar photocurrent–potential curves, but most of the natural samples are highly n-type and only the positive branch is observed. The top curve represents the reduction of residual oxygen in solution. The large Tafel slope was attributed to the partition of the changes in electrode potential between the Helmholtz layer just outside the surface and the space charge layer just beneath the surface, with only changes in the potential across the Helmholtz layer affecting the kinetics of oxygen reduction.

The interfacial capacitance indicated only a small surface charge (much

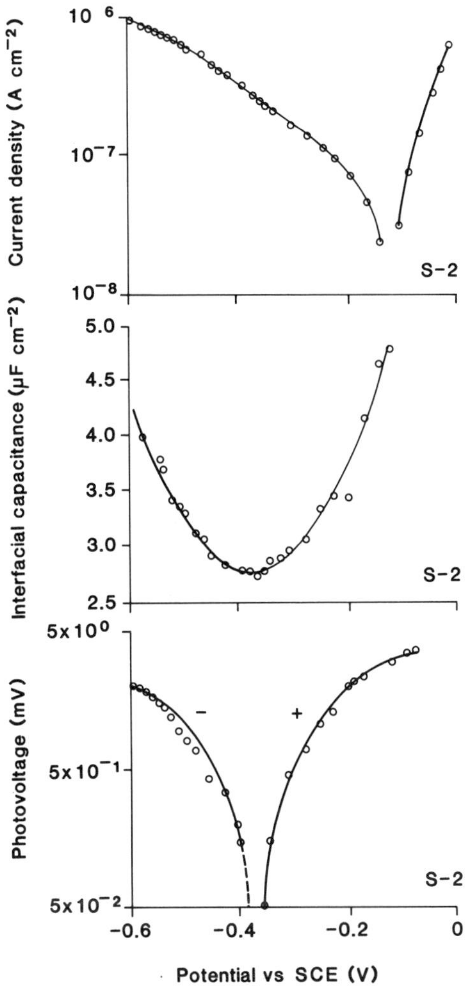

Figure 8.15 The dependence of the photovoltage (bottom), differential capacitance (middle) and interfacial current (top) for a synthetic PbS electrode at pH 9.2. (From Richardson and O'Dell, 1985.)

less than a monolayer) is sufficient to raise the potential of the electrodes so that the surface is p-type, e.g. from approximately −0.6 V to −0.2 V. Figure 8.14 indicates that charge accumulation at the surface occurs in several minutes even in oxygen-purged solutions. Thus, it is doubtful whether the semiconducting properties of galena have a practical influence on its flotation rate.

The results confirm the essential details of Plaksin and Shafeev's model. It is not known whether it is oxygen chemisorption or lead vacancies, produced by reaction (8.4), that is responsible for the upward band bending.

8.6.3 Pyrite

Flotation recovery–potential curves have been determined for the pyrite–ethyl xanthate system at pH 8 (Trahar, 1984) and pH 9.2 (Richardson and Walker, 1985). Figure 8.16 shows that the lower flotation edge corresponds to the reversible potential for the xanthate–dixanthogen couple. Dixanthogen has a small but finite solubility (Hamilton and Woods, 1979), and direct evidence for dixanthogen formation as a result of the oxidation of xanthate on a particle bed of pyrite is shown in the ultraviolet spectra of Figure 8.17. Dixanthogen has weak ultraviolet adsorption bands but the isosbestic points near 235 and 270 nm clearly show the formation of dixanthogen as a soluble oxidation product of xanthate anions.

Trahar (1984) found flotation commencing 300 mV below the potential for dixanthogen formation and maximum recovery of pyrite in the potential region where dixanthogen is formed. This suggests the additional involvement of an iron xanthate compound in rendering pyrite floatable. Voltam-

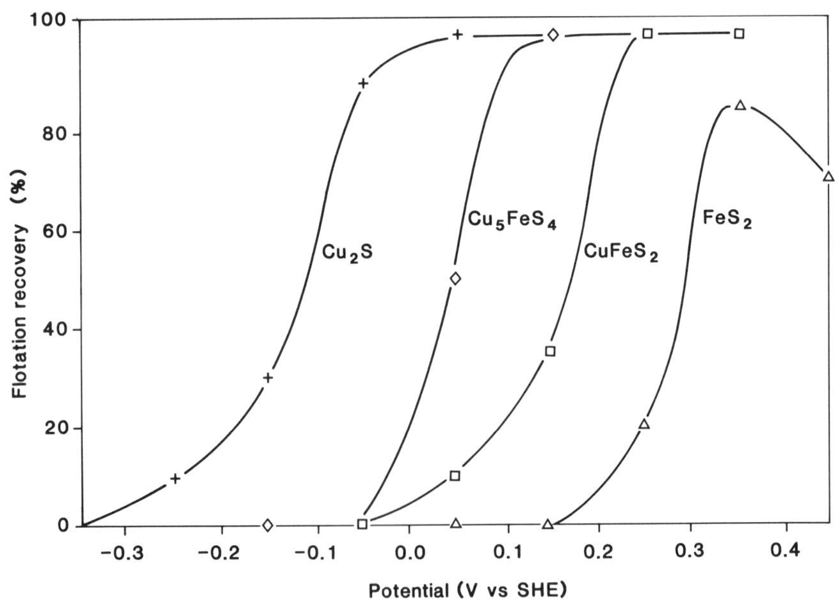

Figure 8.16 Relationship between flotation recovery and conditioning potential for chalcocite, bornite, chalcopyrite and pyrite. Conditioning time = 10 min; flotation time = 2 min; pH = 9.2; ethyl xanthate concentration = 1.44×10^{-5} mol l^{-1} for Cu_2S and $(2 \pm 0.1) \times 10^{-5}$ mol l^{-1} for Cu_5FeS_4, $CuFeS_2$ and FeS_2. (From Richardson and Walker, 1985.)

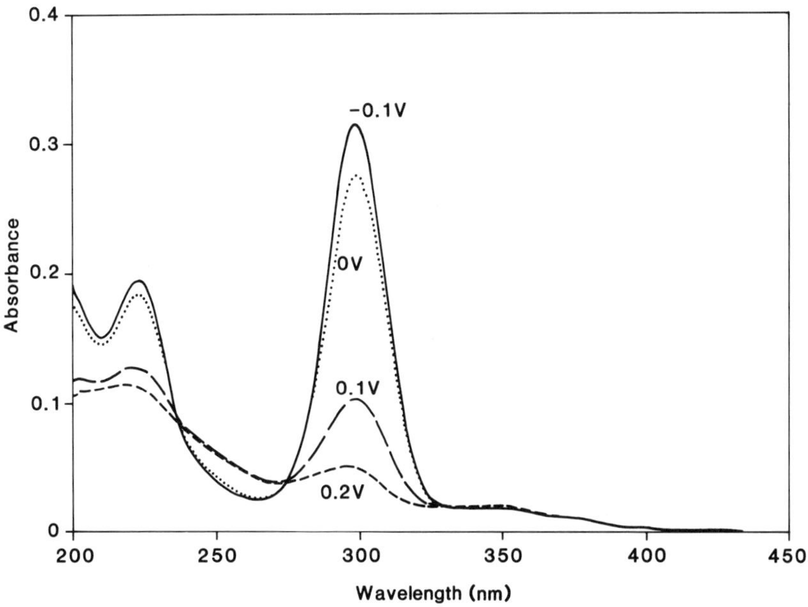

Figure 8.17 Ultraviolet spectra of the solution phase during the electro-oxidation of xanthate to dixanthogen on a pyrite bed electrode. (From Richardson and Walker, 1985.)

metry curves on pyrite (Woods, 1971; Janetski, Woodburn and Woods, 1977), in the vicinity of the flotation edge observed by Trahar do not reveal any additional current due to xanthate. Several iron–hydroxy–xanthate species have been identified and there are suggestions that they may be present on pyrite surfaces (Sheikh and Leja, 1977; Critchley and Hunter, 1986; Wang and Forsberg, 1991). Fornasiero and Ralston (1992) have proposed an iron dissolution–iron hydroxide adsorption mechanism to explain the effect of oxidation and pH on the zeta potential of pyrite. They concluded that in the presence of xanthate, the observed zeta potential–pH and flotation recovery–pH data could be explained only by the formation of ferric and ferrous hydroxide–xanthate species on the pyrite surface. Leppinen, Basilio and Yoon (1988), using *in situ* FTIR, reported that only dixanthogen is observed on the surface of pyrite; however, the spectra were different from that of bulk dixanthogen, which they suggest indicates that dixanthogen is tightly bound to the surface.

Photocurrent measurements have been carried out on fresh surfaces of pyrite produced by fracture under electrolytes (Richardson, Li and Yoon, 1992). Similar to galena, the instantaneous potential of pyrite at fracture decreases by several hundred millivolts towards its field of stability (Figure 8.18). In contrast to galena, an anodic photocurrent is observed on pyrite the instant after fracture, indicating the bands are bent upwards.

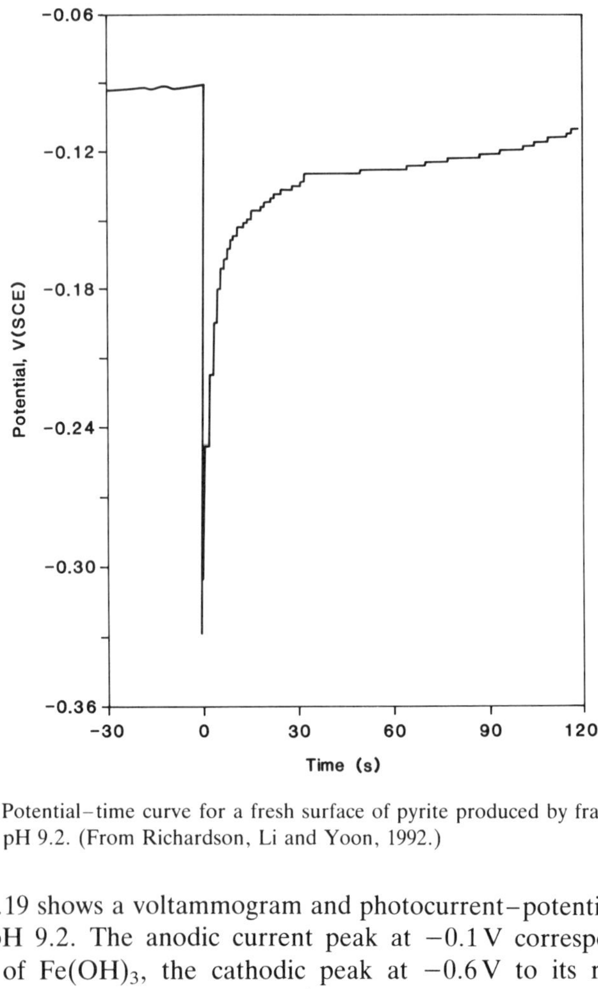

Figure 8.18 Potential–time curve for a fresh surface of pyrite produced by fracture under an electrolyte at pH 9.2. (From Richardson, Li and Yoon, 1992.)

Figure 8.19 shows a voltammogram and photocurrent–potential curve for pyrite at pH 9.2. The anodic current peak at $-0.1\,V$ corresponds to the formation of $Fe(OH)_3$, the cathodic peak at $-0.6\,V$ to its reduction to $Fe(OH)_2$ (Buckley and Woods, 1987). To obtain the flat band potential, pyrite must be strongly reduced at potentials $< -0.9\,V$. The reduction reactions may be

$$FeS_2 + H^+ + 2e^- = FeS + HS^- \tag{8.25}$$

$$FeS_2 + 2H^+ + 4e^- \rightarrow Fe + 2HS^- \tag{8.26}$$

In contrast to galena, the space charge layer on pyrite electrodes is not readily polarized. It is believed that either an intrinsic acceptor-like surface state, possibly associated with cleavage of S_2 bonds at the surface, or the rapid adsorption of hydroxyl ions is responsible for the development of the space charge layer.

During the negative going scan, there is an anodic photocurrent peak near

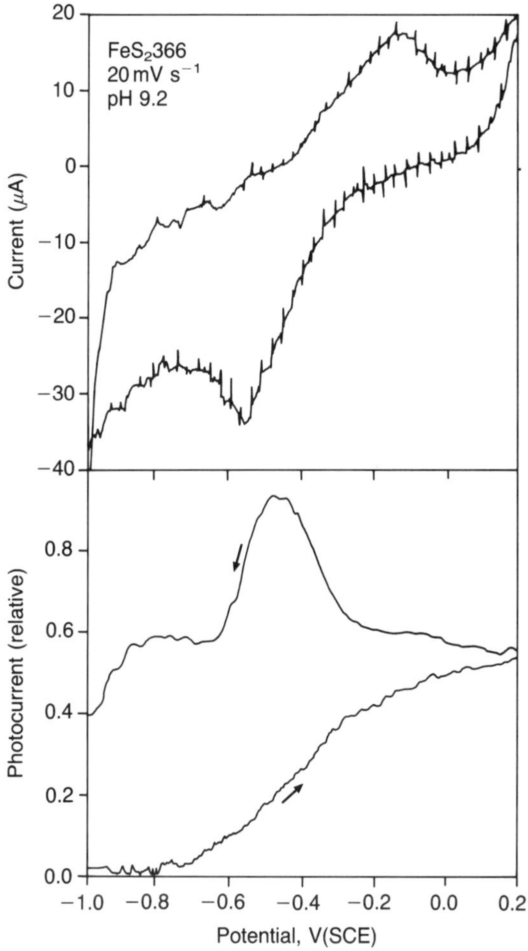

Figure 8.19 Voltammogram (top) and photocurrent–potential curve (bottom) for a fractured pyrite electrode at pH 9.2. (From Richardson, Li and Yoon, 1992, with permission of The Electrochemical Society, Inc.)

−0.55 V. This is attributed to the photo-induced re-oxidation of $Fe(OH)_2$ to $Fe(OH)_3$ by the holes generated by illumination.

The existence of a negative surface charge on pyrite is consistent with zeta potential measurements, which show pyrite is negatively charged in alkaline solutions (Fuerstenau, Kuhn and Elgillani, 1968). It is apparent that the negative surface charge is compensated for in both the semiconductor space charge layer and the Helmholtz layer. The effect of this negative surface charge on reactions with thiol collectors is not known, though it may repel negatively charged xanthate ions.

8.6.4 Copper–iron–sulfides

Figure 8.16 compares the flotation response of chalcopyrite and bornite with that of chalcocite and pyrite. The flotation edge of these Fe-containing minerals occurs at potentials negative to that for pyrite and negative of the reversible potential for the xanthate–dixanthogen couple. This suggests that a xanthate species adsorbs on these minerals prior to dixanthogen formation. The anodic currents observed on voltammetry curves on chalcopyrite (Gardner and Woods, 1979b) and bornite (Buckley, Hamilton and Woods, 1984; Richardson and Walker, 1985) were the same in the absence and presence of ethyl xanthate, as long as the potential sweep remained negative of the xanthate–dixanthogen couple. Thus, no evidence for the formation of a xanthate surface product other than dixanthogen was obtained in these studies. However, currents due to the oxidation of these minerals (reactions (8.9) and (8.10)) do occur in aqueous solution in the potential region where flotation commences, indicating that a metal xanthate could be formed by the interaction of the collector with an oxidation product. The failure to detect the formation of a metal xanthate electrochemically may be due to the reaction occurring by an ion-exchange process involving an oxide product. This requires that the anodic currents, in the presence and absence of xanthate, represent the oxidation of the mineral and that the currents are not changed in the presence of xanthate. Ultraviolet spectroscopy of the solution phase of a bornite bed electrode during linear potential sweeps showed that ethyl xanthate is abstracted from solution in the potential region where mineral oxidation and flotation occur, i.e. at potentials below the xanthate–dixanthogen couple (Richardson and Walker, 1985). However, Leppinen, Basilio and Yoon (1988) have used an ATR cell for *in situ* FTIR measurements of ethyl xanthate on chalcopyrite and concluded that the only xanthate species on the surface is dixanthogen.

8.7 Mineral mixtures

8.7.1 Chalcopyrite–galena

There have been only a few studies of the effects caused by mineral mixtures. The effect of cations derived from the oxidation of one component of a mixture on another component has been discussed by Guy and Trahar (1985). Their studies on single component systems indicated that galena could be floated from chalcopyrite at reducing potentials, while the converse flotation of chalcopyrite from galena would be possible at oxidizing potentials. The two top curves of Figure 8.20 show the flotation of the individual minerals from a mixture with quartz. When the minerals are both present with quartz, they interact with each other in such a manner that the inherent differences in the flotation response is diminished, as indicated by the

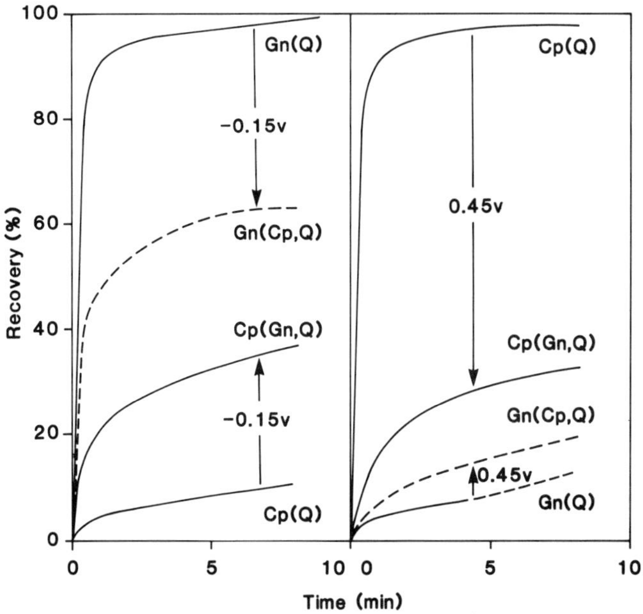

Figure 8.20 Floatabilities under reducing (left) and oxidizing (right) conditions of chalcopyrite, Cp, and galena, Gn, from quartz, Q, and from chalcopyrite, galena, quartz mixtures. Collector, potassium ethyl xanthate. (Adapted from Guy and Trahar, 1985.)

arrows. The reduction in selectivity was explained by a modification of the flotation response of galena (towards chalcopyrite) by Cu ions derived from the oxidation of chalcopyrite and a modification of the flotation of chalcopyrite by replacement of the Fe in chalcopyrite by Pb ions from galena.

8.7.2 Pyrite and chalcocite

Gebhardt and Richardson (1987) obtained similar results of diminished selectivity in mixtures of pyrite and chalcocite. Recovery–potential curves, such as those in Figure 8.16, suggest that chalcocite would be readily floated from pyrite at potentials slightly below that at which dixanthogen is formed. When mixtures of the two minerals are conditioned at the open-circuit potential where some oxidation can occur, Cu ions from chalcocite modify the flotation curve of pyrite, reducing selectivity, as shown in Figure 8.21. The flotation of chalcocite is also reduced, and this is believed to be due to enhanced oxidation by galvanic coupling with pyrite. When the minerals are held at a reducing potential when first introduced into the cell (dot-dash curves), the loss of selectivity is greatly reduced. In this case, chalcocite recovery resembles single-mineral flotation, but pyrite remains depressed even at potentials above the xanthate–dixanthogen couple.

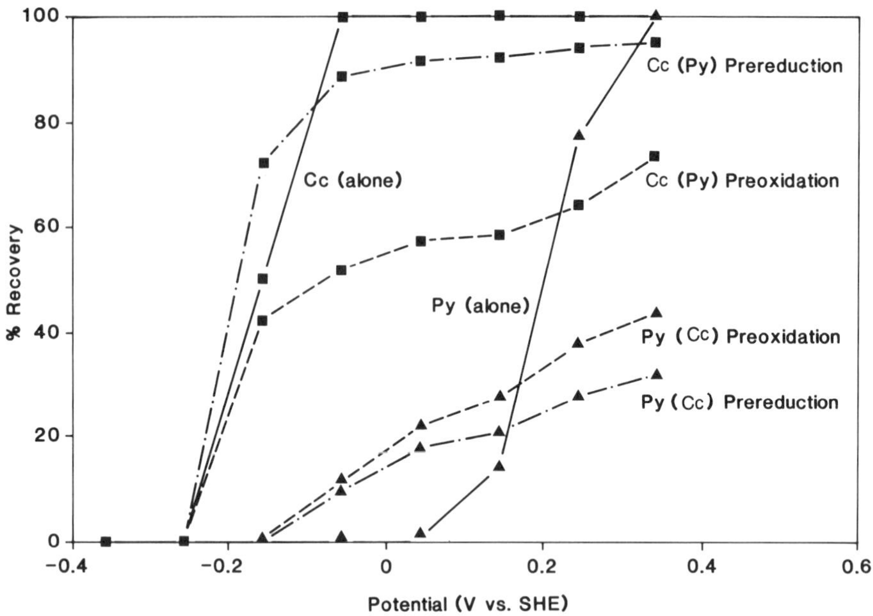

Figure 8.21 Floatabilities of chalcocite, Cc, and pyrite, Py, alone and from mixtures which were pre-reduced at −0.36 V or pre-oxidized at open-circuit (0.2 V) prior to flotation. Collector, ethyl xanthate. (From Gebhardt and Richardson, 1987.)

8.8 Galvanic coupling between minerals

Differences are found in the flotation of sulfides when they are ground under oxidizing or reducing conditions. For galena, the effects are pronounced, as shown in Figure 8.13 (Guy and Trahar, 1985). When ground in non-reducing conditions (ceramic or stainless steel mill), xanthate flotation is rapid over a wide range of potentials and terminates when the pulp becomes sufficiently oxidizing. However, when ground under reducing conditions in a stainless steel mill with sodium sulfide, the recovery–potential curves are much narrower.

Results such as those in Figure 8.13 can be interpreted as modifications in the surface chemistry of minerals due to the oxidizing and reducing conditions imposed during grinding. Grinding is usually carried out with steel grinding media. Iron introduces additional galvanic coupling that can affect mineral and collector oxidation rates and introduce corrosion and wear products in the flotation liquors. Iwasaki and co-workers (Learmont and Iwasaki, 1984; Pozzo, Malisci and Iwasaki, 1990; Ahn and Gebhardt, 1991) have studied the effect of grinding media wear and corrosion products on sulfide flotation. They have established that galvanic coupling between minerals and iron grinding media is usually such that the iron grinding media acts as the anode and the more noble sulfides as the cathode. This can accelerate corrosion of the grinding media and possibly limit oxidation of the sulfides. Ferrous- and ferric-ion corrosion products and iron debris from wear has a pronounced effect on flotation. Pozzo, Malisci and Iwasaki (1990) proposed a model in which oxygen is reduced at the surface of the more noble minerals, producing hydroxyl ions, while iron ions are released by the dissolution of the steel media and more active minerals. This leads to iron hydroxide precipitates. As discussed previously, hydroxides are known to be hydrophilic and their precipitation on mineral surfaces is believed to interfere with collector adsorption, impairing flotation. Alternatively, they may not interfere with collector adsorption, but a larger concentration of collector may be required to overcome the hydrophilic hydroxides to provide a net hydrophobic surface.

8.9 Depression of sulfide minerals

Many of the reagents used as flotation modifiers can be classified as reducing or oxidizing. Two frequently employed reducing reagents are HS^- and CN^- ions, which lower the potential and also adsorb on sulfide minerals to compete with collector adsorption. Sulfide ion is added to a number of flotation circuits as a reducing reagent, usually in the form Na_2S or $NaSH^-$. It is usually a depressant and destroys the hydrophobicity of sulfides. Several mechanisms may be involved, though the major one is believed to be the desorption of xanthate. Castro (1976) and Castro and Larrondo (1981) have proposed the following reactions whereby HS^- displaces chemisorbed xanthate and dixanthogen from copper sulfides:

$$2CuX_{\text{ads}} + SH^- + OH^- \rightleftharpoons 2X^- + Cu_2S + H_2O \tag{8.27}$$

$$4X_2 + 2SH^- + 3H_2O \rightleftharpoons 8X^- + S_2O_3^{2-} + 8H^+ \tag{8.28}$$

$$3X_2 + SH^- + 3H_2O \rightleftharpoons 6X^- + SO_3^{2-} + 7H^+ \tag{8.29}$$

The displacement reactions are illustrated in Figure 8.22, where first xanthate and then sodium sulfide were added to chalcocite at pH 9.2. Xanthate was added in three successions to reach 100% flotation. The top curve shows that sulfide ion lowers the potential of a chalcocite electrode inserted in the cell, the middle curve the accompanying desorption of xanthate, and the bottom curve the decrease in recovery with sulfide addition. Similar studies have recently been reported by Gebhardt and Kotlyar (1991) for the chalcopyrite–xanthate and chalcopyrite–thionocarbamate systems.

Redfearn (1981, 1983) has shown that even with flotation in a nitrogen atmosphere to minimize HS^- consumption, 10 to 100 times the stoichiometric amounts suggested by reactions (8.27)–(8.29) are required. This is

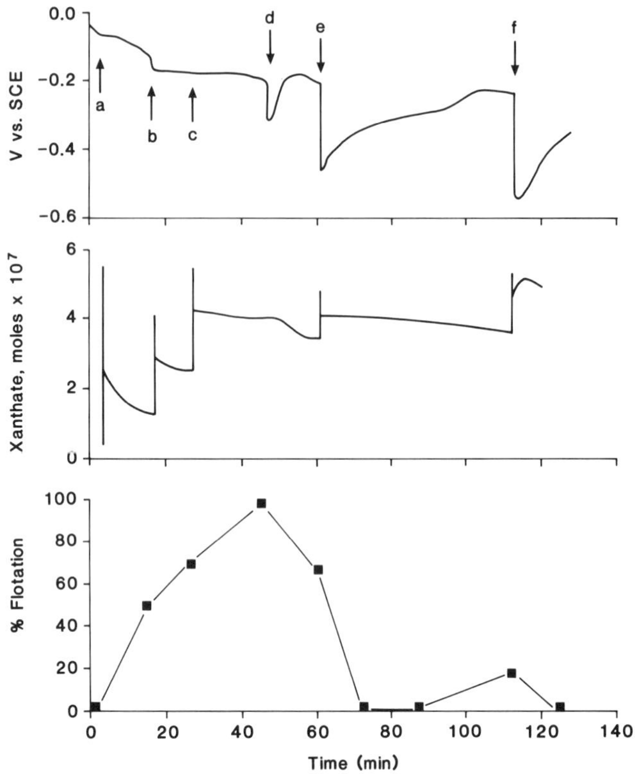

Figure 8.22 Potential, xanthate concentration and flotation response vs. time for chalcocite in 0.05 M sodium borate with xanthate total additions of (a) 6×10^{-7}, (b) 9×10^{-7} and (c) 1.2×10^{-6} moles, followed by sodium sulfide additions of (d) 1.5×10^{-7}, (e) 3×10^{-7} and (f) 4.5×10^{-7} moles.

due to several additional reactions that occur in flotation circuits when HS^- is added. The sulfide ion reacts with dissolved heavy metal ions in solution, precipitating them as the sulfides, which may be beneficial to flotation; it cleanses sulfides of surface oxidation products, reforming the sulfide surfaces, which is beneficial in the treatment of oxidized ores; it can form elemental sulfur on the surfaces of some sulfides, such as chalcopyrite and pyrite (as shown earlier (Figure 8.4)), inducing collectorless flotation, which may be detrimental to grades; it reacts with oxygen, a process that lowers the dissolved oxygen concentration and consumes sulfide ion; and it lowers the redox potential of the solution, depressing sulfides by desorbing collector, as shown above. With chalcocite, the sulfide ion is also believed to react strongly to form a sulfide with a lower copper content, such as covellite. The active collector for covellite is dixanthogen (Allison *et al.*, 1972). Hence, if the surface of chalcocite is converted to a covellite-like surface, then the active collector must switch from chemisorbed xanthate to dixanthogen, a reaction that requires a more oxidizing potential.

Cyanide ions are also reducing reagents that act in a similar fashion to HS^-. In addition to desorbing collector, CN^- may also form hydrophilic metal cyanide complexes on mineral surfaces. These complexes may reinforce the net hydrophilicity produced by the displacement of collector from the surface.

Woods (1984) has summarized the mechanisms by which oxidizing conditions can depress sulfides. These include enhancement of the anodic oxidation of the mineral so that it occurs more rapidly than collector adsorption, and the introduction of another reaction that is more favourable than collector oxidation.

8.10 Summary

This chapter has briefly highlighted recent advances in understanding the chemical reactions that occur at sulfide surfaces relevant to their flotation. Because of the complexity of real flotation systems, research that has most advanced the understanding of the chemistry has used simplified and idealized systems, often single component minerals and single component reagents. Although this led to considerable progress in understanding the basic chemistry of flotation, the complexities of real systems have neither allowed this knowledge to be exploited to develop new flotation technologies from basic principles nor allowed it to optimize flotation by controlling the chemistry. Research on the chemistry of flotation is beginning to move from ideal systems to more realistic systems and to in-plant measurements. This offers some promise that flotation research may some day lead, rather than follow, flotation technology.

References

Ahmed, S.M. (1978a) Electrochemical studies of sulphides, I. The electrocatalytic activity of galena, pyrite and cobalt sulphide for oxygen reduction in relation to xanthate adsorption and flotation. *Int. J. Miner. Process.*, **5**, 163–74.

Ahmed, S.M. (1978b) Electrochemical studies of sulphides, II. Measurement of the galvanic currents in galena and the general mechanism of oxygen reduction and xanthate adsorption on sulphides in relation to flotation. *Int. J. Miner. Process.*, **5**, 175–82.

Ahn, J.H. and Gebhardt, J.E. (1991) Effect of grinding media-chalcopyrite interaction on the self-induced flotation of chalcopyrite. *Int. J. Miner. Process.*, **33**, 243–62.

Allison, S.A. and Finkelstein, N.P. (1971) Study of the products of reaction between galena and aqueous xanthate solutions. *Trans. Instn. Min. Metall. (Sect. C: Min. Proc. Extr. Metall.)*, **80**, C235–9.

Allison, S.A., Gold, L.A., Nicol, M.J. and Granville, A. (1972) A determination of the products of reaction between various sulfide minerals and aqueous xanthate solution, and a correlation of the products with electrode rest potentials. *Metall. Trans.*, **3**, 2613–18.

Barsky, G. (1934) Discussion on 'principles of flotation' – an experimental study of the effect of xanthates on contact angles at mineral surfaces. *Trans Amer. Inst. Min. Metall. Engrs*, **112**, 326.

Basilio, C., Pritzker, M.D. and Yoon, R.-H. (1985) Thermodynamics, electrochemistry and flotation of the chalcocite–potassium ethyl xanthate system, Society of Mining Engineers, preprint 85-89, 10 pp.

Biegler, T. (1976) Oxygen reduction on sulphide minerals Part II. Relation between activity and semiconducting properties of pyrite electrodes. *J. Electroanal. Chem. Interfacial Electrochem.*, **70**, 265–75.

Biegler, T., Rand, D.A.J. and Woods, R. (1975) Oxygen reduction on sulphide minerals Part I. Kinetics and mechanism at rotated pyrite electrodes. *J. Electroanal. Chem. Interfacial Electrochem.*, **60**, 152–62.

Biegler, T., Rand, D.A.J. and Woods, R. (1977) Oxygen reduction on sulphide minerals, in *Trends in Electrochemistry*. (eds J.O'M. Bockris, D.A.J. Rand and B.J. Welch), Plenum Press, New York, pp. 291–302.

Buckley, A. and Woods, R. (1981) Investigation of the surface oxidation of sulfide minerals via ESCA and electrochemical techniques, in *Interfacial Phenomena in Mineral Processing* (eds B. Yarar and D.J. Spottiswood), Engineering Foundation, New York, pp. 3–17.

Buckley, A.N. and Woods, R. (1984) An X-ray photoelectron spectroscopic study of the oxidation of galena. *Appl. Surf. Sci.*, **17**, 401–14.

Buckley, A.N. and Woods, R. (1987) The surface oxidation of pyrite. *Appl. Surf. Sci.*, **27**, 437–52.

Buckley, A.N. and Woods, R. (1991) Adsorption of ethyl xanthate on freshly exposed galena surfaces. *Colloids and Surfaces*, **53**, 33–45.

Buckley, A.N., Hamilton, I.C. and Woods, R. (1984) Investigation of the surface oxidation of bornite by linear potential sweep voltammetry and X-ray photoelectron spectroscopy. *J. Appl. Electrochem.*, **14**, 63–74.

Buckley, A.N., Hamilton, I.C. and Woods, R. (1985) Investigation of the surface oxidation of sulphide minerals by linear potential sweep voltammetry and X-ray photoelectron spectroscopy, in *Flotation of Sulfide Minerals* (ed. K.S.E. Forssberg), Elsevier Science Publishers BV, Amsterdam.

Buckley, A.N., Woods, R. and Wouterlood, H.J. (1988) The deposition of sulfur on pyrite and chalcopyrite from sodium sulfide solutions, *Aust. J. Chem.*, **41**, 1003–11.

Castro, S. (1976) Selective flotation of molybdenite from copper porphyric ores, in *Investigaciones Mineras y Metallurgy*, CIMM, Santiago, pp. 94–110.

Castro, S. and Larrondo, J. (1981) An electrochemical study of depression of flotation of chalcopyrite by cyanide and iron–cyanide complexes. *J. Electroanal. Chem. Interfacial Electrochem.*, **118**, 317–26.

Chander, S. and Fuerstenau, D.W. (1975a) Electrochemical reaction control of contact angles on copper and synthetic chalcocite in aqueous potassium diethyldithiophosphate solutions. *Int. J. Miner. Process.*, **2**, 333–52.

Chander S. and Fuerstenau, D.W. (1975b) On the floatability of sulfide minerals with thiol collectors: the chalcocite/diethyldithiophosphate system, in *Proceedings Eleventh International Mineral Processing Congress*, Cagliari (ed. Ente Minerario Sardo), Instituto di Arte Minerari e Preparazioni di Minerali, Universita di Cagliari, Cagliari, pp. 583–604.

Cook, M.A. and Nixon, J.C. (1950) Theory of water-repellent films on solids formed by adsorption from aqueous solutions of heteropolar compounds. *J. Phys Colloid Chem.*, **54**, 445–59.

Cook, M.A. and Wadsworth, M.E. (1957) Hydrolytic and ion pair adsorption processes in flotation, ion exchange and corrosion, in *Proceedings of the 2nd International Congress on Surface Activity*, Vol. 3 (ed. J.H. Schulman), Butterworth, London, pp. 228–42.

REFERENCES

Critchley, J.K. and Hunter, C.J. (1986) Solubility products of ferric xanthates. *Trans. Instn. Min. Metall. (Sect. C: Min. Proc. Extr. Metall.)*, **95**, 173–5.

Finkelstein, N.P., Allison, S.A., Lovell, V.M. and Stewart, B.V. (1975) Natural and induced hydrophobicity in sulfide mineral systems, in *Advances in Interfacial Phenomena of Particulate/Solution/Gas Systems, Applications to Flotation Research, Am. Inst. Chem. Eng. Symp. Ser.*, No. 150, 71, pp. 165–75.

Fornasiero, D. and Ralston, J. (1992) The interaction of ethyl xanthate with pyrite, in *Proc. Electrochemistry in Mineral and Metal Processing, III* (eds R. Woods and P.E. Richardson), The Electrochemistry Society, 92-17, pp. 191–220.

Fowkes, F.M. (1967) Attractive forces at solid liquid interfaces, in *Wetting*, Soc. Chem. Industry Monograph No. 25, Royal Society of Chemistry, London, pp. 3–30.

Frankl, D.R. and Ulmer, E.A. (1966) Theory of the small-signal photovoltage at semiconductor surfaces. *Surf. Sci.*, **6**, 115–23.

Fuerstenau, M.C., Kuhn, M.C. and Elgillani, D. (1968) Role of dixanthogen in xanthate flotation of pyrite. *Trans Amer. Inst. Min. Metall. Engrs.*, **241**, 148.

Gardner, J.R. and Woods, R. (1979a) A study of the surface oxidation of galena using cyclic voltammetry. *J. Electroanal. Chem. Interfacial Electrochem.*, **100**, 447–59.

Gardner, J.R. and Woods, R. (1979b) An electrochemical investigation of the natural floatability of chalcopyrite. *Int. J. Miner. Process.*, **6**, 1–16.

Gaudin, A.M., Miaw, H.L. and Spedden, H.R. (1957) Native floatability and crystal structure, in *Proceedings of the 2nd International Congress on Surface Activity*, Vol. 3 (ed. J.H. Schulman), Butterworth, London, pp. 202–19.

Gebhardt, J.E. and Kotlyar, D.G. (1991) Hydrosulfide depression of copper-sulfide minerals floated by xanthate and thionocarbamate collectors, in *Proceedings of the Copper 91-Cobre 91 International Symposium*, Vol. II (eds C.S. Dobby, S.A. Argyropoulos and S.R. Rao), Pergamon Press, Oxford.

Gebhardt, J.E. and Richardson, P.E. (1987) Differential flotation of a chalcocite–pyrite particle bed by electrochemical control. *Min. Metal. Proc.*, Aug., 140–5.

Guy, P.J. and Trahar, W.J. (1984) The influence of grinding and flotation environments on the laboratory batch flotation of galena. *Int. J. Miner. Process.*, **12**, 15–38.

Guy, P.J. and Trahar, W.J. (1985) The effects of oxidation and mineral interactions on sulfide flotation, in *Flotation of Sulfide Minerals* (ed. K.S.E. Forssberg), Elsevier, Amsterdam, pp. 61–79.

Hamilton, I.C. and Woods, R. (1979) The effect of alkyl chain length on the aqueous solubility and redox properties of symmetrical dixanthogens. *Aust. J. Chem.*, **32**, 2171–9.

Hamilton, I.C. and Woods, R. (1981) An investigation of surface oxidation of pyrite and pyrrhotite by linear potential sweep voltammetry. *J. Electroanal. Chem. Interfacial Electrochem.*, **118**, 327–43.

Hamilton, I.C. and Woods, R. (1984) A voltammetric study of the surface oxidation of sulfide minerals, in *Proc. Electrochemistry in Mineral and Metal Processing* (eds P.E. Richardson, S. Srinivasan and R. Woods). The Electrochemical Society, 84-10, pp. 259–85.

Heyes, G.W. and Trahar, W.J. (1977) The natural flotability of chalcopyrite. *Int. J. Miner. Process.*, **4**, 317–44.

Heyes, G.W. and Trahar, W.J. (1979) Oxidation–reduction effects in the flotation of chalcocite and cuprite. *Int. J. Miner. Process.*, **6**, 229–52.

Heyes, G.W. and Trahar, W.J. (1984) The flotation of pyrite and pyrrhotite in the absence of conventional collectors, in *Proc. Electrochemistry in Mineral and Metal Processing* (eds P.E. Richardson, S. Srinivasan and R. Woods). The Electrochemical Society, 84-10, 219–32.

Janetski, N.D., Woodburn, S.I. and Woods, R. (1977) An electrochemical investigation of pyrite flotation and depression. *Int. J. Miner. Process.*, **4**, 227–39.
Jones, M.H. and Woodcock, J.T. (1978) Perxanthates – a new factor in the theory and practice of flotation. *Int. J. Miner. Process.*, **5**, 285–96.

Koch, D.F.A. and McIntyre, F.J. (1976) The application of reflectance spectroscopy to a study of the anodic oxidation of cuprous sulphide. *J. Electroanal. Chem. Interfacial Electrochem.*, **71**, 285–96.
Kowal, A. and Pomianowski, A. (1973) Cyclic voltammetry of ethyl xanthate on a natural copper sulphide electrode. *J. Electroanal. Chem. Interfacial Electrochem.*, **46**, 411–20.

Lamache, M., Bauer, D. and Pegouret, J. (1981) Comportement electrochimique de la galene (PbS) dans les conditions de pH proches de celles de la flottation, *Electrochim. Acta*, **26**, 1845–50.
Laskowski, J. and Kitchener, J.A. (1969) The hydrophilic/hydrophobic transition of silica. *J. Colloids Interfacial Sci.*, **29**, 670.
Learmont, M.E. and Iwasaki, I. (1984) The effect of grinding media on flotation of galena, Society of Mining Engineers, preprint 84-84, pp. 1–13.
Leja, J. (1982) *Surface Chemistry of Flotation*. Plenum Press, New York.
Leppinen, J.O., Basilio, C.I. and Yoon, R.-H. (1988) Spectroelectrochemical study of ethyl xanthate adsorption on sulfide minerals, in *Proc. Electrochemistry in Mineral and Metal Processing II* (eds P.E. Richardson and R. Woods), The Electrochemical Society, 88-21, pp. 49–65.
Luttrell, G.H. and Yoon, R.H. (1983) Surface chemistry of collectorless flotation of chalcopyrite, preprint no. 83-196, 112th SME-AIMe Annual Meeting, Atlanta, Georgia.

Many, A., Goldstein, Y. and Grover, N.B. (1971) *Semiconductor Surfaces*, North-Holland Publishing Co., New York.

O'Dell, C.S., Dooley, R.E., Walker, G.W. and Richardson, P.E. (1984) Chemical and electrochemical reactions in the chalcocite–ethyl xanthate system, in *Proc. Electrochemistry in Mineral and Metal Processing* (eds P.E. Richardson, S. Srinivasan and R. Woods), The Electrochemical Society, 84-10, pp. 81-95.

Plaksin, I.N. and Bessonov, S.V. (1957) Role of gases in flotation reactions, in *Proceedings of the 2nd International Congress on Surface Activity*, Vol. 3 (ed. H.H. Schulman), pp. 361–7.
Plaksin, I.N. and Shafeev, R.Sh. (1960) The influence of certain semiconductor properties of the surface on the reaction between xanthate and galenite, in *Academy of Sciences of the USSR Proceedings, Physical Chemistry Section*, **132**, pp. 421–3.
Poling, G.W. (1976) Reactions between thiol reagents and sulphide minerals, in *Flotation, A.M. Gaudin Memorial Volume*, Vol. 1 (ed. M.C. Fuerstenau), American Institute of Mining and Metallurgical Engineers, New York, pp. 334–63.
Pozzo, R.L., Malisci, A.S. and Iwasaki, I. (1990) Pyrite–pyrrhotite–grinding media contact and its effect on flotation. *Min. Metall. Proc.*, pp. 15–21.
Pritzker, M.D. and Yoon, R.-H. (1984) Thermodynamic calculations and electrochemical studies on the galena ethyl-xanthate system, in *Proc. Electrochemistry in Mineral and Metal Processing* (eds P.E. Richardson, S. Srinivasan and R. Woods), The Electrochemical Society, 84-10, pp. 26–53.

Ralston, J. and Healy, T.W. (1980a) Activation of zinc sulfide with Cu_{II}, Cd_{II}, and Pb_{II}: I. Activation in weakly acidic media. *Int. J. Miner. Process.*, **7**, 175–201.
Ralston, J. and Healy, T.W. (1980b) Activation of zinc sulfide with Cu_{II}, Cd_{II}, and Pb_{II}: II. Activation in neutral and weakly alkaline media. *Int. J. Miner. Process.*, **7**, 203–17.

Rand, D.A.J. (1977) Oxygen reduction on sulphide minerals. Part III. Comparison of the activity of various copper, iron, lead, and nickel mineral electrodes. *J. Electroanal. Chem. Interfacial Electrochem.*, **83**, 19–32.

Redfearn, M.A. (1981) The use of nitrogen as a replacement for flotation air in the Gibraltar molybdenum circuit, 8 pp. paper presented at Canadian Mineral Processors B.C. Meeting.

Redfearn, M.A. (1983) The role of nitrogen in the flotation of by-product molybdenite at Gibraltar Mines, paper no. 83-64, 112th SME-AIME Annual Meeting, Atlanta, Georgia.

Richardson, P.E. and Maust, E.E. Jr (1976) Surface stoichiometry of galena in aqueous electrolytes and its effect on xanthate interactions, in *Flotation, A.M. Gaudin Memorial Volume*, Vol. 1 (ed. M.C. Fuerstenau) American Institute of Mining and Metallurgical Engineers, New York, pp. 364–92.

Richardson, P.E. and O'Dell, C.S. (1985) Semiconducting characteristics of galena electrodes. *J. Electrochem. Soc.*, **132**, 1350–6.

Richardson, P.E. and Walker, G.W. (1985) The flotation of chalcocite, bornite, chalcopyrite, and pyrite in an electrochemical flotation cell, paper presented at XV International Mineral Processing Congress, Cannes.

Richardson, P.E., Li, Y. and Yoon, R.-H. (1992) The photoelectrochemistry of in situ fractured pyrite electrodes, in *Proc. Electrochemistry in Mineral and Metal Processing, III* (eds R. Woods and P.E. Richardson), The Electrochemistry Society, 92-17, 342–54.

Richardson, P.E., Stout, J.V. III, Proctor, C.L. and Walker, G.W. (1984) Electrochemical flotation of sulfides: chalcocite–ethylxanthate interactions. *Int. J. Miner. Process.*, **12**, 73–93.

Salamy, S.G. and Nixon, J.C. (1954) Reaction between a mercury surface and some flotation reagents: an electrochemical study. *Aust. J. Chem.*, **7**, 146–56.

Scanlon, W.W. (1959) Polar semiconductors, in *Solid State Physics*, Vol. 9 (eds F. Seitz and D. Turnbull), Academic Press, New York, pp. 83–137.

Senior, G.D. and Trahar, W.J. (1991) The influence of metal hydroxide and collector on the flotation of chalcopyrite. *Inter. J. Mineral Process.*, **33**, 321–41.

Shannon, L.K. and Trahar, W.J. (1986) The role of collector in sulfide ore flotation, in *Advances in Mineral Processing* (ed. P. Somasundaran), Society of Mining Engineers, Colorado, pp. 408–25.

Sheikh, N. and Leja, J. (1977) Mössbauer spectroscopy of Fe-xanthates. *Separation Science*, **12**, 529–40.

Sutherland, K.L. and Wark, I.W. (1955) *Principles of Flotation*, Australasian Institute of Mining and Metallurgy, Melbourne, 489 pp.

Termes, S.C., Buckley, A.N. and Gillard, R.D. (1987) 2p electron binding energies for the sulfur atoms in metal polysulfides. *Inorg. Chim. Acta*, **126**, 79–82.

Tolun, R. and Kitchener, J.A. (1964) Electrochemical study of the galena–xanthate–oxygen flotation system. *Bull. Instn. Min. Metall.*, **687**, 313–22.

Toperi, D. and Tolun, R. (1969) Electrochemical study and thermodynamic equilibria of the galena–oxygen–xanthate flotation system. *Trans. Instn. Min. Metall. (Sect. C: Min. Proc. Extr. Metall.)*, **78**, C191–7.

Trahar, W.J. (1983) A laboratory study of the influence of sodium sulphide and oxygen on the collectorless flotation of chalcopyrite, *Int. J. Miner. Process.*, **11**, 57–74.

Trahar, W.J. (1984) The influence of pulp potential in sulfide flotation, in *Principles of Mineral Flotation* (eds M.H. Jones and J.T. Woodcock), The Australasian Institute of Mining and Mineral Processing, Victoria, pp. 117–36.

Walker, G.W., Walters, C.P. and Richardson, P.E. (1986) Hydrophobic effects of sulfur and xanthate on metal and mineral surfaces. *Inter. J. Miner. Process.*, **18**, 119–37.

Wang, X.-H. and Forsberg, K.S.E. (1991) Mechanism of pyrite flotation with xanthates. *Int. J. Miner. Process.*, **33**, 275–91.

Wark, I.W. and Cox, A.B. (1935) The physical chemistry of flotation, V. Flotation of graphite and sulfur by collectors of the xanthate type. *J. Phys. Chem.*, **39**, 551–9.

Woods, R. (1971) The oxidation of ethyl xanthate on platinum, gold, copper, and galena electrodes: relation to the mechanism of mineral flotation. *J. Phys. Chem.*, **75**, 354–62.

Woods, R. (1972) The anodic oxidation of ethylxanthate on metal and galena electrodes. *Aust. J. Chem.*, **25**, 2329–35.

Woods, R. (1976) Electrochemistry of sulfide flotation, in *Flotation, A.M. Gaudin Memorial Volume*, Vol. 1 (ed. M.C. Fuerstenau), American Institute of Mining and Metallurgical Engineers, New York, pp. 298–333.

Woods, R. (1984) Electrochemistry of sulfide flotation, in *Principles of Mineral Flotation, The Wark Symposium* (eds M.H. Jones and J.T. Woodcock), Australasian Institute of Mining and Metallurgy, Victoria, pp. 91–116.

Woods, R. and Richardson, P.E. (1986) The flotation of sulfide minerals – electrochemical aspects, in *Advances in Mineral Processing* (ed. P. Somasundaran), Society of Mining Engineers, Colorado, pp. 154–70.

Woods, R., Young, C.S. and Yoon, R.-H. (1990) Ethyl xanthate chemisorption isotherms and E_h–pH diagrams for the copper/water/xanthate and chalcocite/water/xanthate systems. *Int. J. Miner. Process.*, **30**, 17–33.

Woods, R., Basilio, C.I., Kim, D.S. and Yoon, R.-H. (1992) Ethyl xanthate chemisorption isotherms and E_h–pH diagrams for the silver/water/ethyl xanthate system. *J. Electroanal. Chem.*, **328**, 179–94.

Yoon, R.-H. (1981) Collectorless flotation of chalcopyrite and sphalerite ores by using sodium sulfide. *Int. J. Miner. Process.*, **8**, 31–48.

Young, C.A., Woods, R. and Yoon, R.-H. (1988) A voltammetric study of chalcocite oxidation to metastable copper sulfides, in *Proc. Electrochemistry in Mineral and Metal Processing II*, (eds P.E. Richardson and R. Woods), The Electrochemical Society, 88-21, pp. 1–17.

CHAPTER NINE

Clay mineral surfaces

R.A. Schoonheydt

9.1 Introduction

Clay minerals are defined here as micron- and submicron-sized crystalline phyllosilicates. Ideally, they consist of two-dimensional sheets of SiO_4 tetrahedra with an overall composition $Si_2O_5^{2-}$. The individual tetrahedra are linked with neighbouring tetrahedra by sharing three corners. All of the unshared corners point in the same direction and form part of sheet of linked octahedra with, as cations, Al^{3+}, Mg^{2+}, Fe^{3+}, Fe^{2+}. An example of a mineral with such a two-layer arrangement is kaolinite. If the octahedral sheet is sandwiched between two $(Si_2O_5^{2-})^n$ sheets, 2:1 clay minerals are obtained. They are divided into subgroups according to the degree of isomorphous substitution as shown in Table 9.1.

The terms 'dioctahedral' and 'trioctahedral' refer to the main cation in the octahedral sheet, being Al^{3+} in the first case and Mg^{2+} in the second. Because of the charge difference, two out of three octahedral positions are occupied in dioctahedral clays, while all three are occupied in trioctahedral clays. The idealized octahedral sheet of Figure 9.1 shows the two types of octahedral cationic sites and the hydroxyl groups, called inner hydroxyls. The idealized tetrahedral sheet (Figure 9.2) exposes hexagonal arrays of oxygens. A full 2:1 clay structure is shown in Figure 9.3.

Isomorphous substitution occurs in the octahedral and in the tetrahedral sheets, but mainly in the former. A negative lattice charge is created, compensated by exchangeable cations in the interlayer region (Figure 9.3), ideally above the centre of the hexagonal rings of oxygens of the tetrahedral sheet. When the negative lattice charge is in the range 0.25–0.6 e per formula unit, the clays are known as smectites or **swelling clays**. This is a group of particular relevance to this chapter.

Swelling is the process of increasing the distance between two successive clay sheets in a stack by a swelling agent, which in most cases is water. The degree of swelling depends on the size and charge of the exchangeable cations, on the location and extent of isomorphous substitution, on the polarity of the swelling agent, and on the amount of clay in the suspension. A combination of small monovalent cations (Li^+, Na^+) and small polar molecules (H_2O) is ideal for swelling. Theoretically, in an extreme case of swelling, clay sheets would become independent of each other in the

Mineral Surfaces. Edited by D.J. Vaughan and R.A.D. Pattrick. Published in 1995 by Chapman & Hall, London. ISBN 0 412 56340 1

Table 9.1 Classification of 2:1 clay minerals

Charge per unit formula	Group	Subgroup	Name
0	pyrophyllite	dioctahedral	pyrophyllite
	talc	trioctahedral	talc
0.25–0.60	smectite	dioctahedral	montmorillonite beidellite nontronite
		trioctahedral	saponite hectorite
0.60–0.90	vermiculite	dioctahedral	vermiculite
		trioctahedral	vermiculite
1.00	mica	dioctahedral	muscovite
		trioctahedral	biotite

After Schoonheydt, 1991.

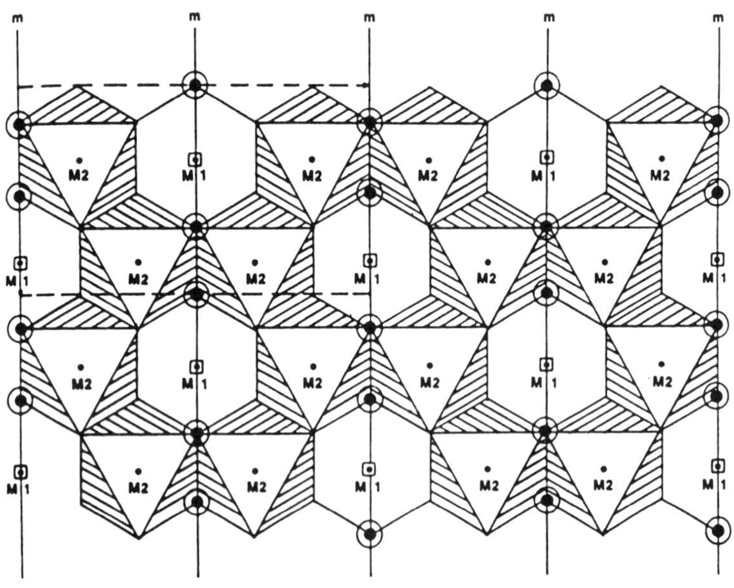

Figure 9.1 Idealized octahedral sheet showing mirror planes (m) and the two types of octahedral sites M_1 and M_2. ⦿ designates hydroxyl groups. (After Güven, 1988.)

suspension, but in practice this situation is probably never fully attained. In any case, the interlamellar or planar surface becomes available for reaction (Figure 9.4) and this is far more important than the edge surface, giving almost infinite possibilities for chemical interactions in two dimensions.

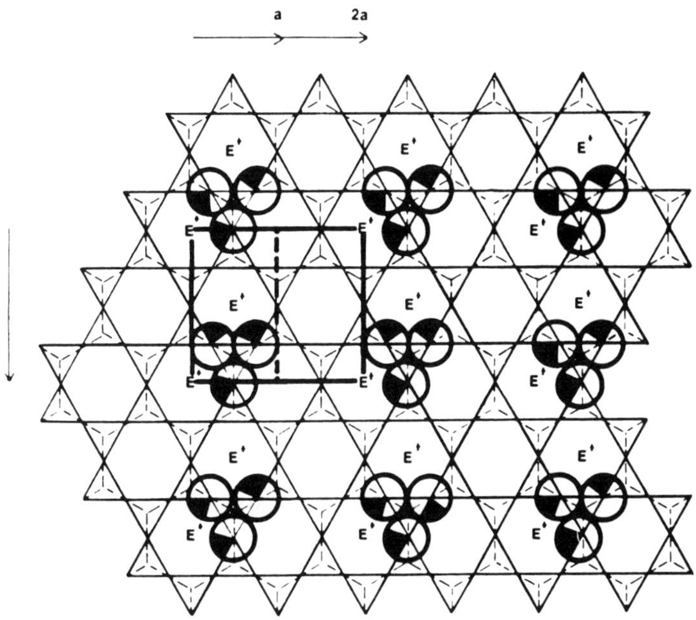

Figure 9.2 The idealized tetrahedral sheet with basal oxygens. E^+ is the favourable position of the octahedral cation. (After Güven, 1988.)

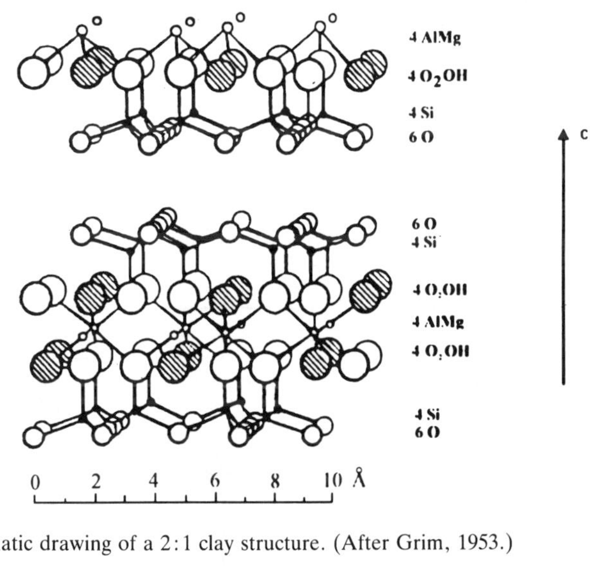

Figure 9.3 Schematic drawing of a 2:1 clay structure. (After Grim, 1953.)

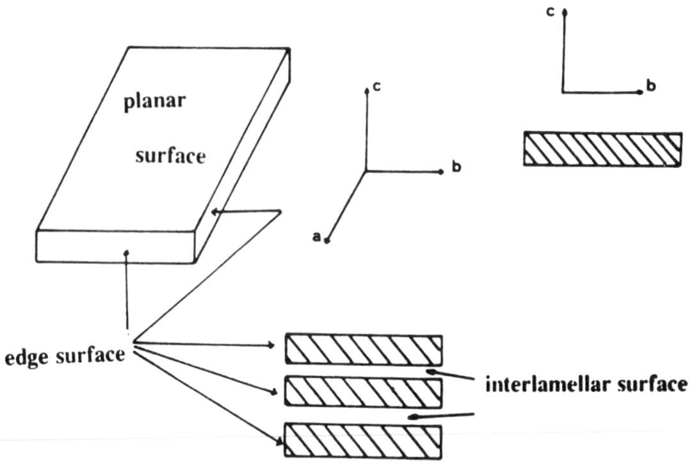

Figure 9.4 Schematic illustration of interlamellar or planar surfaces and edge surfaces in clays.

9.2 Smectites

9.2.1 Chemical composition

Dioctahedral smectites with predominantly octahedral substitution include the montmorillonites, which have a typical structural formula of

$$(Al_{2-x}Mg_xFe_z)^{VI}(Si_{4-y}Al_y)^{IV}O_{10}(OH)_2M^+{}_{x+y}.n\ H_2O$$

and traces of Fe in the octahedral layers. The exchangeable cation is represented by M^+ and VI and IV refer to the octahedral and tetrahedral layers, respectively. The montmorillonites can be subdivided into four groups, based on chemical composition and thermal behaviour (Güven, 1988):

Montmorillonite-type	x	z
Wyoming	0.20–0.38	0.18–0.30
Cheto (Chambers)	0.38–0.50	0.06–0.21
Otay	0.48–0.66	0–0.11
Tatatilla	0.26–0.45	0–0.11

If the predominant isomorphous substitution is in the tetrahedral layer, the mineral is called beidellite and has a typical structural formula of

$$(Al_{2-x}Fe^{3+}{}_yMg_x)^{VI}(Si_{4-z}Al_z)^{IV}O_{10}(OH)_2M^+{}_{x+z}.n\ H_2O$$

Nontronite is an Fe-smectite with Fe^{3+} mainly in the octahedral layer, where it is partially replaced by Al and Mg. Its structural formula is

$$(Si_{4-x-y}Al_xFe_y)^{IV}(Fe_{2-z-w}Al_zMg_w)^{VI}O_{10}(OH)_2M^+_{x+y+z}.n\ H_2O$$

The two most important representatives of the trioctahedral montmorillonites are hectorite and saponite. Hectorite is characterized by isomorphous substitution of Mg^{2+} by Li^+ in the octahedral layer, while for saponite the origin of the layer charge is mainly the isomorphous substitution of Si^{4+} by Al^{3+} in the tetrahedral sheets. The typical structural formulae of saponite and hectorite are respectively:

$$(Mg_{3-x-y}Al_xFe_y)^{VI}(Si_{4-z}Al_z)^{IV}O_{10}(OH)_2M^+_{z-x-y}.n\ H_2O \quad \text{and}$$
$$(Mg_{3-x}Li_x)^{VI}(Si_4)^{IV}O_{10}(OH)_2M^+_x.n\ H_2O.$$

9.2.2 Morphology and surface area

Ideally, montmorillonites occur in the form of hexagonal lamellae, but rhombohedral forms, lamellae with irregular outline, laths and fibres also occur and are a consequence of the crystallization conditions (Figure 9.5). The individual platelets are usually aggregated into particles of various forms and shapes.

Nontronites occur mostly in the form of laths and ribbons, intergrown into aggregates. Hectorite occurs in the form of extremely thin but well-defined laths. These thin laths easily break into minute fragments which aggregate

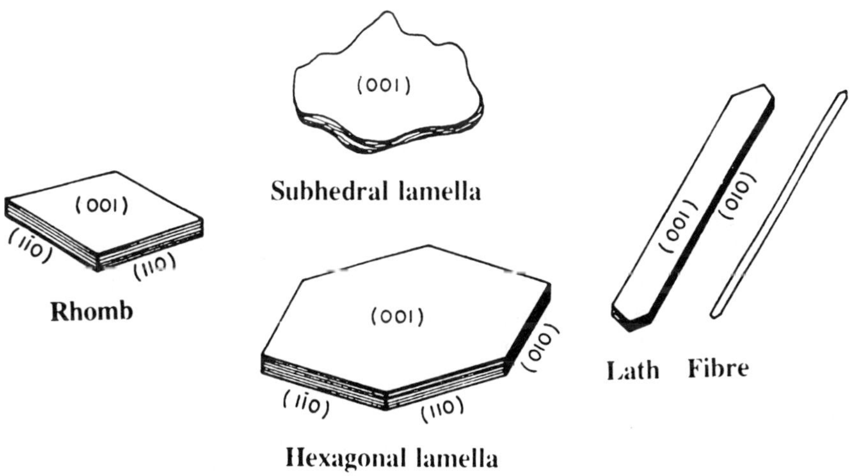

Figure 9.5 Typical crystal forms of smectites. (After Güven, 1988.)

Table 9.2 N_2-BET surface areas (S) and pore volumes (V) of various montmorillonites in their Na^+ forms (size fraction <0.5 μm)

Montmorillonite type	S ($m^2 g^{-1}$)			V ($\mu L g^{-1}$)		
	total	micropores	mesopores	total	micropores	mesopores
Moosburg	124	69	54	116	24	76
Otay	144	82	57	142	33	88
Camp Berteau	81	41	38	95	22	54
Marnia	89	58	32	112	32	55
Greek Yellow	70	48	20	57	22	24
Wyoming Bentonite	44	32	10	37	12	19
Greek White	48	35	14	46	13	26

After Stul and Van Leemput, 1982a, b.

by end-to-end or edge-to-edge associations. Saponites occur as large foliated aggregates of thin films and flakes.

The size, shape, number and stacking of clay platelets in the aggregates, the size of the aggregates, and the stacking of the aggregates themselves are all extremely important factors that determine the external surface area. The surface area of the aggregates is usually determined on a dehydrated sample (393 K); it is controlled not only by the type of smectite and the nature of the exchangeable cation but also by the drying method and the method of determination of the surface area.

Table 9.2 lists N_2-BET (Brunauer Emmett Teller) surface areas and pore volumes of various freeze-dried montmorillonites from different localities. All samples were dehydrated *in vacuo* for 24 h at 393 K prior to the measurement (Stul and Van Leemput, 1982a, b). The surface areas (S) and the pore volumes (V) recorded in Table 9.2 are reasonably well correlated. However, the values of S and V cannot be explained by models of regular stacking of elementary clay platelets. Clearly, irregular stacking must be assumed. This irregularity may occur only in the a, b plane or may be in all three directions, and aggregates of clay platelets are shown in Figure 9.6. Nadeau (1985) measured the physical dimensions of the elementary clay platelet of Wyoming bentonite (size fraction <0.1 μm). As the edge surface was measured to be 7.9 $m^2 g^{-1}$, the data of Table 9.2 show that the majority of the surface area originates from basal planes and from micro- and mesopores formed by the irregular stacking of the elementary particles and of the aggregates. This N_2-BET surface area is also called the external surface area.

The total surface area of an elementary particle is the sum of the edge surface area and of the surface area of its two basal planes. It can be accurately calculated only if the size of the elementary particle is known

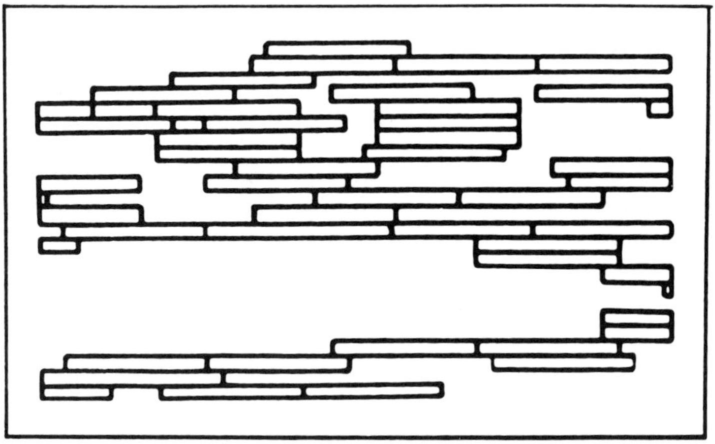

Figure 9.6 Idealized aggregation of elementary clay sheets. (After Fripiat, 1990.)

exactly, which is usually not the case. Experimentally, probe molecules must be adsorbed which fulfil the following conditions:

1. they are able to penetrate between the clay particles in the aggregates;
2. they form a true monolayer on the surface;
3. the interactions between the molecules are negligible with respect to the molecule–surface interaction.

These conditions are never fulfilled. Indeed, condition (1) requires polar molecules such as water and glycols, for which conditions (2) and (3) are not really valid. The theoretical total surface area for 2 μm square elementary particles is 800 $m^2 g^{-1}$ and for 0.05 μm particles it is 832 $m^2 g^{-1}$, the thickness being 0.96 nm (Schoonheydt, 1991). For Wyoming bentonite, Nadeau (1985) obtained S ($m^2 g^{-1}$) = 825/T (nm), where T is the particle thickness, from electron microscopic measurements. This leads to $S = 660\ m^2 g^{-1}$.

9.2.3 Cation exchange capacity

The cation exchange capacity (CEC) is the quantity of exchangeable cations measured in mmole g^{-1}. Defined in this way, it is a purely operational quantity, which depends on the measurement technique, on the conditions of the measurement and on the definition of the unit weight of clay (air-dry, oven-dry . . .). Whatever the method used, the CEC consists of two parts: a pH-independent part, which is due to isomorphous substitution of a tetrahedral or octahedral cation by a cation of lower valency, and a pH-dependent part, which is ascribed to 'broken bonds' at the crystal edges.

The pH-independent part can be calculated if the structural formula of the smectite is known. It can also be measured experimentally with the now classical alkylammonium method, first developed by Lagaly and Weiss (1975). This method is based on the observation of a monolayer-to-bilayer transition in the X-ray diffraction (XRD) patterns of smectites. This transition occurs as soon as the average planar area of the interlayer alkylammonium cation exceeds the average planar area per negative lattice charge. There are three drawbacks to the method:

1. the average planar area of an alkylammonium cation is a theoretical quantity, which depends on the **adopted** zigzag structure of the C-chain, on the assumed orientation of the chain with respect to the plane of the silicate layers, and on the amount of unoccupied space between adjacent cations;
2. at the edges of the aggregates the alkylammonium cations may protrude from the interlamellar space, causing a particle size effect;
3. the charge density of smectites is heterogeneous, which makes the monolayer-to-bilayer transition diffuse in the XRD patterns.

Corrections for the particle size effect have been made (Stul and Mortier, 1974). While addressing the problem of charge heterogeneity, Laird, Scott and Fenton (1989) proposed the empirical relationship

$$\sigma = \frac{38.88}{5.67n_c + 14} - 0.078$$

between layer charge σ and alkylammonium chain length n_c, which correlates well with the layer charge as determined from structural formulae.

The charge heterogeneity is an inherent property of the smectites and up to 10 charge classes in the range 0.20–0.46 e per $Si_4Al_2O_{10}(OH)_2$ have been proposed (Stul and Mortier, 1974; Maes, Stul and Cremers, 1979; Stul and Van Leemput, 1982c; Talibudeen and Goulding, 1983). However, the charge density decreases slightly and is more homogeneous for the smallest size fraction (<0.1 μm) (Rengasamy, Van Assche and Uytterhoeven, 1976).

The pH-dependent part is apparent whenever the CEC is determined with classical techniques such as $^{22}Na^+$, NH_4OOCCH_3 and Ba^{2+}. Maes, Peigneur and Cremers (1975) proposed the following two equations for monovalent and divalent exchange, respectively:

$$\text{CEC (meq g}^{-1}) = 79.9 + 5.04 \text{ pH}$$

$$\text{CEC (meq g}^{-1}) = 96.1 + 3.93 \text{ pH}$$

These equations explain the increase of the CEC with pH, and also indicate that this pH-dependent CEC is significant. It is due to edge sites that

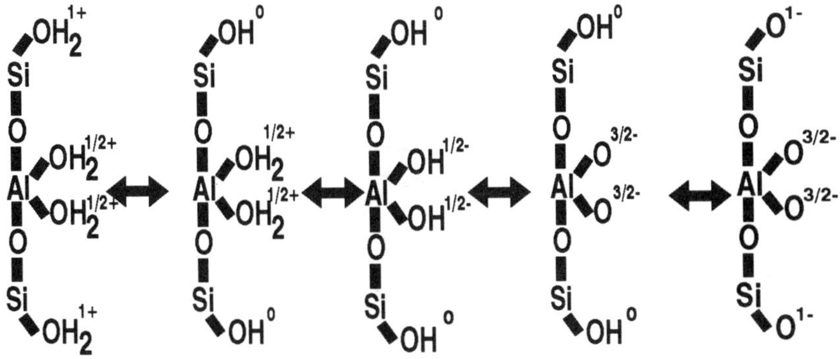

Figure 9.7 Idealized pictures of hydroxyl groups at edges of elementary clay particles. (After White and Zelazny, 1988.)

become deprotonated, and therefore negatively charged upon pH increase. The diagrammatic representation of these sites in Figure 9.7 is from White and Zelazny (1988). These authors estimated the amount of edge sites as 5.32 μmole m^{-2}.

The CEC is different for monovalent and divalent cations. This is ascribed by Peigneur, Maes and Cremers (1975) to the higher affinity of divalent cations for edge sites. Thus, edge sites are important in both CEC mechanisms.

Metal uncharged ligand complexes may also be used for the determination of the CEC of clays. Their advantage is that they are very selectively adsorbed, especially the planar complexes with chelating amines, such as ethylenediamine (en) (Maes, Marijnen and Cremers, 1977; Maes and Cremers, 1978; Maes, Rasquin and Cremers, 1982). The selectivity comes from the rule that 'planar complexes prefer planar surfaces': a geometrical factor and the replacement of two axially coordinated water molecules by the clay surface are the driving forces for the selectivity (Maes *et al.*, 1980). A method has been developed, based on the adsorption of transition metal ion (TMI) complexes, to remove the TMI from waste waters down to ppb levels. In addition, Chabra, Pleysier and Cremers (1975) have shown that $Ag(SC(NH_2)_2)_x^+$ can be used for rapid determination of the CEC of soils.

Non-expandable clays or partially collapsed clays, i.e. clays with a fraction of 1 nm layers, such as micas, illite, charge-reduced montmorillonites or K^+-montmorillonite subjected to wetting–drying cycles, have a very small number of exchange sites, which show an extremely high selectively for Cs^+ (Brouwer *et al.*, 1983; Maes, Verheyden and Cremers, 1985). The nature of these sites is unknown. One hypothesis is that they are frayed-edge sites, as shown in Figure 9.8. They amount to 0.1–0.5% of the CEC. Illitic and

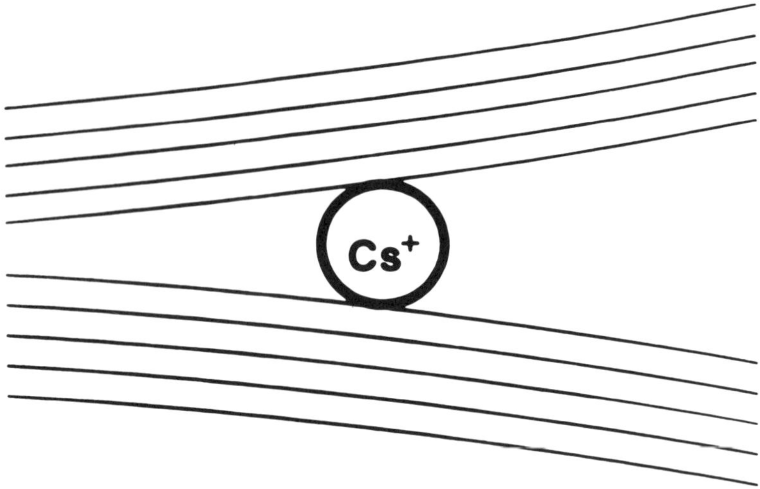

Figure 9.8 Idealized picture of a frayed edge site.

mica-type materials are important components of bricks, roof tiles, subsurface clay layers and shales. Radioactive Cs^+ (in the atmosphere due to a nuclear accident, or in subsurface repositories) may come into contact with these clay-type materials and be very strongly and selectively adsorbed at these high selectivity sites. The mobility of Cs^+ will be reduced to practically zero and it will be impossible to wash it out of these illitic and mica-type materials.

9.3 Clay surfaces in aqueous suspension

Smectites swell in aqueous suspension as a result of the uptake of water in the interlamellar regions of the aggregates. The aggregates may break down and ultimately may be completely dispersed (i.e. a random distribution of elementary particles is present in the suspension, and these particles move independently of each other). The extent of dispersion and the structure and properties of the water molecules near the surface of clays have been the most important research topics related to clay–water suspensions. The properties of the surface sites (exchangeable cations, basal oxygens and edge sites, or O, OH groups, cations and defects) are closely related to these two subjects.

9.3.1 *Aggregation of elementary particles*

There is general agreement that swelling of smectites is not only dependent on the CEC and the nature of the exchangeable cation, but also on the size, shape and history of the clay particles. In clay pastes, the organization of the elementary clay platelets is described at three levels (Figure 9.9): tactoids or micro-aggregates of the elementary sheets with predominantly face-to-face aggregation, aggregates of these tactoids, and flocs. There are also three types of water: floc water in the pores of the flocs, aggregate water in the pores of the aggregates, and interlamellar water (Ben Rhaïem, Pons and Tessier, 1987; Touret *et al.*, 1990). For low charge Na-smectites, the micro-aggregates consist of about four elementary sheets with spacings of 3.5 nm or more. For Ca-smectites the aggregates consist of face-to-face aggregation of 50 elementary sheets, which are arranged in tactoids of 7–11 elementary sheets (Figure 9.10). The interlamellar space in the tactoids is 1.8 nm, while pores of up to 5 nm can be formed in the aggregates. For Mg-smectites, the number of elementary sheets per tactoid is in the range 4–19, depending on

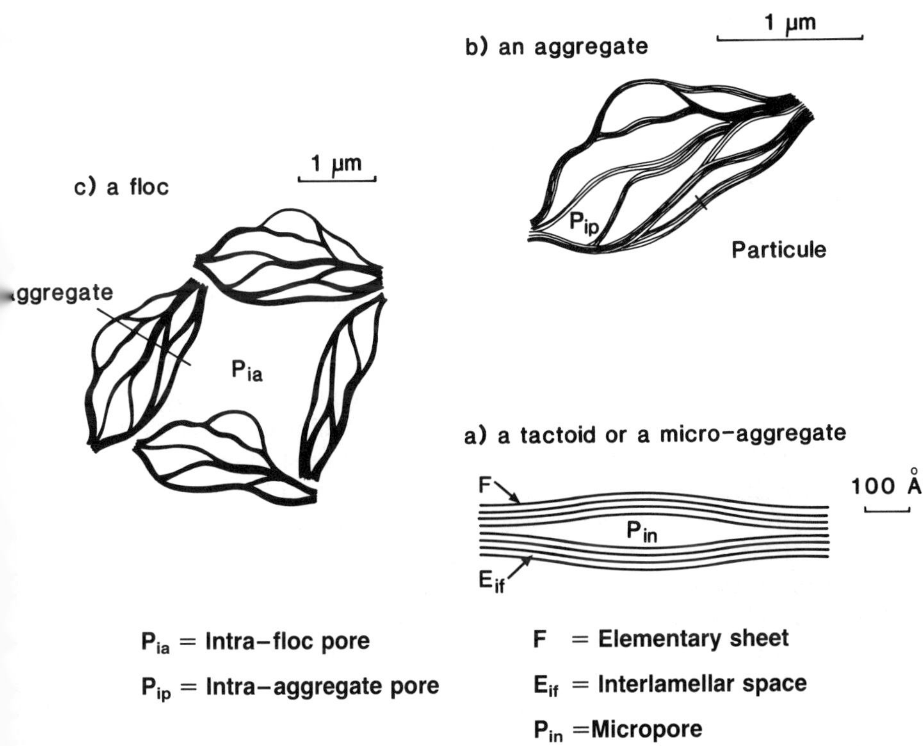

Figure 9.9 Levels of aggregation of elementary clay particles. (After Touret *et al.*, 1990.)

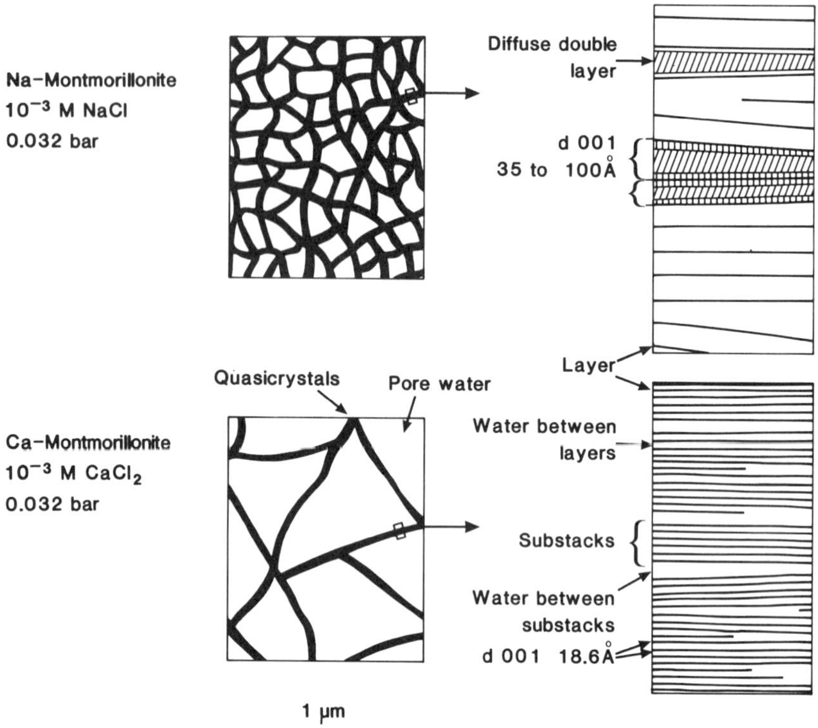

Figure 9.10 Schematic drawing of the microstructure of Na^+- and Ca^{2+}-smectite. (After Ben Rhaïem, Pons and Tessier, 1987.)

the size (length and width) of the sheets, and the interlamellar distance is 2.2 nm.

Whether swelling upon dilution of the suspensions goes to the limit of separation into individual particles is a matter of debate. Small angle neutron scattering data seem to indicate that it does, at least for Li^+-montmorillonite (Sposito and Prost, 1982). However, there are at least three reports in the literature which indicate that, as the concentration of the suspension is decreased to very small values (such as 0.1 wt%), the micro-aggregates or tactoids of elementary platelets do not disintegrate, but continue to swell (Roberson, Weir and Woods, 1968). In other words, the average number of elementary platelets per tactoid does not change with concentration, only the number of tactoids (Fripiat *et al.*, 1982; Vali and Bachmann, 1988).

Table 9.3 gives the number of elementary platelets per tactoid or micro-aggregate for typical smectites. The history of the samples and the measurement technique must be indicated in order for the value of these measurements to be assessed. For dilute suspensions, it seems safe to con-

Table 9.3 Tactoids in smectite suspensions

Montmorillonite	Technique	wt% in suspension	Tactoid
Li^+	SANS	–	single platelets
K^+	SANS	–	two platelets per tactoid; two water layers between adjacent platelets
Cs^+	SANS	–	three platelets per tactoid; two water layers between adjacent platelets
Na^+	SAXS	20	single platelets and tactoids with 4–5 platelets and two water layers between adjacent platelets
K^+	SAXS	20	aggregates with and without water layers between adjacent sheets
Cs^+	SAXS	20	thick aggregate with zero and one water layer between adjacent sheets
Ca^{2+}	SAXS	20	4–5 platelets per tactoid; 3 water layers between adjacent sheets
	EM	–	3 platelets per tactoid

SANS, small angle neutron scattering; SAXS, small angle X-ray scattering; EM, electron microscopy.

clude that mixtures of single platelets and tactoids composed of 2–5 elementary platelets exist in suspension.

The number of water layers between adjacent sheets in micro-aggregates or tactoids is very much dependent on the size and charge of the cation. This has important consequences for adsorption, because in aqeous suspension, adsorption is proportional to the available surface. The state of aggregation of the clay platelets and the ability of the adsorbing molecule to enter into the aggregates and the interlamellar space are important factors. This can be investigated by spectroscopy of adsorbed cationic dyes at small loadings. The small loadings serve not only to keep the clay suspension stable during the measurement, but also to minimize the disturbance of the clay aggregation by the adsorbed dyes. Methylene blue (MB^+) and proflavine (PF^+) are the ideal probe molecules because of their tendency to aggregate in aqueous solution above a critical concentration with formation of dimers, trimers, or even higher aggregates, which can be distinguished spectroscopically. What is measured are essentially π–π^* transitions, the energy of which depends on the polarity of the environment and on the aggregation of the molecules on the clay surface.

The tendency of the dyes to aggregate is enhanced by adsorption on clay (Cenens and Schoonheydt, 1988). That is, the dye molecules are concentrated in a small volume around the clay particles in suspension. It

Table 9.4 Effective surface areas (m^2g^{-1}) of smectites in aqueous suspension from fluorescence spectroscopy of proflavine

Clay	External surface	Total surface
Na^+-H	125	775
Ca^{2+}-H	165	725
Cs^+-H	–	580
K^+-H	135	485
Ca^{2+}-L	–	840
Na^+-BS	110	110

From Cenens, Schoonheydt and De Schryver, 1990.

is possible to describe the aggregation reaction quantitatively in solutions containing clays saturated with Na^+ and Ca^{2+} (not for K^+ and Cs^+) (Cenens and Schoonheydt, 1990):

$$n\,MB^+ \leftrightharpoons (MB^+)_n \qquad \text{with} \qquad K_n = [(MB^+)_n]/[MB^+]^n.$$

This means that the surface of Na^+ and Ca^{2+} clays is freely available to MB^+. That is, these suspensions contain individual clay sheets or aggregates which are swollen to such an extent that free diffusion of MB^+ in the interlamellar space is possible. In the case of K^+ and especially Cs^+ clays, MB^+ diffuses into the interlamellar space but is not allowed to dimerize because two adjacent sheets in the tactoids are separated by one water layer only.

One can go one step further with a quantitative interpretation of the fluorescence intensity of adsorbed PF^+ (Cenens, Schoonheydt and De Schryver, 1990). The fluorescence is quenched by dimerization due to a static quenching mechanism (Perrin mechanism). A quenching volume can be defined, which is interpreted in terms of an effective surface area of clays, assuming that the reaction takes place in a 1 nm thick layer around the particles. These surface areas (Table 9.4) are the first to be published for clays in aqueous suspension. Although their absolute value can be questioned, because they depend on the assumed layer thickness, the trend is evident, and in line with the qualitative ideas about the cation-dependence of smectite aggregation. Again, this analysis of fluorescence data is not applicable to all types of clay suspensions and more research needs to be done in order to analyse all the factors that influence the data, and to refine the effective surface areas.

9.3.2 The structure of water near the surface

The question of whether or not, and how far, water molecules 'feel' the surface of clays has given rise to two schools of thought. For a detailed discussion the reader is referred to Sposito and Prost (1982). The thermodynamic school (Low, 1979, 1980; Low and Margheim, 1979) gives a macroscopic description of the clay–water system. The clay phase and the water phase are separated and all observed changes with respect to bulk water are attributed to the water phase only. As a result, the properties of the water phase are changed with respect to those of bulk water up to several nanometres from the surface, and this for water contents up to 28 kg per single kg of clay.

The spectroscopic school, on the other hand, gives a microscopic picture of the adsorbed water phase. Indeed, any spectroscopy determines the nature of the molecule only in its immediate environment and long distance perturbation can only be indirectly examined. A comparison of the molecular properties of adsorbed water with those of bulk water suggests that the influence of the surface extends, at most, to three water layers. With quasi-elastic neutron scattering (QENS) experiments the molecular mobility on a time-scale of 10^{-10}–10^{-13} s can be examined for one and two layers of water molecules in the interlamellar space. For Li^+, proton correlation times for the rotation of the water molecule around the O–Li^+ axis, and of the Li^+–hydrate as a whole, were determined to be $1.3–2.2 \times 10^{-12}$ s and $2.8–3.8 \times 10^{-12}$ s, respectively. The diffusion coefficient of this interlayer water was determined to be $4.5 \times 10^{-7}–6.7 \times 10^{-8}\ cm^2\ s^{-1}$, increasing with the degree of hydration (Poinsignon *et al.*, 1987). For the two-layer hydrate of Ca^{2+}-montmorillonite, the hydration shell water is immobile on the QENS time-scale, i.e. motions are slower than 10^{-10} s (Tuck *et al.*, 1984) and thus amenable to study by nuclear magnetic resonance (NMR). This is the approach of Delville, Grandjean and Laszlo (1991) and Grandjean and Laszlo (1990), who used oriented clay sheets in a magnetic field. They found preferred orientations of water molecules with respect to the surface and the cations with two re-orientation possibilities. These are rotations around the O–H axis, and rotations around the M^{n+}–O axis; the former is preferred with Na^+, the latter with Ca^{2+}. The interlamellar water that is not in the first coordination sphere of the cations is liquid-like, with a diffusion coefficient about half that of liquid water.

On the NMR time-scale, two types of water can be identified: **phase a** or bulk-like water and **phase b** water, three layers thick, on the surface of micro-aggregates and aggregates. Presumably there is no fast exchange (on the NMR time-scale) between the latter phase and interlamellar water (Fripiat *et al.*, 1982; Fripiat, Letellier and Levitz, 1984). For smectites, phase b water is on the **external** surface of tactoids, showing that tactoids are present even in dilute suspensions.

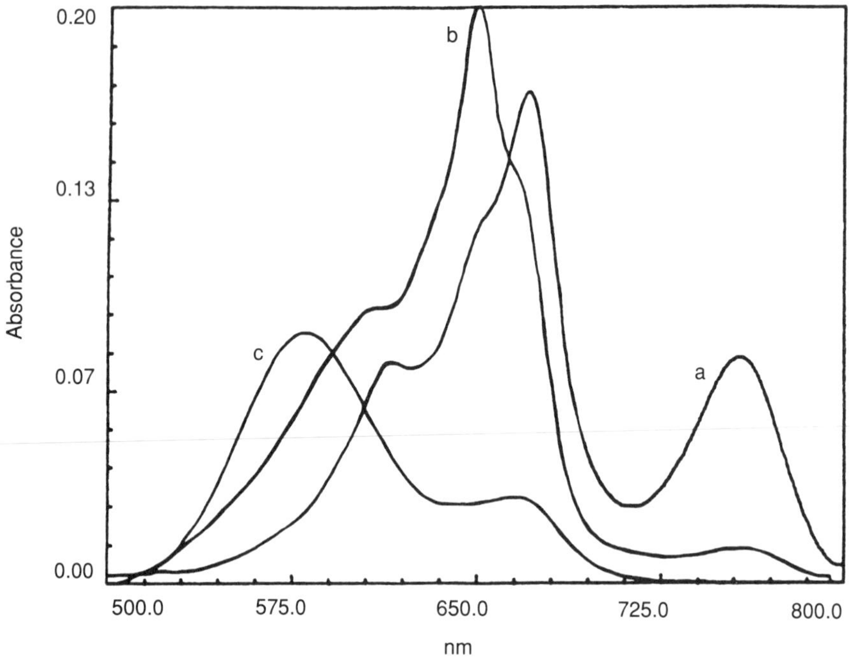

Figure 9.11 Spectra of methylene blue on clays in aqueous suspension: (a) K^+-laponite (loading is 0.2% of CEC); (b) Na^+-hectorite (loading is 0.2% of CEC); (c) Na^+-laponite (loading is 20% of CEC).

In aqueous suspensions, there are two sources of acid sites: OH groups at the edges, and water molecules in the coordination sphere of cations. The acidity of residual water in air-dry or vacuum-dry smectites has been known for more than 20 years (Mortland *et al.*, 1963). Protonation of ammonia, pyridine and basic dye molecules and demetallation of adsorbed porphyrins are typical results of the interaction of this water with adsorbed bases. It is now clear that even in dilute aqueous suspensions, acid sites may be present on the surface. At pH 4, the synthetic mica-type montmorillonite carries only doubly protonated methylene blue at low loadings. This is due to 1–5 μmole g^{-1} of strong acid sites, mainly on the edge surface (Cenens and Schoonheydt, 1988). These are called 'strong' acid sites because, in solution, the double protonation of MB requires pH <1. Such sites are, however, also present on alkali-exchanged smectites such as hectorite and its synthetic analogue laponite. Moreover, there is a definite cation effect as the number of these sites increases from Na^+ to K^+ to Cs^+ (Figure 9.11). This is in reverse order of the polarizing power of the cations. Two explanations are possible:

1. there are OH groups at the edges, so there must be more such sites on a

highly aggregated smectite, such as Cs^+-exchanged hectorite, than on a swollen, Na^+-exchanged hectorite;
2. Cs^+-exchanged clays have collapsed regions in their aggregates with few, but strongly polarized, interlamellar water molecules able to protonate incoming MB molecules.

In these regions, only monomeric MB can be adsorbed because of lack of space for dimerization. Thus the ratio of monomer: dimer is higher on Cs^+- than on Na^+-hectorite. This is also observed experimentally (Cenens and Schoonheydt, 1988).

9.4 Clay materials science

Clay materials science is the study of the chemistry underlying the development of clays and clay minerals into materials for industrial or possible industrial use. For instance, a classical example is the chemistry involved in heat treatment in the production of ceramics. At the laboratory level, clay minerals have been modified so as to obtain heterogeneous catalysts, selective adsorbents and supramolecular structures. Recent achievements in the first two areas will be highlighted.

9.4.1 Heterogeneous catalysts

Three types of heterogeneous catalysts have been developed based on smectites: acid-activated clays, pillared clays, and heterogenized homogeneous catalysts. The first group has reached the commercial stage.

Acid activation consists of treating the smectites with strong mineral acids such as HCl, H_2SO_4 and HNO_3. Exchangeable cations and octahedral cations are leached out of the lattice and SiO_2 is precipitated on the surface. At low levels of lattice attack, the reaction is diffusion-controlled (diffusion of protons towards the reaction front, or diffusion of cations towards the solution). As the reaction proceeds, the chemical reaction itself becomes rate-determining (Corma, Mifsud and Saur, 1990). The result of the acid activation is a material that barely shows the characteristic X-ray pattern of the original clay, with an enhanced surface area and enhanced acidity. It is widely used as decolourizer. Historically, it was the first commercial cracking catalyst. Nowadays, the active phase in cracking catalysts is a zeolite which is mixed with treated kaolinite and alumina (Hettinger, 1991).

Commercial acid-activated clays, such as K10 derived from a montmorillonite-type of clay, are also used as supports for reagents or catalysts. Laszlo (1987, 1990) has promoted the use of anhydrous $Fe(NO_3)_3$/K10 (clayfen) and anhydrous $Cu(NO_3)_2$/K10 (claycop) as reagents for aromatic nitration reactions. Smectites exchanged with multivalent cations are acid catalysts, even without acid-activation, because of the strong acidity of the residual water in the coordination sphere of the multivalent cations (Adams, 1987; Ballantine *et al.*, 1987). This is especially true of Al^{3+}, Fe^{3+}, Zr^{4+} and Cr^{3+}. Diels-Adler additions and Friedel-Crafts reactions are catalysed by Lewis acid clays, especialy K10, impregnated with transition metal chlorides and carefully dehydrated (Laszlo, 1987). These materials are analogous to the impregnated nitrates. The question then is, are they real catalysts or simply reagents? Secondly, they may suffer from possible loss of metal in the solution phase, thus potentially contaminating the reaction products.

Acid-activation is dependent on particle size and chemical composition. With synthetic clays, these factors can be controlled. In order to understand all the factors that determine the acid-activation process and the catalytic function of acid-activated clays, synthetic clays offer the best avenue for progress (Urabe, 1990).

The heterogenization of homogeneous catalysts by simple ion exchange of cationic complexes or *in situ* synthesis in the interlamellar space was an area of active research in the 1980s (Rupert, Granquist and Pinnavaia, 1987). Recently, this research area has emerged again, thanks to the work of Choudhary, Subba Rao and Prasad (1991). Here, an acid clay is derived with silane compounds on which the active site is subsequently grafted.

In the past ten years, however, most efforts have been put into the pillared clays. Pillaring can be defined as the process of propping the elementary clay sheets apart by a pillaring agent, such that permanent porosity is created (Schoonheydt, 1991). In the narrow sense, the pillaring agents are polynuclear inorganic cations, which upon calcination form robust, thermally stable pillars between clay sheets. The height of the pillars is at least 0.9 nm or four oxygen layers. It is believed that the polynuclear cations are adsorbed in the interlamellar space via a cation-exchange process. The three relevant parameters are:

1. at least a tenfold excess of pillaring cations in solution is required (excess with respect to the CEC);
2. partially neutralized solutions;
3. washing after the exchange.

In the case of Al-pillaring, it has been hypothesized that the Keggin ion $[Al_{13}O_4(OH)_{24}(H_2O)_{12}]^{7+}$ is the main pillaring agent. This is based on two observations: 1. it is the only polynuclear species detectable in solution by ^{27}Al NMR; 2. the spacing of 1.8 nm agrees exactly with the size of the Al_{13} cluster. Against the Al_{13} hypothesis are two observations: 1. after exchange and before washing the spacing is 1.25 nm; it increases to the typical 1.8 nm after washing only; 2. a broad 1.8 nm spacing is obtained after exchange with concentrated Al solutions in which it is impossible to detect the Al_{13} by NMR. This spacing decreases to 1.5 nm after the first washing, and reaches the typical 1.8–1.9 nm value upon further washing (Schutz *et al.*, 1987;

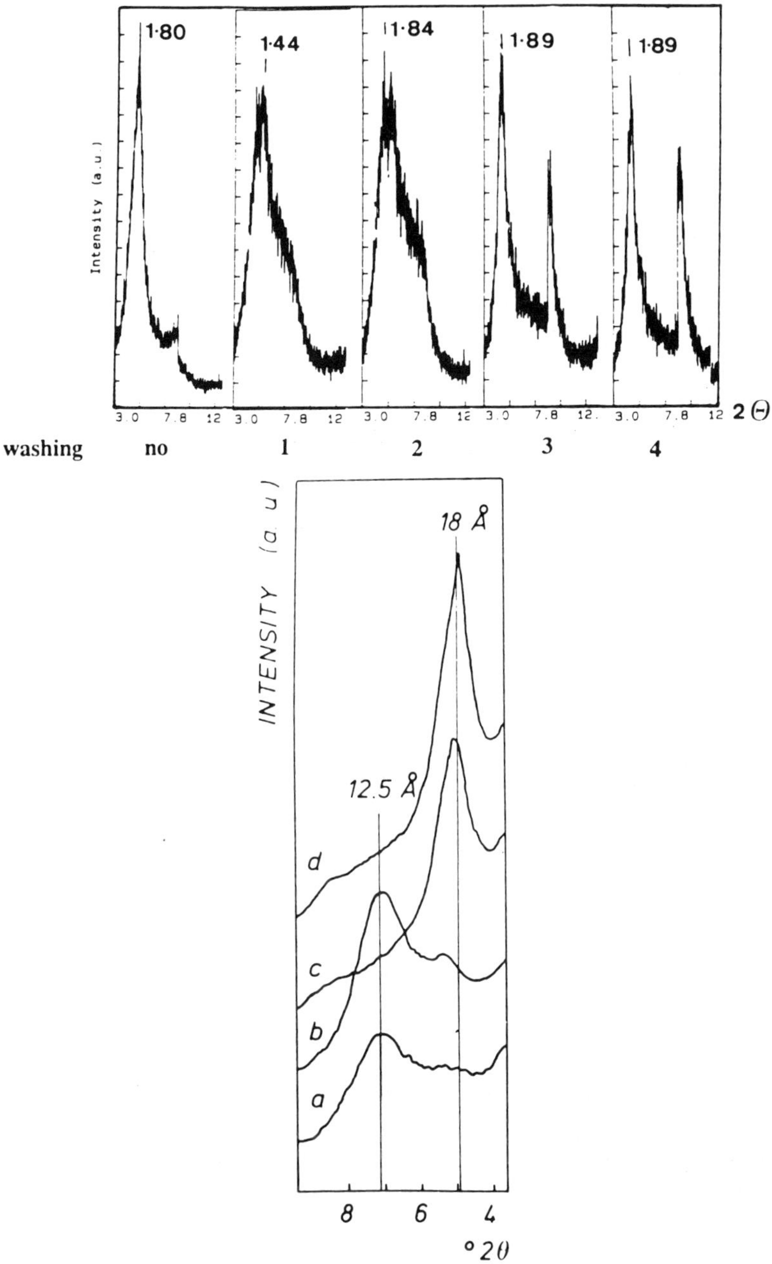

Figure 9.12 Evolution of d_{001} line in the XRD spectra of pillared clays with washing; upper part, saponite; lower part, beidellite (a, b, c and d indicate successive washings).

Figueras *et al.*, 1990; Schoonheydt and Leeman, 1992). These two observations are shown in Figure 9.12.

It appears therefore that the pillars are formed upon washing and the question of the nature of the pillaring agent cannot be answered and may even be irrelevant. It is probable that a range of polynuclear Al complexes are capable of giving a 1.8 nm spacing, including the Keggion ion. Despite these difficulties in understanding the pillaring process, new developments have continued to emerge. For instance: pillaring with GaAl polynuclear species (Bradley and Kydd, 1991; Gonzalez *et al.*, 1991, 1992); the competitive ion exchange method (Figueras *et al.*, 1990); the rare earth pillars, which give spacings up to 2.6 nm (Tokarz and Shabtai, 1985; McCauley, 1990; Sterte, 1990).

A second problem of pillared clays concerns the chemistry of the calcination process. In the case of Al pillars, the most widely studied, calcination can be represented, assuming Al_{13} pillars, by the reaction:

$$[Al_{13}O_4(OH)_{24}(H_2O)_{12}]^{7+} \rightarrow 6.5Al_2O_3 + 7H^+ + 20.5H_2O$$

Protons are generated which are either removed as water or react with structural oxygens to form an acidic clay. A discussion point is whether there is a chemical interaction between the Al_2O_3 pillar and the surface oxygens, leading to anchored pillars. To the author's knowledge, three such cases have been reported in the literature. The first is the pillared beidellite which after calcination gives a weak and broad ^{27}Al line at 56.5 ppm, distinctly different from that of tetrahedral Al in the lattice (67.5 ppm) and in the Keggin ion (62.5 ppm) (Plee *et al.*, 1985). The suggested explanation is a chemical anchoring of the pillar, thus:

$$\text{Si—OH—Al + HO—Al—pillar} \rightarrow \text{(Si,Al)—O—Al-pillar.}$$

The second case is that of fluor-hectorite (Pinnavaia *et al.*, 1985). The evidence is a shift of the −93.8 ppm ^{29}Si NMR line of the original clay to −96.7 ppm upon pillaring. This suggests inversion of Si-tetrahedra and linkage of these inverted tetrahedra to the pillar via a common oxygen atom. A clay structure of inverted Si-tetrahedra was proposed by Edelman and Favejee a long time ago (Grim, 1953). The third piece of evidence is from Bukka, Miller and Shabtai (1992). These authors suggested that the higher thermal stability of pillared Cheto-type smectites compared with pillared Wyoming montmorillonite is due to the high temperature anchoring of the Al-pillar via an oxygen atom of an inverted Si-tetrahedron, analogous to fluor-hectorite.

It is clear from these studies that the chemical composition of the smectite has an, as yet unknown, influence on the pillaring process. The observed NMR shifts in fluor-hectorite may be caused by at least three factors:

calcination; formation of acidic clays; and interaction with the pillars, which may include chemical anchoring. However, as long as contributions from these three factors remain uncertain, the assignment of the shifts to chemical anchoring only must be considered as hypothetical. In any case, an acidic clay is produced pillared with Al_2O_3 pillars, the nuclearity of which is unknown. The acidity is due to the clay and to the pillar. The latter introduces a strong component of Lewis acidity.

9.4.2 *Selective adsorbents*

The newest developments are in the areas of removal of non-ionic organics with organoclays and the enantio-selective clays for separation of optical isomers.

Non-ionic organics, such as aromatics and chlorinated hydrocarbons, are not adsorbed significantly by 'normal' smectites. This is because the smectites are hydrophilic. Thus, water fills the interlamellar space and the organics can be adsorbed only on the external surface, if they are adsorbed at all. If the clay is made organophilic, efficient adsorption can occur, owing mainly to van der Waals interactions, though specific adsorption forces cannot be excluded (Mortland, 1970; McBride, Pinnavaia and Mortland, 1977). The latter will depend on the type of organoclays and the type of organic compound. Two types of organoclays have been developed: those obtained by exchange of tetra-alkylammonium cations (Wolfe, Demirel and Bauvaux, 1985; Lee *et al.*, 1990), and hydroxy-Al clays (or La-clays) exchanged with surfactants such as cetylpyridinium cations (Srinivasan and Fogler, 1990a, b). These clays are called inorgano-organoclays or IOCs.

In the case of tetra-alkylammonium clays, the adsorption capacity for gas phase adsorption is mainly determined by the length of the alkyl chains in the order $TMA^+ > TEA^+ > TPA^+$, respectively tetramethyl ammonium, tetraethyl ammonium and tetrapropyl ammonium cations. This is the order of available space and indicates that the adsorption occurs both on the external surface and in the interlamellar space. In the case of adsorption from aqueous solution, a competition between water and the organic is involved, which is strongly dependent on the charge density of the clay and the properties of the organic (Lee *et al.*, 1990). Thus, benzene is adsorbed from aqueous solution on TMA^+-smectite both on the external surface and in the interlamellar space, but to a lesser extent than from the vapour phase. Toluene and o-xylene, on the other hand, are significantly adsorbed only on low charge smectite from aqueous solution; whereas on high charge smectites the TMA^+ molecules take in more space and water apparently blocks some sites, thus inducing a shape-selectivity into the system.

In the IOCs, the amount of surfactant must be such that they are adsorbed with their tails perpendicular to the clay surface, and the charged **head** groups point away from the surface. This prevents the IOC from flocculation in aqueous suspension, and the amount of organic is sufficient to render the clay organophilic, and thus an efficient adsorber of organic pollutants (Srinivasan and Fogler, 1990a, b). That the positive head of the surfactant points away from the surface means that the latter is positively charged, owing to the hydroxy-Al (La) compounds. Materials (inorgano-organo-clays) are obtained which are competitive with carbon in their ability to adsorb organic pollutants.

In the early 1980s, Yamagishi and Soma (1981), Yamagishi (1982a, b, 1983), Yamagishi and Ohnishi (1982) and Yamagishi and Fujita (1984) developed chirally modified smectite materials for optical resolution of an adsorbent. Thus, montmorillonite loaded with the Λ form of $M(phen)_3^{2+}$

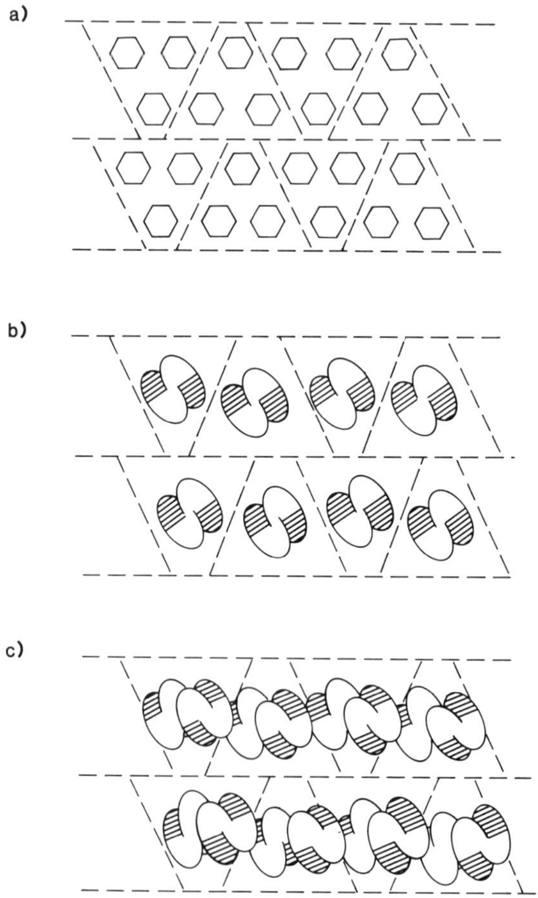

Figure 9.13 Adsorption model of tris (1, 10-phenanthroline) complexes: (a) ditrigonal holes of the clay surfaces; (b) adsorption model of one optical isomer, each complex occupying 3 ditrigonal holes; and (c) adsorption model of the two optical isomers.

(phen = 1,10-phenanthroline, M = Ni^{2+}, Fe^{2+}, Ru^{2+}) preferentially adsorbs the mirror image form (Δ) of a range of analogous complexes. Such modified clays are also able to separate optical isomers of organic compounds (Nakamura, Yamagishi and Iwamoto, 1990). The obvious consequence is that the racemic mixture of these phen complexes is adsorbed at up to twice the CEC, that is with co-adsorption of the anion. The enantiomeric forms, on the other hand, are adsorbed up to the CEC. The picture that emerges is that the racemic mixture is adsorbed in pairs of the two optical isomers with an exact geometrical fit (Figure 9.13).

However, the situation is the reverse for $Ru(bipy)_3^{2+}$ (bipy = 2,2′ bipyridine): the enantiomers are adsorbed at up to twice the CEC and the racemic mixture up to the CEC. The latter can be brought to the level of the enantiomeric adsorption in the presence of NaCl. This indicates that not only geometrical considerations of matching the shape and size of the complexes with the ditrigonal holes on the surface, but also the aggregation of the clay particles, and therefore also the size of the clay particles, is important (Villemure, 1990; Villemure and Bard, 1990).

There is clear spectroscopic evidence that shows this. Joshi and Ghosh (1987) and Kamat *et al.* (1991) have given subtle but clear spectral differences between adsorbed Λ- and Δ-$Ru(bipy)_3^{2+}$ on one hand, and between $Ru(bipy)_3^{2+}$ and $Ru(phen)_3^{2+}$ on the other hand, and this at small loadings. Clearly there are small differences in adsorption depending on the optical isomer and the nature of the ligands. In fact, this is not surprising at all. These types of aromatic complexes are extremely selectively adsorbed by smectites (Schoonheydt *et al.*, 1978) in excess of the CEC. This means that when a solution is presented to the clay, the complexes do not discriminate between the various sites but adsorb on the first site they collide with, which very probably is on the external surface of a more or less expanded aggregate. This adsorbed complex creates a somewhat less hydrophilic local environment and the next incoming complex will preferentially be adsorbed in the neighbourhood of the first one. Thus, a non-random distribution is created which is not the most thermodynamically stable one. Therefore, a slow rearrangement of the elementary clay sheets will occur, together with a redistribution of the complexes. The measurements on such systems are probably undertaken on systems in pseudo-equilibrium and will strongly depend on the experimental conditions.

9.5 Future directions in clay materials research

The production of clay minerals as materials for modern applications requires the scientific development of custom-made clays. These are materials designed for a specific application. Clay minerals, and smectites in particular, are particularly well suited for such developments. Indeed, a whole range of smectites from hydrophobic to hydrophilic can be produced by adsorption of suitable (organic) molecules; surface areas, pore volumes and pore diameters can be designed by pillaring; and particle size and shape and the degree of aggregation can now be monitored for use of smectites in suspension.

For advanced applications, chemically well-characterized and pure smectites are needed. The necessary degree of purity is usually not available from natural clays. To avoid the very expensive and time-consuming purification procedures, smectite synthesis seems to be the only valid alternative. Research on smectite synthesis and, indeed, clay synthesis in general, is at a low level but slowly expanding. Sakurai, Urabe and Izumi (1990, 1991) have synthesized a tetrasilicic mica-type of material (TSM) for development as a heterogeneous catalyst. Transition metal ions, substituting for Al^{3+} or Mg^{2+} in the octahedral layer, are known in natural smectites (Güven, 1988), and it is of no surprise that a substantial effort has been made on the isomorphous substitution of transition metal ions in the octahedral layer (Mosser *et al.*, 1990; Petit and Decarreau, 1990). Others have developed Ni-containing clays as heterogeneous catalysts (Urabe, Koga and Izumi, 1989). Templated synthesis of smectites has been known for a long time (Barrer and Dicks, 1967), but has been revitalized by the recent, exciting work of Carrado (1992).

The development of clay materials also requires a detailed knowledge of the organization of the elementary clay sheets to form micro-aggregates and larger associations. This is the study of the aggregation phenomenon, both in (aqueous) suspension and in air-dry materials. The measurement of the rheological properties of aqueous suspensions gives numbers that are difficult to interpret in terms of the fundamental properties of the clay particles (size, shape, chemical composition and interparticle interaction energy). It is therefore necessary to develop new methods of investigation of the smectite surface in aqueous suspension. One possibility, developed in our group, is the adsorption of cationic dyes and the study of the spectroscopic properties of these adsorbed dyes as a function of suspension properties and loading (Cenens and Schoonheydt, 1988, 1990; Cenens, Schoonheydt and De Schryver, 1990).

The advantages of these dye probes is their strong adsorption and their high extinction coefficient. The latter allows the investigation of extremely small loadings, which do not disturb the clay suspension. The time-dependence of the absorption as well as the emission spectra of the adsorbed dyes (Viaene *et al.*, 1987; Schoonheydt and Heughebaert, 1992) show that clay suspensions are quasi-equilibrium systems which continuously change over periods of hours and even days.

An interesting application is the photostabilization of photosensitive pesticides in the presence of clay-adsorbed dyes (Margulies, Rozen and Cohen, 1985). The idea is that the dye captures the (sun)light, protecting the pesticide from degradation. The efficiency of the process depends on the organization of the two molecules on the surface: is it a random distribution, an aggregation, or does the pesticide dissolve in the admicelles of the dye on the surface? This aggregation of light-adsorbing molecules on surfaces can conveniently be studied by spectroscopic means (Margulies and Rozen, 1986; Margulies, Rozen and Nir, 1988; Rytwo *et al.*, 1991).

A third area of research that deserves more attention in the future is the theoretical modelling of clay minerals. The ultimate goal is an understanding of the relations between structure and chemical composition on the one hand, and properties on the other hand. Electrostatic potential calculations have been performed by Giese (1984) and Bleam (1990a, b). Systematic theoretical studies, analogous to those on zeolites, are lacking, but two new developments are worth mentioning. Hess and Saunders (1992) have done crystal orbital calculations on kaolinite, and Baekelandt *et al.* (1991) have developed the electronegativity equalization method. The latter is a semi-empirical approach to density functional theory, with the advantage that the structure is fully taken into account. As a consequence, the calculated properties really reflect the effect of structure, i.e. of the specific arrangement of the atoms in space.

I have stressed these three areas of research on clay minerals (synthesis, aggregation and theoretical modelling) because they require the development and application of advanced experimental and theoretical techniques and will surely advance our knowledge of these exciting materials.

Acknowledgements

The author acknowledges his PhD and engineering students for their continuous outstanding experimental and theoretical work. Six months' collaboration with Prof. Cliff Johnston is highly appreciated. Financial support from the Belgian Fund for Scientific Research and the Belgian Government (Concerted Research Action, Inter-University Poles of Attraction) is acknowledged.

References

Adams, J.M. (1987) Synthetic organic chemistry using pillared, cation exchanged and acid-treated montmorillonite catalysts: a review. *Appl. Clay Sci.*, **3**, 309–42.

Baekelandt, B.G., Mortier, W.J., Lievens, J.L. and Schoonheydt, R.A. (1991) Probing the reactivity of different sites within a molecule or solid by direct computation of molecular sensitivities via an extension of the electro-negativity equalization method. *J. Am. Chem. Soc.*, **113**, 6730–4.

Ballantine, J.A., Graham, P., Patel, I. *et al.* (1987) New differential thermogravimetric method using cyclohexylammine for measuring the concentration of interlamellar protons in clay catalyst, in *Proceedings International Clay Conference 1985* (eds L.G. Schultz, H. van Olphen and F.A. Mumpton), The Clay Minerals Society, Bloomington, pp. 311–18.

Barrer, R.M. and Dicks, L.W.R. (1967) Chemistry of soil minerals. Part IV. Synthetic alkyl-ammonium montmorillonites and hectorites. *J. Chem. Soc. (A), Inorg. Phys. Theor.*, pp. 1523–9.

Ben Rhaïem, H., Pons, C.H. and Tessier, D. (1987) Factors affecting the microstructure of smectites: role of cations and history of applied stress, in *Proceedings International Clay Conference 1985* (eds L.G. Schultz, H. van Olphen and F.A. Mumpton), The Clay Minerals Society, Bloomington, pp. 292–7.

Bleam, W.F. (1990a) Electrostatic potential at the basal (001) surface of talc and pyrophillite as related to the tetrahedral sheet distortions. *Clays and Clay Min.*, **38**, 522–6.

Bleam, W.F. (1990b) The nature of cation-substitution sites in phyllosilicates. *Clays and Clay Min.*, **38**, 527–36.

Bradley, S.M. and Kydd, R.A. (1991) A comparison of the thermal stability of Ga_{13}, $GaAl_{12}$ and Al_{13}-pillared clay minerals. *Catalysis Lett.*, **8**, 185–92.

Brouwer, E., Baeyens, B., Maes, A. and Cremers, A. (1983) Cesium and rubidium ion equilibria in illite clay. *J. Phys. Chem.*, **87**, 1213–19.

Bukka, K., Miller, J.D. and Shabtai, J. (1992) FTIR study of deuterated montmorillonites: structural features relevant to pillared clay stability. *Clays and Clay Min.*, **40**, 92–102.

Carrado, K. (1992) Preparation of hectorite clays utilizing organic and organometallic complexes during hydrothermal crystallization. *Ind. Eng. Chem. Res.*, **31**, 1654–9.

Cenens, J. and Schoonheydt, R.A. (1988) Visible spectroscopy of methylene blue on hectorite, laponite B and barasym in aqueous suspension. *Clays and Clay Min.*, **36**, 214–24.

Cenens, J. and Schoonheydt, R.A. (1990) Quantitative absorption spectroscopy of cationic dyes on clays, in *Proceedings 9th International Clay Conference 1989* (eds V.C. Farmer and Y. Tardy), *Sci. Géol. Mém.*, **85**, 15–23.

Cenens, J., Schoonheydt, R.A. and De Schryver, F.C. (1990) Probing the surface of clays in aqueous suspension by fluorescence spectroscopy of proflavine, in *Spectroscopic Characterization of Minerals and their Surfaces* (eds L.M. Coyne, S.W.S. McKeever and D.F. Blake), *ACS Symp. Ser.*, **415**, American Chemical Society, Washington, DC, pp. 378–95.

Chabra, R., Pleysier, P. and Cremers, A. (1975) The measurement of the cation exchange capacity and exchangeable cations in soils: a new method, in *Proceedings International Clay Conference 1975* (ed. S.W. Bailey), Applied Publishing, Wilmette, pp. 439–49.

Choudhary, B.M., Subba Rao, Y.V. and Prasad, B.P. (1991) New triphase catalysts from montmorillonite. *Clays and Clay Min.*, **39**, 329–32.

Corma, A., Mifsud, A. and Saur, E. (1990) Kinetics of the acid leaching of palygorskite: influence of the octahedral sheet composition. *Clay Min.*, **25**, 197–205.

Delville, A., Grandjean, J. and Laszlo, P. (1991) Order acquisition by clay platelets in a magnetic field. NMR study of the structure and microdynamics of the adsorbed water layer. *J. Phys. Chem.*, **95**, 1383–92.

Figueras, F., Klapyta, Z., Massiani, P. *et al.* (1990) Use of competitive ion exchange for intercalation of montmorillonite with hydroxi-aluminium species. *Clays and Clay Min.*, **38**, 257–64.

Fripiat, J.J. (1990) Surface activities of clays, in *Spectroscopic Characterization of Minerals and their Surfaces* (eds L.M. Coyne, S.W.S. McKeever and D.F. Blake). *ACS Symp. Ser.*, **415**, American Chemical Society, Washington, DC, pp. 360–77.

Fripiat, J.J., Letellier, M. and Levitz P. (1984) Interaction of water with clay surfaces, in *Clay Minerals: Their Structure, Behaviour and Use* (eds L. Fowden, R.M., Barrer and P.B. Tinken), The Royal Society, London, pp. 67–79.

Fripiat, J., Cases, J., François, M. and Letellier, M. (1982) Thermodynamics and microdynamic behavior of water in clay suspensions and gels. *J. Colloid and Interface Sci.*, **89**, 378–400.

Giese, R.F. (1984) Electrostatic models of micas. *Rev. Mineral.*, **13**, 105–44.
Gonzalez, F., Pesquera, C., Benito, I. and Mendioroz, S. (1991) Aluminium–gallium pillared montmorillonite with high thermal stability. *J. Chem. Soc., Chem. Comm.*, pp. 587–8.
Gonzalez, F., Pesquera, C., Blanco, C. *et al.* (1992) Synthesis and characterisation of Al–Ga pillared clays with high thermal and hydrothermal stability. *Inorg. Chem.*, **31**, 727–31.
Grandjean, J. and Laszlo, P. (1990) Multinuclear magnetic resonance studies of structure and dynamics at the interface of clay materials, in *Spectroscopic Characterisation of Minerals and their Surfaces* (eds L.M. Coyne, S.W.S. McKeever and D.F. Blake), *ACS Symp. Ser.*, **415**, American Chemical Society, Washington, DC, pp. 396–406.
Grim, R.E. (1953) *Clay Mineralogy*, McGraw-Hill, New York.
Güven, N. (1988) Smectites, in *Hydrous Phyllosilicates* (ed. S.W. Bailey), Mineralogical Society of America, Washington, DC, pp. 497–559.

Hess, A.C. and Saunders, V.R. (1992) Periodic ab initio Hartree-Fock calculations of the low symmetry mineral kaolinite. *J. Phys. Chem.*, **96**, 4367–74.
Hettinger, W.P. Jr. (1991) Contribution to catalytic cracking in the petroleum industry. *Appl. Clay Sci.*, **5**, 445–68.

Joshi, V. and Ghosh, P.K. (1987) Luminescence spectroscopy as a probe of binding state differences between enantiomeric and racemic poly(pyridyl)-ruthenium(II) chelates intercalated in smectite clays. *J. Chem. Soc., Chem. Comm.*, pp. 789–91.

Kamat, P.V., Gopidas, K.R., Mukherjee, T. *et al.* (1991) Spectral differences between enantiomeric and racemic $Ru(bipy)_3^{2+}$ on layered clays: probable causes. *J. Phys. Chem.*, **95**, 10009–18.

Lagaly, G. and Weiss, A. (1975) The layer charge of smectitic layer silicates, in *Proceedings International Clay Conference 1975* (ed. S.W. Bailey), Applied Publishing Ltd., Wilmette, pp. 157–72.
Laird, D.A., Scott, A.D. and Fenton, T.E. (1989) Evaluation of the alkylammonium method of determining layer charge. *Clays and Clay Min.*, **37**, 41–6.
Laszlo, P. (1987) Chemical reactions on clays. *Science*, **235**, 1473–7.
Laszlo, P. (1990) Les Argiles en Chimie. *La Recherche*, **219**, 314–23.
Lee, J.F., Mortland, M.M., Chian, C.T. *et al.* (1990) Adsorption of benzene, toluene and xylene by two tetramethylammonium smectites having different charge densities. *Clays and Clay Min.*, **38**, 113–20.
Low, Ph.F. (1979) Nature and properties of water in montmorillonite–water systems. *Soil Sci. Soc. America J.*, **43**, 651–8.
Low, Ph.F. (1980) The swelling of clay: II. Montmorillonites. *Soil Sci. Soc. America J.*, **44**, 667–76.
Low, Ph.F. and Margheim, J.F. (1979) The swelling of clay: I. Basic concepts and empirical equations. *Soil Sci. Soc. America J.*, **43**, 473–81.

Maes, A. and Cremers, A. (1978) Stability of metal uncharged ligand complexes in ion exchangers. Part 3. Complex ion selectivity and stepwise stability constants. *J. Chem. Soc., Faraday Trans. I*, **74**, 2470–80.
Maes, A., Marijnen, P. and Cremers, A. (1977) Stability of metal uncharged ligand complexes in ion exchangers. *J. Chem. Soc., Faraday Trans. I*, **73**, 1297–301.
Maes, A., Peigneur, P. and Cremers, A. (1975) Thermodynamics of transition metal ion exchange in montmorillonite, in *Proceedings International Clay Conference 1975* (ed. S.W. Bailey), Applied Publishing, Wilmette, pp. 319–29.
Maes, A., Rasquin, E. and Cremers, A. (1982) Thermodynamic study of complexation on

exchange equilibria in Wyoming bentonite clay. *J. Chem. Soc., Faraday Trans. I*, **78**, 2041–9.

Maes, A., Stul, M.S. and Cremers, A. (1979) Layer charge – cation exchange capacity relationships in montmorillonite. *Clays and Clay Min.*, **27**, 387–92.

Maes, A., Verheyden, D. and Cremers, A. (1985) Formation of highly selective cesium exchange sites in montmorillonites. *Clays and Clay Min.*, **33**, 251–7.

Maes, A., Schoonheydt, R.A., Cremers, A. and Uytterhoeven, J.B. (1980) Spectroscopy of $Cu(en)_2^{2+}$ on clay surfaces. Surface and charge density effects. *J. Phys. Chem.*, **84**, 2795–9.

Margulies, L. and Rozen, H. (1986) Adsorption of methylgreen on montmorillonite. *J. Molec. Struct.*, **141**, 219–26.

Margulies, L., Rozen, H. and Cohen, E. (1985) Energy transfer at the surface of clays and protection of pesticides from photodegradation. *Nature*, **315**, 658–9.

Margulies, L., Rozen, H. and Nir, S. (1988) Model for competitive adsorption of organic cations on clays. *Clays and Clay Min.*, **36**, 270–6.

McBride, M.B., Pinnavaia, T.J. and Mortland, M.M. (1977) Adsorption of aromatic molecules by clays in aqueous suspension, in *Fate of Pollutants in the Air and Water Environments*, Part I, Vol. 8 (ed. I.H. Suffet), Wiley, New York, pp. 145–54.

McCauley, J.R. (1990) Stable intercalated clays and preparation method, US Patent 4,957,889.

Mortland, M.M. (1970) Clay–organic complexes and interactions. *Adv. Agronomy*, **22**, 73–117.

Mortland, M.M., Fripiat, J.J., Chaussidon, J. and Uytterhoeven, J.B. (1963) Interaction between ammonia and the expanding lattices of montmorillonite and vermiculite. *J. Phys. Chem.*, **17**, 248–58.

Mosser, C., Mestdagh, M., Decarreau, A. and Herbillon, A.J. (1990) Spectroscopic (ESR, EXAFS) evidence of copper for (aluminium–magnesium) substitution in octahedral sheets. *Clay Min.*, **25**, 271–82.

Nadeau, P.H. (1985) The physical dimensions of fundamental clay particles, *Clay Min.*, **20**, 499–514.

Nakamura, Y., Yamagishi, A. and Iwamoto, T. (1990) Clay columnn chromatography for optical resolution: partial resolution of 1,1′-binaphtol on optically active $[Co(phen)_{3-x}am_x]^{n+}$–montmorillonite columns. *Clay Sci.*, **8**, 17–24.

Peigneur, P., Maes, A. and Cremers, A. (1975) Heterogeneity of charge density distribution in montmorillonite as inferred from cobalt adsorption. *Clays and Clay Min.*, **23**, 71–5.

Petit, S. and Decarreau, A. (1990) Hydrothermal (200 °C) synthesis and crystal chemistry of iron-rich kaolinites. *Clay Min.*, **25**, 181–96.

Pinnavaia, T.J., Landau, S.D., Tzou, M.S. and Johnson, I.D. (1985) Layer cross-linking in pillared clays. *J. Am. Chem. Soc.*, **107**, 7222–4.

Plee, D., Borg, L., Gatineau, L. and Fripiat, J.J. (1985) High-resolution solid-state ^{27}Al and ^{29}Si nuclear magnetic resonance study of pillared clays. *J. Am. Chem. Soc.*, **107**, 2362–9.

Poinsignon, C., Estrade-Szwarckopf, J., Conard, J. and Dianoux, A.J. (1987) Water dynamics in the clay–water system: a quasi-elastic neutron-scattering study, in *Proceedings International Clay Conference 1985* (eds L.G. Schultz, H. van Olphen and F.A. Mumpton), The Clay Minerals Society, Bloomington, pp. 284–91.

Rengasamy, P., Van Assche, J.B. and Uytterhoeven, J.B. (1976) Particle size of Wyoming bentonite and its relation to the cation exchange capacity and the homogeneity of the charge density. *J. Chem. Soc., Faraday Trans. I*, **72**, 376–81.

Roberson, H.E., Weir, A.H. and Woods, R.D. (1968) Morphology of particles in size-fractionated Na-montmorillonites. *Clays and Clay Min.*, **16**, 239–47.

Rupert, J.P., Granquist, W.T. and Pinnavaia, T.J. (1987) Catalytic properties of clay minerals, in *Chemistry of Clays and Clay Minerals* (ed. A.C.D. Newman), Mineralogical Society, London, pp. 275–318.

Rytwo, G., Serban, G., Nir, S. and Margulies, L. (1991) Use of methylene blue and crystal violet for determination of exchangeable cations in montmorillonite. *Clays and Clay Min.*, **39**, 551–5.

Sakurai, H., Urabe, K. and Izumi, Y. (1990) Pillared tetrasilicic mica catalysts modified by fixed interlayer cations. Classification of fixation mode by cation. *Bull. Chem. Soc. Jpn*, **63**, 1389–95.

Sakurai, H., Urabe, K. and Izumi, Y. (1991) Pillared tetrasilicic mica catalysts having fixed interlayer Ca ions. Comparison with other clays. *Bull. Chem. Soc. Jpn*, **64**, 227–35.

Schoonheydt, R.A. (1991) Clays: from two to three dimensions, in *Introduction to Zeolite Science and Practice* (eds H. van Bekkum, E.M. Flanigen and J.C. Jansen), Elsevier, Amsterdam, pp. 201–39.

Schoonheydt, R.A. and Heughebaert, L. (1992) Clay adsorbed dyes: methylene blue on laponite. *Clay Min.*, **27**, 91–100.

Schoonheydt, R.A. and Leeman, H. (1992) Pillaring of saponite in concentrated medium. *Clay Min.*, **27**, 249–52.

Schoonheydt, R.A., Pelgrims, J., Heroes, Y. and Uytterhoeven, J.B. (1978) Characterisation of tris(2,2′-bipyridyl)ruthenium(II) on hectorite. *Clay Min.*, **13**, 435–8.

Schutz, A., Stone, W.E.E., Poncelet, G. and Fripiat, J.J. (1987) Preparation and characterisation of bidimensional zeolite structures obtained from synthetic beidellite and hydroxylaluminium solution. *Clays and Clay Min.*, **35**, 251–68.

Sposito, G. and Prost, R. (1982) Structure of water adsorbed on smectites. *Chem. Rev.*, **82**, 553–73.

Srinivasan, K.R. and Fogler, H.S. (1990a) Use of inorgano–organo-clays in the removal of priority pollutants from industrial wastewater: structural aspects. *Clays and Clay Min.*, **38**, 277–86.

Srinivasan, K.R. and Fogler, H.S. (1990b) Use of inorgano–organo-clays in the removal of priority pollutants from industrial wastewater: adsorption of benzo(a)pyrene and chlorophenols from aqueous solution. *Clays and Clay Min.*, **38**, 287–93.

Sterte, J. (1990) Preparation and properties of large pore RE/Al-pillared montmorillonites. A comparison of RE-cations, in *Preparation of Catalysts V* (eds P. Grange, P.A. Jacobs and G. Poncelet), Elsevier, Amsterdam, pp. 113–22.

Stul, M.S. and Mortier, W. (1974) The heterogeneity of the charge density in montmorillonites. *Clays and Clay Min.*, **22**, 391–6.

Stul, M.S. and Van Leemput, L. (1982a) The texture of montmorillonites as influenced by the exchangeable cations and the drying methods. I: External surface area related to the stacking units of the aggregates. *Surf. Technol.*, **16**, 89–100.

Stul, M.S. and Van Leemput, L. (1982b) The texture of montmorillonites as influenced by the exchangeable cations and the drying methods. II: A comparative porosity study. *Surf. Technol.*, **16**, 101–12.

Stul, M.S. and Van Leemput, L. (1982c) Particle size distribution, cation exchange capacity and charge density of deferrated montmorillonites. *Clay Min.*, **17**, 209–15.

Talibudeen, O. and Goulding, K.W.T. (1983) Charge heterogeneity in smectites. *Clays and Clay Min.*, **31**, 37–42.

Tokarz, M. and Shabtai, J. (1985) Cross-linked smectites IV. Preparation and properties of hydroxylaluminium-pillared Ce- and La-montmorillonites and fluorinated NH_4^+-montmorillonites. *Clays and Clay Min.*, **37**, 89–98.

Touret, O., Pons, C.H., Tessier, D. and Tardy, Y. (1990) Etude de la répartition de l'eau dans des argiles saturées en Mg^{2+} aux faibles teneurs en eau. *Clay Min.*, **25**, 217–33.

Tuck, J.J., Hall, P.L., Hayes, M.H.B., Ross, D.K. and Poinsignon, C. (1984) Quasi-elastic neutron-scattering studies of the dynamics of intercalated molecules in charge-deficient layer silicates. *J. Chem. Soc., Faraday Trans. I*, **80**, 309–24.

Urabe, K. (1990) Synthetic clays as catalysts – fascinating swellable crystals. *Nendo Kagaku*, **30**, 148–56.
Urabe, K., Koga, M. and Izumi, Y. (1989) Synthetic Ni-substituted saponite as a catalyst for selective dimerisation of ethene. *J. Chem. Soc., Chem. Comm.*, pp. 807–8.

Vali, H. and Bachmann, L. (1988) Ultrastructure and flow behavior of colloidal smectite dispersions. *J. Colloid Interface Sci.*, **126**, 278–91.
Viaene, K., Caigui, J., Schoonheydt, R.A. and De Schryver, F.C. (1987) Study of the adsorption on clay particles by means of a fluorescent probe. *Langmuir*, **3**, 107–11.
Villemure, G. (1990) Effect of negative surface charge densities of smectite clays on the adsorption isotherms of racemic and enantiomeric tris(2,2′-bipyridine)ruthenium(II)chloride. *Clays and Clay Min.*, **38**, 623–30.
Villemure, G. and Bard, A.J. (1990) Clay modified electrodes. Part 10. Studies of clay adsorbed $Ru(bipy)_3^{2+}$ enantiomers by UV-visible spectroscopy and cyclic voltammetry. *J. Electroanal. Chem.*, **283**, 403–20.

White, G.N. and Zelazny, L.W. (1988) Analysis and implications of the edge structure of dioctahedral phyllosilicates. *Clays and Clay Min.*, **36**, 141–6.
Wolfe, T.A., Demirel, T. and Bauvaux, E.R. (1985) Interaction of aliphatic amines with montmorillonite to enhance adsorption of organic pollutants. *Clays and Clay Min.*, **33**, 301–11.

Yamagishi, A. (1982a) Racemic adsorption of dicyanobis(1,10-phenanthroline)iron(II) in colloidally dispersed sodium montmorillonite. *Inorg. Chem.*, **21**, 1778–82.
Yamagishi, A. (1982b) Racemic adsorption of iron(II)tris(1,10-phenanthroline)chelate on a colloidal clay. *J. Phys. Chem.*, **86**, 2472–9.
Yamagishi, A. (1983) Chirality recognition of a clay surface modified by an optically active metal chelate. *J. Chem. Soc., Dalton Trans.*, 679–81.
Yamagishi, A. and Fujita, N. (1984) Racemic adsorption of a bis(chelate)cobalt(III) complex by colloidally dispersed sodium montmorillonite. *J. Colloid Interface Sci.*, **100**, 136–42.
Yamagishi, A. and Ohnishi, R. (1982) Clay column chromatography for optical resolution: initial resolutions of bis(acetylacetonate)(glycinato)cobalt(III) and (acetylacetonato)-bis(glycinato)cobalt(III) on a tris(1,10-phenanthroline)nickel(II)-montmorillonite column. *Inorg. Chem.*, **21**, 4233–6.
Yamagishi, A. and Soma, M. (1981) Improvement of the optical purity of metal tris-(1,10-phenanthroline)chelates by use of adsorption on a colloidal clay. *J. Chem. Soc., Chem. Comm.*, 539–80.

CHAPTER TEN

Zeolite surfaces and reactivity

A. Dyer

10.1 Introduction

The word **zeolite** is derived from two Greek words meaning **boiling stones**. This was the name given (in 1756) by a Swedish mineralogist, Cronstedt, to a new mineral that he had observed to lose water very easily when heated. This mineral was later named stilbite.

There are now about 46 zeolites known as natural minerals, typified by unique structural arrangements of $(SiO_4)^{4-}$ and $(AlO_4)^{5-}$ tetrahedra joined at all four corners to produce three-dimensional frameworks. In many zeolites these frameworks are open so cations and water molecules, contained in cavities and channels within the frameworks, have freedom of movement (Dyer, 1988). This facile movement is recognized as promoting ease of cation exchange and water loss – the latter being responsible for the intumescence noted by Cronstedt.

The regularity and openness of zeolitic structures confers properties that have become of considerable industrial and fundamental research interest, as will be described later. This interest has led to laboratory and plant synthesis of many other materials which are structurally related to the natural zeolite minerals. The consequence of this is that the strict definition of zeolites as aluminosilicates has been extended to include frameworks that contain other tetrahedrally coordinated atoms such as Fe, Co and P. In fact these **zeotypes** are also known in nature, as will be described later. The Structure Commission of the International Zeolite Association now includes a variety of other species in its definitive *Atlas of Zeolite Structure Types* (Meier and Olson, 1992). These include:

- silica phases with frameworks such as the natural mineral melanophlogite and the synthetic dodecasils;
- materials with frameworks containing some OH species completing the tetrahedral units, e.g. partheite ($Ca_8[Al_{16}Si_{16}O_{60}(OH)_8]16H_2O$);
- natural and synthetic zeotypes;
- structural variants based on cancrinite and sodalite, i.e. feldspathoids.

The justification for the inclusion of feldspathoids is that the cages joined in a space-filling way in these minerals are often the cages that form parts of zeolite frameworks.

Mineral Surfaces. Edited by D.J. Vaughan and R.A.D. Pattrick. Published in 1995 by Chapman & Hall, London. ISBN 0 412 56340 1

10.2 Zeolite structures

Figure 10.1 shows the basic tetrahedral building units and an element of structure showing how they can link to create a chain. Chains like this can be assembled to form zeolite frameworks such as in the mineral natrolite.

Identification of repeating units of linked tetrahedra (Si and Al) provides a way of classifying zeolite structures and several secondary building units are used in this way. These are shown in Figure 10.2. In some cases, even more complex structural units can be recognized and help us to understand the zeolite structures. A well-known example of this can be seen in the commercially useful zeolites A, X and Y, which can be regarded as based upon assemblages of sodalite units. Sodalite units are a basket of Si and Al tetrahedra (Figure 10.3) which, in the feldspathoid mineral of the same name, are joined in a space-filling manner, as mentioned earlier (Figure 10.4). In the synthetic species zeolite A, the units are linked by all their square faces in the manner illustrated in Figure 10.5. It can be seen that this creates a cubic structure, with each sodalite unit (also known as 'β cages') occupying positions analogous to those assumed by Na and Cl in the sodium chloride lattice.

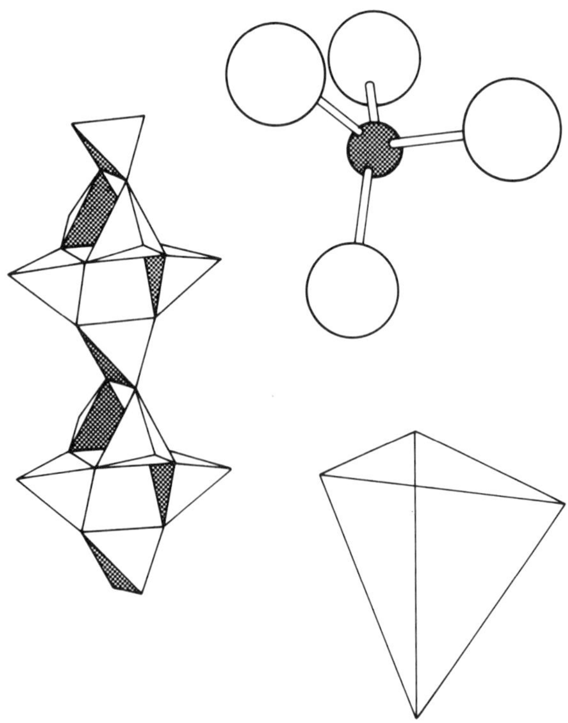

Figure 10.1 $[SiO_4]^{4-}$ tetrahedron (black circle Si) and linked chain of tetrahedra.

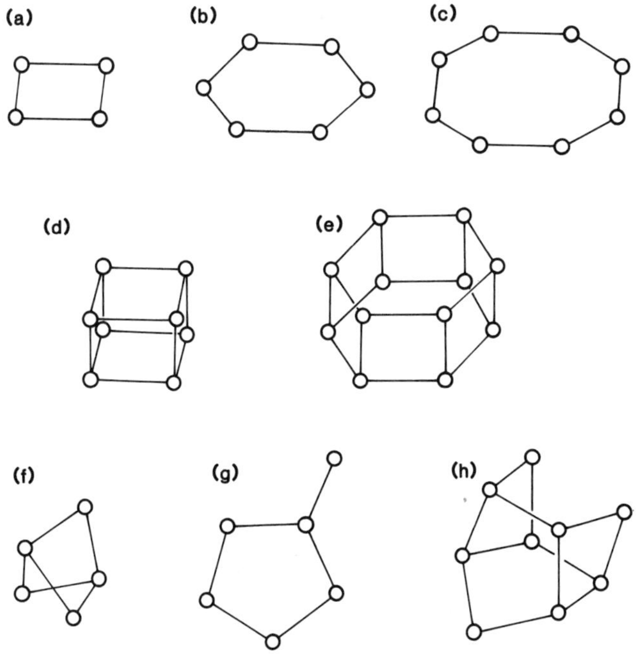

Figure 10.2 Secondary building units of zeolite frameworks (circle is Si, solid line represents O – compare 2(e) with similar constituent building units in Figure 10.7 for an example of their use).

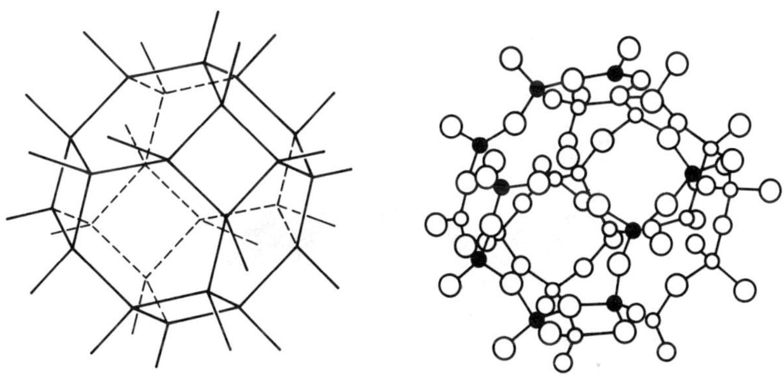

Figure 10.3 Sodalite unit, line drawing and ball and stick representations.

The structures of zeolites X and Y, which differ in the Si/Al ratios of their frameworks, are created in a similar way by linking the units via all their hexagonal faces. This creates a hexagonal structure with the cages linked in the way that the carbon atoms form a diamond lattice (Figure 10.6). Note

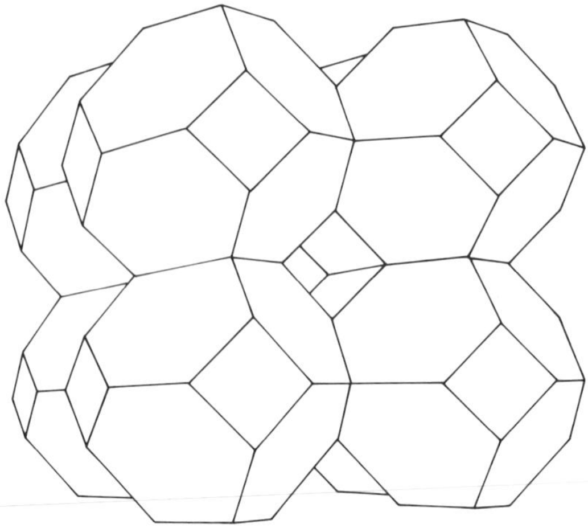

Figure 10.4 Space-filling arrangement of units in sodalite structure.

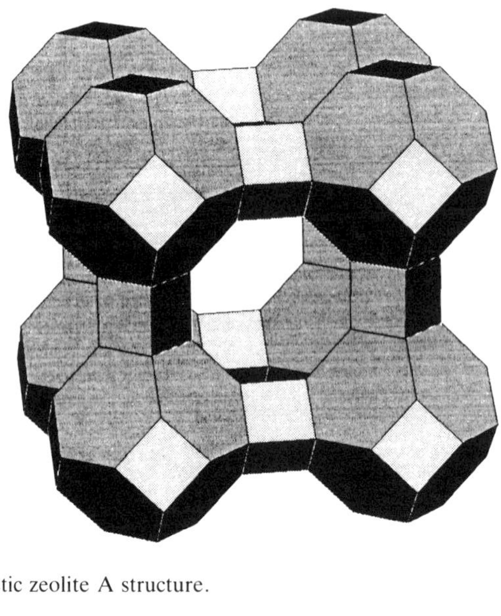

Figure 10.5 Synthetic zeolite A structure.

that zeolite A has no known natural counterpart, but zeolites X and Y are synthetic isomorphs of the natural mineral faujasite. Another example of this identification of useful zeolite structural units is in the chabazite, gmelinite and erionite minerals, which are formed from double hexagonal rings linked as illustrated for chabazite in Figure 10.7.

The *Atlas of Zeolite Structure Types* (Meier and Olson, 1992) contains stereopairs of all known zeolite frameworks as listed in Table 10.1.

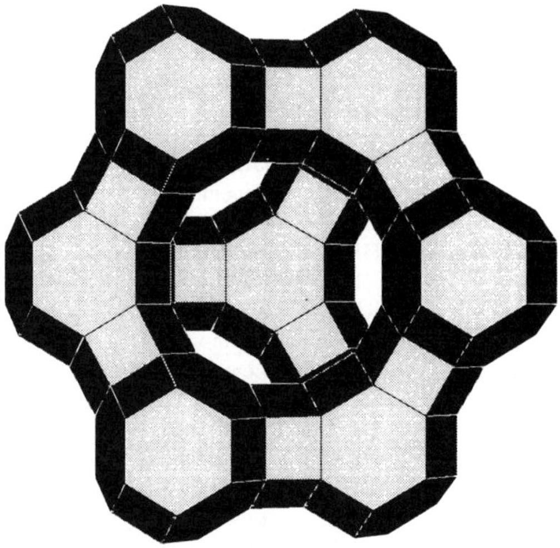

Figure 10.6 Synthetic faujasite (X, Y) structure.

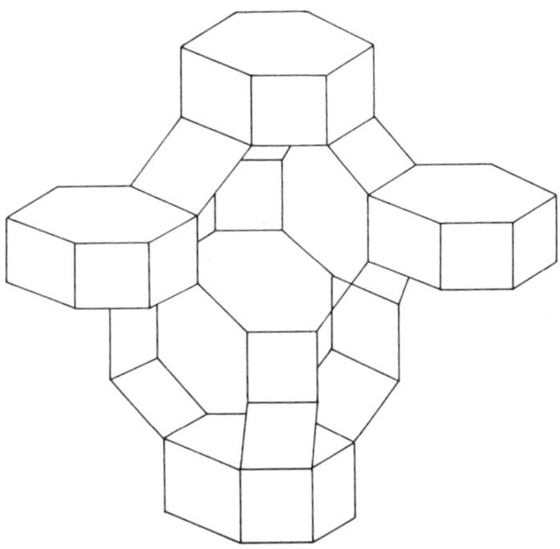

Figure 10.7 Chabazite structure.

10.3 Creation of natural and synthetic zeolites and zeotypes

10.3.1 Natural zeolites

Every geological museum has prized zeolite specimens, valued because of their beauty, and originating from basalts or rocks of similar volcanic origin – long regarded as the traditional route of zeolite genesis. This view changed some 30 years ago with the discovery of vast worldwide deposits of zeolites

Table 10.1 Zeolite frameworks

IUPAC code	Name	IUPAC code	Name	IUPAC code	Name
ABW*	Li-A(BW)	GIS	Gismondine	MOR	Mordenite
ANA	Analcime	GMR	Gmelinite	MTN*	ZSM-39
BEA	Beta (Tschernichite)	GOO	Goosecreekite	MTT*	ZSM-23
		HEU	Heulandite[3]	MTW*	ZSM-12
BIK	Bikitaite	JBW*	Na J	NAT	Natrolite
BOG	Boggsite	KFI*	ZK-5	NES*	NU-87
BPH*	Beryllophosphate H[1]	LAU	Laumontite	NON*	Nonasil[4]
BRE	Brewsterite	LEV	Levyne	OFF	Offretite
CHA	Chabazite	LOS*	Losod	PAR†	Partheite
CHI†	Chiavennite[2]	LOV	Lovdarite[2]	PAU	Paulingite
DAC	Dachiardite	LTA*	Linde type A	PHI	Phillipsite
EAB*	TMA-E(AB)	LTL	Linde type L	RHO*	RHO (Pahasapaite)
EDI	Edingtonite	LTN*	Linde type N		
EMT*	EMC-2 (Hexagonal Faujasite)	MAZ	Mazzite	ROG†	Roggianite
		MEI*	ZSM-18	STI	Stilbite
EPI	Epistilbite	MEL*	ZSM-11	THO	Thomsonite
ERI	Erionite	MER	Merlionite	TON	Theta-1
EUO*	EU-1	MFI*	ZSM-5	WEN†	Wenkite
FAU	Faujasite	MFS*	ZSM-57	YUG	Yugawaralite
FER	Ferrierite	MON	Montesommaite		

* Synthetic, no known natural counterpart
† Incomplete framework
1 Isotypic with synthetic zeolite Q
2 Beryllosilicate
3 Isotypic with clinoptilolite
4 Isotypic with ZSM-51

Listing excludes felspathoid frameworks and synthetic zeotypes but includes those natural zeotypes (CHI, LOV) with unique frameworks.

After Meier and Olson (1992)

in sedimentary and low grade metamorphic rocks. These zeolites can be of >90% purity and are mined commercially in USA, Cuba, Hungary, Bulgaria, Japan, Slovakia, South Africa, Italy, Russia, Indonesia and Korea, though some deposits have <60% zeolite content. It is now recognized that zeolites form readily under a variety of conditions, and Gottardi (1989) has categorized them as forming via various 'routes'.

DIAGENETIC ROUTES

- in soils where a high pH has arisen from arid or semi-arid conditions when evaporation produces a high concentration of sodium carbonate, which reacts with clay minerals to produce zeolites. The analcime deposits of the San Joaquin Valley, California are a typical example of this.
- in hydrologically open systems where percolation of meteoric water through tuff layers creates a series of vertical zones. The top zones

contain glass that has been altered to clay minerals, but the deeper zones contain clinoptilolite, chabazite and phillipsite and, ultimately, analcime. This phenomenon has been seen in the John Day Formation in Oregon and at Oahü, in Hawaii.

- in hydrologically closed systems, e.g. a saline lacrustrine environment in a closed basin under arid, or semi-arid, conditions. Lakewaters of high pH and saline composition arise from evaporation, and they interact with silica sources (clays, tuffs, plagioclase or even biogenic silica) to form zeolites. Lake Tecopa in California and Lake Magadi in Kenya contain zeolites created in this way.
- in marine sediments, where zeolites are formed largely by pore fluids interacting with glasses of submarine volcanic origin. This has happened in the Pacific, Indian and Atlantic Oceans, i.e. in deep-sea environments. Shallow sea sediments that contain zeolites also are known and are widespread, e.g. the andesitic tuffite of the Italian Northern Apennines.
- in burial diagenesis, in which zones at increasing depth show different zeolite contents, like those previously mentioned for hydrologically open systems. This may also involve very low grade metamorphic processes. This diagenetic route is illustrated by deposits in Japan.

HYDROTHERMAL ROUTES

- in geothermal occurrences typified by those in Iceland, Yellowstone Park (USA) and at Wairakei (New Zealand). At these locations, zeolites arise from hydrothermal circulation on existing volcanic ash. Often, well-defined zones can be identified with clinoptilolite and mordenite at shallow depths in 'cooler' zones, and with the deeper (hotter) environments containing analcime, heulandite, laumontite and wairakite. The assemblages feature a continuous transition to burial diagenesis.
- in hydrothermal ore deposits, which can contain local zeolites. Examples are harmotome found in the sulfide deposits at Andreasburg in Germany and in the lead-barite deposits in the northern Pennines, England (e.g. Settlingstones).
- in late hydrothermal deposition in pegmatites, such as can be seen in the Elba (Italy) granodiorite pegmatite, which contains a suite of zeolites.
- in hydrothermal veins and geodes in feldspathic rocks, such as the chabazite and stilbite in the Adamello granite in Italy, caused by hydrothermal fluids.

ZEOLITES IN BASALTS

These are likely to be the result of thermal water dissolving glass from the basalt flow, with subsequent deposition in regions of lower temperature, though some of the deposits have been explained using a diagenetic model. Many examples are known of this genesis, and they are often prized as museum specimens.

OTHER ROUTES

- genesis in 'geoautoclaves'; this has been suggested as a description for conditions of high temperature and vapour pressure existing in an ignimbrite immediately after its deposition. Zeolitization has been said to occur under these conditions in locations in Italy, Bulgaria and Japan.
- magmatic genesis is said to be the origin of an analcime in lavas of the Colima volcanic complex in Mexico.

10.3.2 Synthetic zeolites

The general route whereby zeolites and zeotypes can be synthesized on a laboratory or a plant scale is shown schematically in Figure 10.8. The structure-directing agents listed are only examples of a wide range of materials that have been used. Modern thinking suggests that these agents

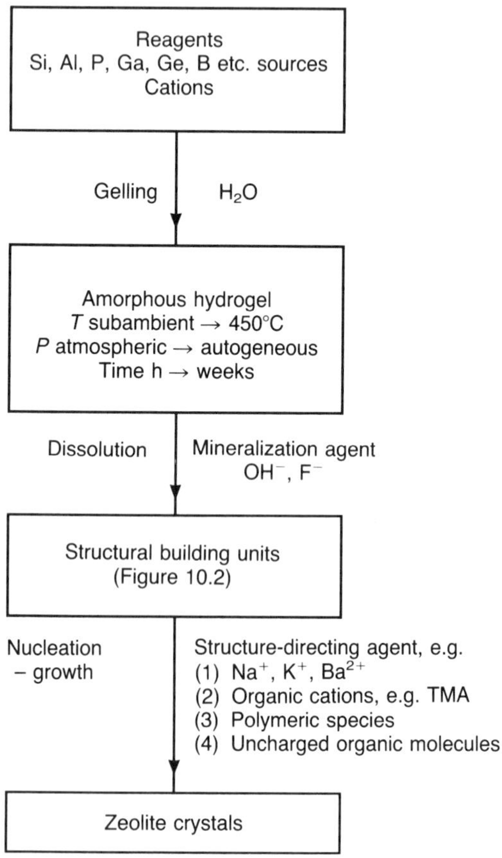

Figure 10.8 Zeolite synthesis.

create structured water, around which the sub-units assemble. This has replaced the earlier concept of the direct templating of zeolite frameworks by organic cations in solution coupled with a void-filling function.

The use of organic reagents as constituents in zeolite-forming gels has been particularly successful in promoting the synthesis of zeolites with very low Al content. These 'high silica' zeolites have created considerable commercial interest. The best known of these products is ZSM-5 (Zeolite Socony Mobil), which can be made with a Si/Al ratio in the range 2–50. A pure silica pseudomorph of ZSM-5 is also well known as 'silicalite'. Further details of zeolite synthesis methodology can be found in the text by Barrer (1982) and in a recent review by Gilson (1992).

10.3.3 Some links between newer synthetic species and natural zeolites/zeotypes

The listings, by Meier and Olson (1992), of framework structures of zeolitic character link natural and synthetic species in a number of ways. The list includes mazzite, a high silica content zeolite found at Mount Semiol, Loire, France, after its synthetic counterparts (omega or ZSM-4) had been structurally characterized from a synthesis using only an organic cation (tetramethyl ammonium – TMA). More recently, natural counterparts of the synthetic zeolites L and β have been found (perlialite and tschernichite, repectively), identified by earlier laboratory characterization of synthetic products from an inorganic or organic cation synthesis mixture.

The proliferation of synthetic zeotype materials to some extent mirrors the range of natural materials. Several natural minerals have the analcime frameworks but with atoms other than Si or Al in tetrahedral interstices. Examples are viseite, a natural SAPO (silico alumino phosphate), and hsianghualite, a beryllosilicate. Kehoeite, suggested as a natural AlPO (aluminophosphate), has now been discredited (McConnell, 1990).

On firmer ground, chiavennite and lovdarite are beryllosilicates with open unique frameworks and a natural beryllophosphate analogue of the synthetic aluminosilicate zeolite RHO has been identified (pahasapaite). Perhaps many more similar species form in nature but with limited stability. Finally, it is fascinating to record that a natural aluminosilicate containing tetramethyl ammonium cations has been reported from the Central Urals (Sokolova, Rybakov and Pautov, 1981). It is structurally related to sodalite and the direct templating of the sodalite cage by the TMA cation is one of the very few cases where the structure-directing role of the cation does not seem to involve the water structure effect (Gilson, 1992).

10.4 Zeolite characterization

Traditionally, zeolites have been identified by X-ray powder diffractometry, but this is not always unequivocal and closely related minerals such as

Table 10.2 Isostructural zeolite minerals

Framework and structure type code	Mineral and major cation content
ANA, analcime	Analcime (Na), leucite (K), pollucite (Cs), wairakite (Ca)
CHA, chabazite	Chabazite (Ca), willhendersonite (K, Ca), (herschelite, Na)
GIS, gismondine	Amicite (Na, K), garronite (Na, Ca), gobbinsite (Na), gismondine (Ca)
HEU, heulandite	Heulandite (Ca), clinoptilolite (Na, K)
NAT, natrolite	Gonnardite (Na, Ca), natrolite (Na), mesolite (Na, Ca), scolecite (Ca)
PHI, phillipsite	Harmotome (Ba, Na), phillipsite (K, Ca)
STI, stilbite	Barrerite (Na), stellerite (Ca), stilbite (Ca, Na)

After Gottardi and Galli (1985); Meier and Olson (1992)

stilbite and stellerite need full chemical analyses to distinguish between them. This raises a contentious point in zeolite mineralogy because stilbite and stellerite have identical Si/Al frameworks, differing only in cation content. Stellerite contains only Ca, whereas stilbite has a 25% Na cation content. They are usually recorded as separate minerals, but the pure Na analogue (originally named 'barrerite') is now discredited. There are other examples of zeolites having the same framework topology recognized as separate mineralogical species and these are listed in Table 10.2. Consultation of this list shows a further inconsistency in that the Na-rich variety of chabazite, named 'herschelite' is discounted by both mineralogists and structural scientists! The framework structures, with the cation and water sites, are defined by single crystal X-ray studies aided by a battery of techniques, such as electron and neutron diffraction, Rietveld analysis of diffraction data, high resolution electron microscopy and synchrotron beam diffraction studies.

Another technique that finds wide employment in zeolite characterization is that of magic angle spinning nuclear magnetic resonance (MASNMR). This enables fine detail of the environments of the atoms in the tetrahedral positions to be clarified, e.g. the respective Al near neighbours to Si tetrahedra, and also the presence of octahedrally coordinated Al species, which are produced by Al displaced from their normal tetrahedral positions. This is a particular adjunct of the methods used to modify synthetic zeolites to improve their catalytic properties. These will be mentioned later. A good general survey of these modern methods has been provided by Van Hooff and Roelofsen (1981). More routine characterizations by thermal analysis, especially differential thermogravimetry (DTG) and infrared spectroscopy, provide useful 'fingerprints'. Gottardi and Galli (1985) show examples of this in their textbook on natural zeolites.

10.5 Zeolite surfaces, internal and external

A key point in the appreciation of zeolite surfaces is that many species have a large **internal** surface available, and it is this surface that controls most of their useful properties.

Zeolites have often been likened to sponges with large internal cavities and channels available for use. The extent of this voidage varies with zeolite molecular architecture and some examples are given in Table 10.3 for both natural and common synthetic species.

Of course, these voids contain cations and water molecules, and those zeolites with a large void volume can be expected to be those that have high mobilities for cations and water – i.e. facile ion exchange and water loss/regain – fundamental properties in their use as ion exchangers, molecular sieves and catalysts, as will be described later in this chapter. Usually, available surfaces are quantified by gas sorption techniques (e.g. N_2 uptake analysed by BET isotherm plots). Zeolites have nitrogen surface areas in the approximate region $100–1000\,m^2\,g^{-1}$, but these must be regarded as estimates because of the assumption of monolayer coverage as required by BET theory. In zeolites of high voidage (e.g. chabazite, A, X and Y) it can be assumed that only 1% is on the **external** crystal surface.

Reference to the illustrations of zeolite structures in Figures 10.5–10.7 shows that access to cavities and channels within the zeolite frameworks is controlled by windows proscribed by oxygen atoms and this is the basis of the zeolite gas sorptive properties.

It should also be noted that –OH groups occur on both internal and external surfaces, normally in low concentration (about $3\,nmol\,g^{-1}$). This is

Table 10.3 Void fractions of some zeolites

Natural	Composition*	Void fraction†
Analcime	$Na_{16}[Al_{16}Si_{32}O_{96}]16H_2O$	0.18
Mordenite	$Na_8[Al_8Si_{40}O_{96}]24H_2O$	0.28
Clinoptilolite	$(Na,K)_6[Al_6Si_{30}O_{72}]20H_2O$	0.34
Gmelinite	$(Na_2,Ca)_4[Al_8Si_{16}O_{48}]24H_2O$	0.44
Chabazite	$Ca_6[Al_{12}Si_{24}O_{72}]40H_2O$	0.47
Synthetic	**Composition***	**Void fraction†**
ZSM-5	$Na_n[Al_nSi_{96-n}O_{192}]16H_2O$	0.10
L	$K_6Na_3[Al_9Si_{27}O_{72}]21H_2O$	0.32
A	$Na_{12}[Al_{12}Si_{12}O_{48}]27H_2O$	0.47
Y	$Na_{56}[Al_{56}Si_{136}O_{384}]250H_2O$	0.48
X	$Na_{87}[Al_{87}Si_{105}O_{384}]234H_2O$	0.50

* Square brackets represent framework constituents
† Assessed from water content

After Breck (1984)

one reason why the current thinking on classification of zeolite structures includes frameworks known to contain terminal −OH moieties, as mentioned earlier. These −OH groups can also be bridging entities – especially in synthetic high silica zeolites such as ZSM-5.

The presence of these −OH groups can have important consequences in zeolite catalytic activity. They also provide ways in which modifications can be carried out which have interesting effects on the useful properties of the synthetic minerals.

10.6 Zeolite properties

10.6.1 Ion exchange

THEORY

The process of zeolite cation exchange follows the classical, reversible, stoichiometric replacement of cations from the internal cavities within the

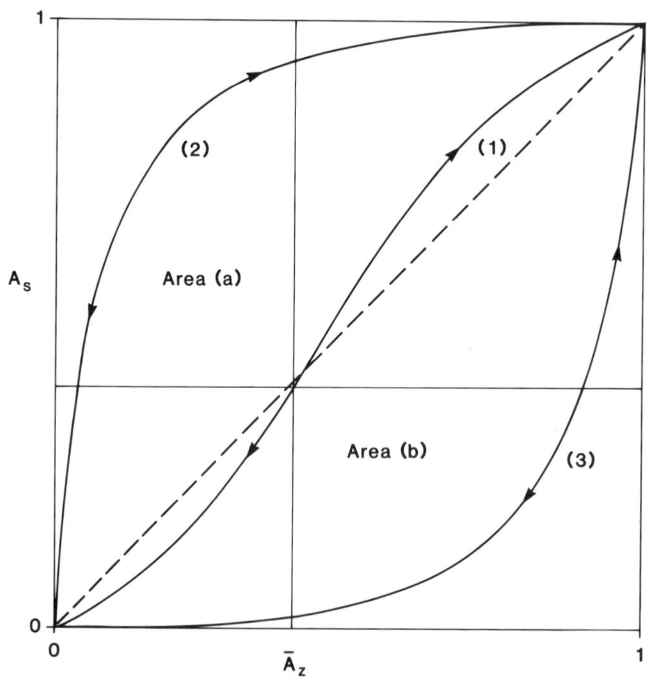

Figure 10.9 Ion exchange isotherm. A_s is the concentration of ion A in solution, $\bar{A}_z$ is the ion A concentration in the zeolite phase. Curve 1 represents equal selectivity solution/solid for A, Curve 2 solid unselective for A, Curve 3 solid selective for A. The ratio of Area (a) to Area (b) quantifies the selectivity, or lack of selectivity of the solid phase for ion A.

zeolite particles and is rate controlled by cation diffusion within the solid phase (**particle diffusion**). For this reason zeolites have been extensively studied as convenient model substances to construct theories of the kinetic and equilibrium aspect of ion-exchange (Dyer, 1981), but, from a practical point of view, the construction of ion-exchange isotherms can give useful indications of selective cation uptake (Figure 10.9). It is known that selectivity can vary with concentration, and the measurement of selectivities at the very low concentrations of interest to environmental control (or indeed the uses mentioned earlier) have seldom been carried out.

The reasons why a zeolite may be selective for a particular cation can be defined only broadly. In general, zeolites with an open structure and high silica/alumina ratio will prefer large unhydrated cations (e.g. Cs uptake in clinoptilolite), whereas those with Si/Al close to unity will often take up the small, more hydrated cations (e.g. Ca into zeolite A). This must be taken as only a rough guide and there clearly are many complex interactions involved. For example, a particular cation might be too large (either as a bare ion or in a hydrated state) to pass into part, or all, of the zeolite internal structure (ion-sieving effect).

DETERGENCY

Most elementary chemistry texts contain reference to the use of zeolites as water softeners. This is misleading because the inorganic materials, pioneered by Gans at the turn of the century and sold as ion exchangers for water treatment, were amorphous aluminosilicates. In practice, no real market currently exists for the use of natural zeolites in this traditional way.

There is, however, a growing use of zeolites in washing powders and other household detergents. Zeolites are added to these products as 'builders', i.e. water softeners, to improve washing efficiency. The most commonly used zeolite builder is the synthetic zeolite A, which has the previously noted high selectivity for Ca^{2+}, hence reducing water hardness. It is less selective for Mg^{2+}, but none the less use in this way accounts for at least 500 000 t per annum and as such represents the major industrial use of zeolites by a wide margin. Natural zeolites also are used in this way, with clinoptilolite being the preferred builder, though chabazite also has been used.

The advantage of zeolite builders comes in their use as a replacement (sometimes only partial) of sodium tripolyphosphate, thus reducing the level of phosphate released into the environment. This provides a good example of a use relying on removal of a specific ion from aqueous media, and there are other circumstances where zeolites are useful in this way.

RADIOISOTOPE CONTROL

Clinoptilolite, having a relatively high Si/Al ratio, has the ability to scavenge Cs present in very low concentrations in nuclear wastes. This has been developed by British Nuclear Fuels in their site ion exchange effluent plant

(SIXEP), at Sellafield, which has greatly reduced the discharge of Cs radioisotopes into the Irish Sea. (It also reduces ^{90}Sr release.) The clinoptilolite used comes from a deposit which is 98% pure, located at Mud Hills, near the Edwards Air Force base in California. Its performance requirement is to take up 20 moles of Cs, and 1 mole of Sr, from competing concentrations of Na (7×10^5 mol), Mg (6×10^3 mol) and Ca (5×10^3 mol).

The Cs selectivity of clinoptilolite has led to its being studied as a possible way of reducing Cs radioisotope body burdens in sheep, and mordenite has been investigated for the same end in reindeer. Mordenite has also been used to treat aqueous nuclear waste, as has an acid washed chabazite (AW500). Finally, clinoptilolite amendments to soil have greatly reduced Cs-uptake by plants and an especially significant experiment demonstrating this has been carried out on Bikini Atoll. Clinoptilolite from the Belli Plest deposit in Bulgaria was dropped by air to help control the Chernobyl accident.

WASTE-WATER TREATMENT

The ubiquitous clinoptilolite also has a selectivity for the NH_4^+ ion and this has promoted its use in municipal sewage treatment, as at Lake Tahoe in the USA and at Vae, near Budapest. These are examples of the use of local mineral resources, and the Japanese have made a similar use, particularly in cleaning up water from fish farms. The Italians have used chabazite-rich tuffs for a variety of water treatments. Details of uses of natural zeolites in this and many other ways have been reviewed by Dyer (1989).

AGRICULTURE and HORTICULTURE

There is a considerable literature describing how natural zeolites (usually) can be fed to various animal species (sheep, pigs, beef cattle, horses, etc.) with advantage. This seems to arise from zeolite ion exchange for ammonium in the gut controlling excess pH, which increases weight gain and reduces certain diseases (e.g. scour in pigs). There also exists the possibility that the high surface area of zeolites scavenges toxins.

The literature also relates the improvement of soil quality by 5–10% zeolite additions, followed by a better crop. The exact mechanism invoked has yet to be defined, but water species of zeolite surfaces are known to dissociate to give hydroxyl ions and protons. The protons produced are selectively taken into the zeolite, as H_3O^+ ions, and this can be expected to be a significant factor at low concentrations of other cations important to plant growth. This may be the reason why NH_4^+ ions, which are normally selectively held in a zeolite, can become available to crops, the zeolite acting as a controlled-release fertilizer. Again, the zeolite most mentioned for these applications is clinoptilolite (Dyer, 1989).

10.6.2 Sorption and separations on zeolites

When the inherent water has been removed from a zeolite, the remaining framework often has a useful capacity to take in guest molecules. Removal of water is described as 'activation' and can be attained by heating (to about 350 °C), usually in a vacuum or in combination with an inert gas purge on an industrial scale.

Figures 10.5–10.7 show that entry to the internal surface is gained by progress through rings of oxygen atoms. Zeolite A (Figure 10.5), for example, has two restricting rings, one a single ring of 6 oxygen atoms (S6R) controlling entry into the sodalite cage, and the second an 8 oxygen ring (S8R) controlling entry to the main cavity. These rings have sizes comparable to molecular dimensions and so can behave as 'molecular sieves'. Table 10.4 shows some examples of typical rings found in zeolite structures, with an indication of their sizes and the types of molecule able to pass through them.

'Fine tuning' of effective window- (ring-) size can sometimes be achieved by cation exchange. The well-known commercial molecular sieves produced by Union Oil Products (UOP) (originally Union Carbide) are Linde MS 3A, 4A and 5A, and these are the K-, Na- and Ca- exchanged forms of zeolite A, respectively, where the numbers represent effective pore sizes measured in Ångströms.

10.6.3 Uses of zeolites as molecular sieves

A detailed account of the uses of zeolites as molecular sieves is given by Breck (1984). Briefly, the major uses are as follows.

DESICCANTS

The ion-exchanged forms of zeolite A are well known as a general laboratory drying agent in the form of extrudates or beads bonded by the addition of

Table 10.4 Molecular sieving, correlation between oxygen ring size and molecules admitted

Crystallographic dimension (Å)	No. of oxygens	Zeolite	Molecules admitted (examples)
2.3	6	A, X	H_2O, NH_3
3.8*	8	Chabazite	He, Ne, Ar, CO_2
4.1	8	A	C_2H_2, CH_3NH_2
4.2, 5.4	10	Ferrierite	as through 8 ring
7.4	12	X, Y	cyclohexane, benzene, CCl_4, B_5H_9

* Distorted

Two zeotypes with larger rings are known – (a) VPI-5 an aluminophosphate with an S18R (12.1 Å) and (b) cloverite, a gallophosphate with a S20R (13.2 × 4.0 Å) approximately 17% clay mineral (e.g. attapulgite). In the same form, they are widely used to scavenge water from air lines, railway hydraulic brakes, refrigeration fluids and transformer oils. They also can be used in sealed double-glazing units. On a larger scale, 3A is used to dry refinery gas streams ('cracked gas') and 4A for drying natural gas, oxygen and hydrogen. The success of these materials as desiccants arises for several possible reasons:

- their ability to exclude other molecules on the basis of size;
- the strong affinity of water for the internal zeolite surfaces;
- accommodation of water molecules in voids containing cations;
- ease of regeneration;
- long useful life (natural gas drying towers have a 5-year life-span).

GAS PURIFICATION

Methods of gas purification are many and varied and use both natural and synthetic zeolites. The natural zeolites most often used are chabazite, clinoptilolite and mordenite. The latter two are useful for dealing with acid gas streams.

Common purifications are the use of 4A to remove CO_2 from natural gas as well as 'sweetening' the gas, i.e. removing H_2S and nitrogen products. Zeolite 13X (synthetic faujasite containing only Na cations) also removes undesirable larger molecules that contain S and N atoms.

GAS SEPARATION

The best known method of gas separation is the use of 5A to separate n- from iso-paraffins by exclusion of the iso-paraffins from the zeolite by molecular sieving. This is the basis of several commercial processes (e.g. Union Carbide – Isosiv, and Exxon – Ensorb). Their purpose is to provide feedstocks for the detergent and chemical industry. Currently there is considerable interest in using zeolites in 'pressure swing applications' (PSAs), especially in the enrichment of gases from the air. Oxygen production in hospitals is an example of this, both in the provision of oxygen lines and in portable units for individual use or use in field hospitals, as in the Gulf War. Again zeolite A is widely used, but clinoptilolite, mordenite and chabazite are important substrates in PSA applications, e.g. the separation of methane and nitrogen mixtures using clinoptilolite. These applications rely on preferential sorptions into the zeolite, based on the quadrupole moment (N_2) or dipolar interactions (H_2O,CO_2).

Finally, synthetic zeolites have long been used in gas–solid chromatography for permanent gas and hydrocarbon analyses. Some modern GC capillaries are coated with zeolites.

LIQUID PHASE SEPARATION

Union Oil Products have pioneered several small scale processes to fractionate mixtures of liquids. Examples of these are listed in Table 10.5.

Table 10.5 UOP processes for liquid phase separations

Process	Separation	Zeolite*	Desorbent
Parex	*p*-xylene from its isomers	Y, X	toluene
Ebex	ethylbenzene from its isomers	Y, X	*p*-diethyl benzene
Molex	*n*-paraffins/iso-paraffins, naphthalenes, aromatics	5 Å	light paraffins
Olex	linear long chain paraffins from other paraffins	X ?	?
Sorbutene	But-1-ene from C_4 olefines and paraffins	?	?
Sarex	Fructose from sucrose and other sugars (corn syrup feed)	Y	aqueous systems

* Often exchanged with K, Ca, Sr, Ba

After Dyer (1988)

Only the Molex process depends on molecular size; the others function on the basis of a liquid chromatographic fractionation. Recent work (Dyer and Emms, unpublished data) shows that high silica zeolites (synthetic) make good high performance liquid chromatographic (HPLC) media. The mechanisms whereby these chromatography separations are achieved await elucidation, but this work shows that both internal and external surface attractions are important, as are zeolite structures and the nature of the cation present. These conclusions offer promise for both analytical and bulk separations.

CARRIERS

The encapsulation of molecules into zeolite voids has been useful in the handling of unpleasant chemicals such as vulcanizing agents, and to carry CO_2 in carbonated drinks. Other prospective uses are to store and transport methane and as drug-delivery systems for veterinary use.

MODIFICATION

In addition to the use of cation exchange to tailor molecular sieving, the presence of $-OH$ groups on the zeolite surface can be used to anchor $-SiH_3$ groups via a silanation process (using silane, SiH_4). These can then be hydrolysed to partially block the zeolite pores with a silica species.

10.6.4 Zeolites as catalysts

Thirty years ago it was discovered that a synthetic faujasite could be used as a cracking catalyst. Since that time there has been an explosion of patents, conference reports and journal publications devoted to reactions catalysed by zeolites. It is not the purpose of this article to cover this literature, especially as there has been an excellent recent review by John, Clark and

Maxwell (1992), which summarizes all the chemical reactions known to be promoted by zeolites. They also discuss likely mechanisms and the nature of the zeolites used.

This chapter will provide an overview of current commercial use of zeolite catalysts together with some indications of their mode of action and ease of modification.

COMMERCIAL PROCESSES USING ZEOLITE CATALYSTS AND THEIR ADVANTAGES

Table 10.6 provides some details of zeolite catalyst usage. The most well known is that of zeolite Y as a fluidized bed catalyst (FCC) to crack crude oil for petrol production. Essentially all the world's motor vehicle fuels are produced this way.

The zeolite is used as a **promoter** and as such comprises up to 50% of a small composite with a clay or silica binder. As such it can remain stable

Table 10.6 Commercial processes using zeolitic materials

Process	Catalyst	Advantages in using zeolite based catalysts
Catalytic cracking	Ultrastable Y* (USY) ZSM-5 Rare earth exchanged Y (REY)	Selectivity and high conversion rates
Hydrocracking	Y, erionite mordenite	High conversion rates
Selecto-reforming	Ni erionite, Ni erionite/ clinoptilolite	Increase in octane number via LPG production
Hydro-isomerism	Pt-mordenite	Converts low octane, pentane and hexane to higher octane yields
Dewaxing	ZSM-5, ZSM-23, mordenite	Improved pour points for heavy oils and fuels
Benzene alkylation	ZSM-5, REY	Ethylbenzene and styrene production with low by-product yield
Xylene isomerization	ZSM-5	Increase in *p*-xylene yield with low by-product yield
Methanol to gasoline conversion	ZSM-5	High gasoline yield with high octane rating
NO_X reduction	H-mordenite	Effluent clean-up in nitric acid and nuclear reprocessing plants
Aromatic hydroxylation/ olefine epoxidation	Titanium Silicalite	Styrene → phenylacetaldehyde, Ethene → ethylene oxide and other reactions
Liquid propane gas (LPG) to aromatics	Ga ZSM-5	Benzene, toluene, xylene production

* USY, zeolite Y treated (e.g. by steaming) to reduce aluminium content and improve thermal and hydrothermal stability.

throughout many cycles in the catalyst riser (at 480–520 °C), where it meets a downstream of crude oil, followed by a stream blast to release the cracking products and then a stream of air in a regenerator (590–730 °C). This regeneration process cleans the catalyst of coke and residual oil products.

Clearly the success of a catalyst depends on its ability to withstand these extreme conditions and this is one of the advantages of the ultrastable synthetic faujasite (USY) and ZSM-5 materials; this is always a consideration in the other applications listed in Table 10.6. It is, however, only one of the known advantages acquired when zeolites are considered as industrial catalysts. Other benefits come from the zeolite properties of high internal surface area and variable pore size, which lead to both reactant and product selectivity via molecular sieving. This is described as **shape-selectivity** and is controlled by **configurational diffusion** – the phrase used to describe a diffusion-controlled regime in which useful catalytic reactions are encouraged by a matching of size, shape and orientation of the product and reactant molecules to the molecular architecture of the zeolite. An additional bonus comes from the ease with which zeolites can be modified. Examples of this have been mentioned earlier, e.g. ion-exchange and the introduction of heteroatoms (other than Si, Al) into the zeolite framework. This latter change can be achieved by post-synthesis modification and an interesting prospect is that of inducing ion-exchange and catalytic properties into the uncharged frameworks of aluminophosphates (AlPO) by inserting Si atoms into the existing AlPO structure. A further modification is that of salt occlusion, i.e. the introduction of a salt molecule into a zeolite cavity to attain improved thermal stability and reduce the deleterious effects of trace elements, such as Ni, Fe and V, which occur in some crude oils.

SOURCE OF ZEOLITE CATALYTIC ACTIVITY

Generally speaking, zeolites have high catalytic activity because their structure can produce a higher proportion of the so-called **active sites** per unit surface area (internal) than competing catalysts. These sites are the traditional Brønsted and Lewis sites.

The Brønsted sites arise from the creation of hydroxyl groups and this can be via one of three routes illustrated below, where Z is a zeolite substrate.

1. Ammonium ion exchange (e.g. as in production of USY)

$$NaZ_{(s)} + NH^+_{4(aq)} \rightleftharpoons NH_4Z_{(s)} + Na^+_{(aq)}$$

$$NH_4Z_{(s)} \xrightarrow{\text{calcine}} HZ_{(s)} + NH_{3(g)}$$

2. Polyvalent ion exchange (e.g. as in rare earth exchanged Y–ReY)

$$NaZ_{(s)} + M(H_2O)^{n+}_{aq} \rightleftharpoons M(H_2O)^{n+}Z_{(s)} + n\,Na^+_{aq}$$

$$M(H_2O)^{n+}Z_{(s)} \xrightarrow{\text{calcine}} MOH^{(n-1)}_{s} + HZ_{s}$$

3. Direct ion exchange with mineral acid (e.g. with ZSM-5).

In all cases the **protonated** zeolite, HZ, is created. Brønsted sites exist on the zeolite aluminosilicate surfaces, which when heated above 550 °C lose water to produce Lewis sites thought to be tri-coordinated Al; shown below as Al*

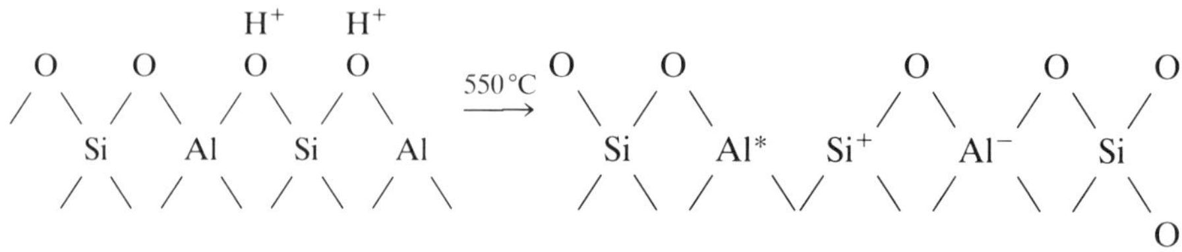

Brønsted site   Lewis site

The Lewis sites are unstable, especially into the presence of water vapour, and an annealing process stabilizes the structure. This creates 'true' Lewis sites by ejecting Al species from the framework *viz.*

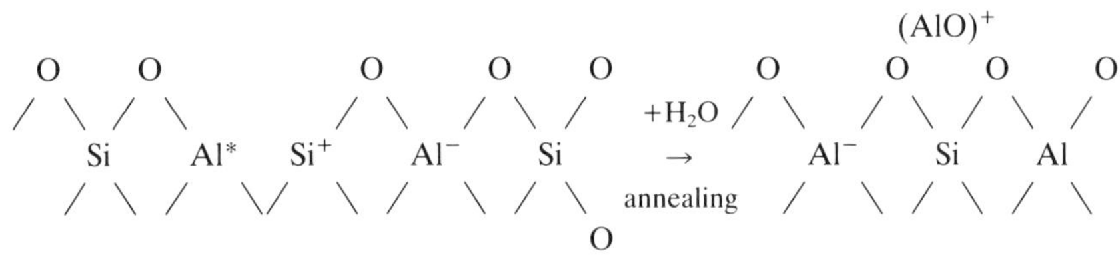

This extra framework Al can be identified by MASNMR but is in at least two forms. It is these Al species that are thought to be the **super acid** sites inducing the high zeolite surface reactivity. Confirmation of this is still in prospect.

Another aspect to zeolite catalytic activity is the use of ion-exchange modification to introduce elements of known catalytic propensity into the zeolite promoter. Examples of this are included in Table 10.6. There has been a considerable amount of work devoted to the investigation of the dispersion of metal particles (e.g. Ni and Fe) on zeolite substrates to enhance specific organic reactions. This can be extended to the **multifunctional catalysts** containing more than one metal.

10.7 The future

10.7.1 Catalysis

The SAPO materials seem likely to provide novel catalysts, especially as the emphasis of zeolite usage turns to chemical intermediates and fine chemical production (Hoelderich, 1989), and processes using the B analogue of ZSM-5 ('boralite') and Ga-L are thought to be under development.

Another advance seems likely through the use of zeolites as enzyme mimics. This centres on the reactions of organometallics with zeolite internal surfaces. The best known example is the production of a $[Co^{2+}$(bipyridyl)(terpyridine)$]^{2+}$ complex inside the main cavity of zeolite Y that can selectively and reversibly absorb oxygen from the air. John, Clark and Maxwell (1992) also review the catalytic potential of a Co phthalocyanine moiety prepared in the Y cavity.

10.7.2 Zeolites as hosts

Materials can be loaded into zeolite voids to endow them with particular properties. Conducting polymers can be formed in channels to create 'molecular wires' and clusters of semiconducting sulfides (PbS, CdS) created in cages ('quantum dots'). The composite materials produced have potential for use in optical switching, optical data storage and microsensors.

Clusters of metals can be established with interesting catalytic properties (e.g. Pt), but higher loading can result in composites with unique magnetic properties, which may provide a route to the manufacture of lightweight magnets.

10.7.3 Zeolite films and membranes

Considerable interest is being devoted at present to the production of zeolite thin films and membranes, particularly for gas separations.

10.8 Conclusions

The interest in zeolites continues to grow. It can be predicted that natural zeolites will be more widely considered for roles in environmental clean-up processes – especially those involving radioisotopes and heavy metals. New zeolite phases appear almost weekly and, to date, very little is known of their sorption and ion-exchange capabilities. It is to be expected that novel applications will arise from such studies. The future may well see 'designer' sorbents and exchangers modelled by a computer, programmed to envisage heteroatom frameworks and effective cation locations. That this is already being actively considered, with designer catalysts, can be seen in a recent publication describing progress in the modelling of structure and reactivity in zeolites (Catlow, 1992).

References

Barrer, R.M. (1982) *Hydrothermal Chemistry of Zeolites*, Academic Press, London.
Breck, D.W. (1984) *Zeolite Molecular Sieves*. Krieger, USA.

Catlow, C.R.A. (1992) *Modelling of Structure and Reactivity in Zeolites*. Academic Press, London.

Dyer, A. (1981) Modern theories of ion exchange and ion exchange selectivity with particular reference to zeolites, in *Inorganic Ion Exchangers in Chemical Analysis* (eds M. Qureshi and K.G. Varshney), CRC Press, USA, pp. 33–56.

Dyer, A. (1988) *An Introduction to Zeolite Molecular Sieves*, John Wiley & Sons, Chichester.

Dyer, A. (1989) Uses of natural zeolites, in *Chemistry and Industry*, Part 6, pp. 241–6.

Gilson, J.-P. (1992) Organic and inorganic agents in the synthesis of molecular sieves, in *Zeolite Microporous Solids: Synthesis, Structure, and Reactivity* (eds E.G. Derouane, F. Lemos, C. Naccache and F.R. Riberro), *NATO ASI Series*, Kluwer Academic Publishers, Dordrecht, pp. 19–48.

Gottardi, G. (1989) The genesis of zeolites. *Eur. J. Mineral.*, **1**, 479–87.

Gottardi, G. and Galli, E. (1985) *Natural Zeolites*. Springer-Verlag, Berlin.

Hoelderich, W.F. (1989) Zeolites: catalysts for the synthesis of organic compounds, in *Zeolites, Facts, Figures, Future* (eds P.A. Jacobs and R.A. van Santen), *Studies in Surface Science and Catalysis*, Vol. 49, Elsevier, Amsterdam, pp. 69–93.

John, C.S., Clark, D.M. and Maxwell, I.E. (1992) New insights into zeolite catalysis, in *Perspectives in Catalysis* (eds J.M. Thomas and K.I. Zamarov), Blackwell Publications (for IUPAC), Oxford, pp. 387–430.

McConnell, D. (1990) Kehoeite is not a valid species. *Mineral. Mag.*, **55**, 657. Meier, W.M. and Olson, D.H. (1992) Atlas of Zeolite Structure Types, *Zeolites*. Spec. Issue, **12**, 1–200.

Sokolova, E.V., Rybakov, V.B. and Pautov, L.A. (1989) Crystal structure of a new natural tetramethyl ammonium aluminosilicate $[N(CH_3)_4][Si_2(Si_{0.5}Al_{0.5})O_6]_2$. *Dokl. Akad. Nauk. SSR*, **317**, 884–7.

Van Hooff, J.H.C. and Roelofsen, J.W. (1991) Techniques for zeolite characterization, in *Introduction to Zeolite Science and Practice* (eds H. van Bekkum, E.M. Flanigen and J.C. Jansen), *Studies in Surface Science and Catalysis*, Vol. 58, Elsevier, Amsterdam, pp. 242–84.

Index

Page numbers appearing in **bold** refer to tables and those in *italic* refer to figures.